Advanced Vector Analysis
for Scientists and Engineers

WITPRESS

WIT Press publishes leading books in Science and Technology.
Visit our website for the current list of titles.
www.witpress.com

WITeLibrary

Home of the Transactions of the Wessex Institute, the WIT electronic-library provides the
international scientific community with immediate and permanent access to individual
papers presented at WIT conferences. Visit the eLibrary at
http://library.witpress.com

Advanced Vector Analysis for Scientists and Engineers

By

M. Rahman

WITPRESS Southampton, Boston

Advanced Vector Analysis
for Scientists and Engineers

M. Rahman

Published by

WIT Press

Ashurst Lodge, Ashurst, Southampton, SO40 7AA, UK
Tel: 44 (0) 238 029 3223; Fax: 44 (0) 238 029 2853
E-Mail: witpress@witpress.com
http://www.witpress.com

For USA, Canada and Mexico

WIT Press

25 Bridge Street, Billerica, MA 01821, USA
Tel: 978 667 5841; Fax: 978 667 7582
E-Mail: infousa@witpress.com
http://www.witpress.com

British Library Cataloguing-in-Publication Data

A Catalogue record for this book is available
from the British Library

ISBN-13: 978-1-84564-093-4

LOC: 2006938300

Contents

Preface

Vector Analysis is one of the most useful branches of mathematics. It is used in practical problems arising in engineering and the applied sciences. The main purpose of this book is to illustrate the application of vector calculus to physical problems. This book is suitable for a one-semester course for senior undergraduate and graduate students in science and engineering. It is also suitable for scientists and engineers working in practical fields.

This book is the outgrowth of class notes used over years of teaching vector analysis in many universities including Dalhousie University. The theory is explained clearly and there is an abundance of worked examples throughout. The beauty of the book is its richness of examples. There are eight chapters, each one containing ample exercises at the end. There are also three important appendices at the end of the book. Solutions to selected exercises have been included in Appendix A. Appendix B contains a quick summary of important vector formulae, and Appendix C contains a brief historical background of vector calculus which was originally introduced in the days of Aristotle during the fifteenth century. A systematic development of the evolution of vectors with basic vector algebra is described. This appendix also describes the modern vector calculus and vector analysis due to the contributions of Josiah Willard Gibbs (1839–1903) and Oliver Heaviside (1850–1925). Later Maxwell contributed heavily in the electromagnetic theory applications.

Chapter 1 introduces the algebra of vectors. This chapter contains vector addition and subtraction, scalar and vector products and vector triple products. Representation of vectors by the coordinate system; equations of lines and planes using vector algebra are demonstrated. The chapter contains many applied exercises.

In Chapter 2, we consider vector functions of one variable to illustrate vector differentiation and integration along with the geometric interpretation of a position vector. Vector integration is applied to determine the length and arc length of a curve. This chapter also contains particle motion on a curve, its velocity and acceleration. Tangential and normal components of acceleration of a particle moving in space are discussed in a lucid manner. The chapter ends with a considerable number of practical exercises.

Partial derivatives of a scalar function for several independent variables are clearly defined in Chapter 3 and the idea is extended to vector functions. The functions of multiple variables and the concept of Jacobian are stated clearly with examples. The chapter is concluded with ample exercises of practical interest.

The Del ∇ operator (gradient operator) plays a very important role in practical problems arising in engineering and physical sciences. Chapter 4 is devoted to the precise definition of this important operator. With the help of this operator, a number of quantities such as gradient, divergence and curl are defined in the light of physical applications. Applied problems such as continuity of a fluid flow in incompressible and compressible fluids and also rotation of a fluid in the viscous case are very elegantly described with physical interpretation. We have also cited some formulae involving the Del operator with proofs. The Laplace operator ∇^2 plays a very significant role in physical problems and this is discussed very thoroughly. The chapter ends with exercises containing a number of important problems.

In Chapter 5, we have discussed the mathematical theory containing line, surface and volume integrals starting with the basics. A conservative vector field has been defined and illustrated with examples. Surface and volume integrals are also discussed with some practical examples. Triple integrals in cylindrical and spherical coordinates are discussed in this chapter. Practical situations are illustrated. Ample problems are included at the end of the chapter.

Chapter 6 contains some important and extremely useful integral theorems such as Green's Theorem, Stokes' Theorem and the Divergence Theorem. Many examples are solved to help the reader understand the practical use of these sophisticated theorems. They are very useful for the further study of research problems. The chapter concludes with ample exercises of academic and physical interest.

Orthogonal curvilinear coordinate systems are discussed in Chapter 7. The development of grad, div, curl are demonstrated clearly using the orthogonal curvilinear coordinate system. These important quantities are derived in the frames of Cartesian, cylindrical polar and spherical polar coordinates. Some practical examples are worked out. This chapter concludes with a number of practical exercises.

The last chapter (Chapter 8) of this book contains the applications of vector analysis. We have selected problems in fluid flow, electromagnetic theory with emphasis on Maxwell's equations. Some solution techniques are considered for Maxwell's wave and heat equations. A brief discussion on nonlinear ocean wave interaction has been presented at the end to show why vector analysis is important to the solution of physical problems. We have briefly demonstrated how to solve the nonlinear wave energy equation. The next section of the chapter deals with the numerical simulation of a vector field using Mathematica 3.0. The last section examines a practical problem of irrotational and inviscid fluid flow. The chapter concludes with a number of exercises of physical interest.

Chapters 1–5 contain elementary vector calculus and chapters 6–8 have dealt with advanced material of vector analysis. The book is developed sequentially starting with the elementary level and ending with the advanced level. It is our hope that the reader will find the book useful.

The Author
Halifax, Canada 2007

Acknowledgements

The Author is thankful to the Natural Sciences and Engineering Research Council of Canada for their financial support. Thanks are also extended to Professor Carlos Brebbia for his constant encouragement to write this book. Mr. Adhi Susilo deserves my appreciations for helping me in designing all the figures contained in this book. I am grateful to WIT Press for their efforts in producing such a superb book.

Chapter 1

The algebra of vectors

1.1 Introduction

The starting point for the definition of a vector will be the intuitive one encountered in elementary physics. From the historical background of vector analysis described in Appendix C, we know that there are physical quantities such as length, mass, density, volume, speed, temperature, concentration, etc., that have only magnitude associated with them. These can be represented by real numbers and are called *scalars*. There are many quantities, however, that have both magnitude and direction associated with them. Such quantities cannot be completely described by a single real number. The most notable concepts in this category are velocity, acceleration and force. To represent such quantities mathematically, we introduce *vectors*. In this chapter, we shall investigate the ideas of vector analysis following the development and notation of Josiah Willard Gibbs (1839–1903), the great American mathematical physicist and thermodynamicist, and Oliver Heaviside (1850–1925), a British electrical engineer. Their work was independent but the end results were effectively the same, apart from differences in notation and terminology.

Definition 1.1 *A vector is defined as a directed line segment.*

The length of a vector will be denoted by the word *magnitude*. Any physical element that has magnitude and direction, and hence can be represented as a vector, will also be designated as a vector. In Fig. 1.1a, b and c, we show two vectors **A** and **B** along a line, three vectors **A**, **B** and **C** in a plane, and three vectors **A**, **B** and **C** in space, respectively. It is customary to place an arrowhead on a vector and call

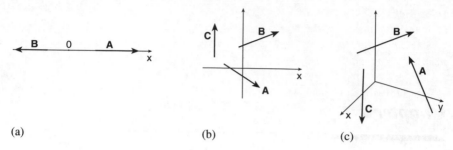

Figure 1.1   Vectors as directed line segments.

this end the *tip* of the vector. The other end is called the *tail* of the vector and the direction of the vector is from tail to tip.

To distinguish vectors from scalars, we shall consistently denote the former by symbols, either in uppercase or lowercase letters, set in boldface type thus, **A**. When written by hand, a vector quantity is indicated by putting an arrow above the appropriate symbol. A scalar quantity can be represented by a mark on a fixed scale. However, to represent a vector, we must use a directed line segment whose direction is the same as that of the vector and whose length is equal to the magnitude of the vector.

For convenience, the representative line segments shall be referred to as the vector itself. The magnitude or length of a vector **A** is also called the *absolute value* of the vector and is indicated either by enclosing the symbol for the vector between ordinary absolute-value bars or simply by setting the symbol for the vector in ordinary rather than boldface type, i.e.

$$A = |\mathbf{A}|.$$

The above represents the magnitude or absolute value of the vector **A**. A unit vector can then be defined in the following manner.

Definition 1.2   *A vector* **A** *is said to be a unit vector if it has length equal to one unit, i.e.* **A** *is a unit vector if*

$$|\mathbf{A}| = 1.$$

Thus, any vector **A** can be made a unit vector if it is divided by its magnitude such that $\mathbf{A}/A = \mathbf{e}$, where **e** is the unit vector. Hence, $\mathbf{A} = A\mathbf{e}$. It will follow that a *zero vector* will be one whose absolute value is zero. The direction of the zero vector is undefined and it is denoted by a zero set in boldface type. From Definition 1.1, there is no restriction on the point of application of a vector. We may thus place the tail of the vector in any spot we wish. This suggests the following definition for equality of vectors.

Definition 1.3   *Two vectors whose magnitudes and directions equal are said to be equal, regardless of the points in space from which they may be drawn. If two vectors have the same length but are oppositely directed, either is said to be the negative of the other.*

From this it can be seen that a vector quantity can be represented equally well by any of infinitely many equivalent line segments, all having the same length and direction. It is, therefore, customary to say that *a vector can be moved parallel to itself without change.* However, in some applications, e.g. in dealing with forces whose points of application or lines of action cannot be shifted, it is necessary to think of a vector as fixed or at least restricted in position. Such vectors are usually said to be *bound* in contrast to unrestricted vectors, which are said to be *free.*

1.2 Addition and subtraction of vectors

Suppose there are two vectors, **A** and **B**, that are drawn from the same point. The *sum* of the two vectors **A** and **B** is defined by the familiar parallelogram law, i.e.

Definition 1.4   *If two vectors **A** and **B** are drawn from the same point, or origin, and if the parallelogram having **A** and **B** as adjacent sides is constructed, then the sum **A** + **B** is the vector represented by the diagonal of the parallelogram which passes through the common origin of **A** and **B** (see Fig. 1.2a).*

From the above definition, it is clear that

$$\mathbf{A} + \mathbf{B} = \mathbf{B} + \mathbf{A},$$

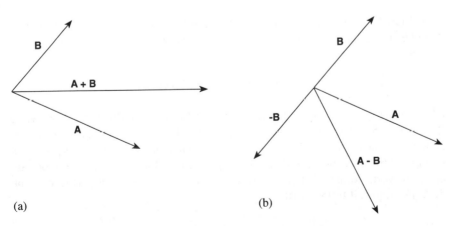

(a)   (b)

Figure 1.2   The addition and subtraction of vectors.

i.e. *vector addition is commutative* and that

$$A + (B + C) = (A + B) + C,$$

i.e. *vector addition is associative.* By the *difference* of two vectors **A** and **B**, we mean the sum of the first and the negative of the second (see Fig. 1.2b), i.e.

$$A - B = A + (-B).$$

To multiply a vector **A** by a scalar α, we will introduce the vector $\alpha A = A\alpha$ whose length is equal to the product of $|\alpha|$ and the magnitude of **A**, and whose direction is the same as the direction of **A** if α is positive and opposite to it if α is negative. It is not difficult to show that the following rules hold for scalar multiplication:

$$\lambda(A + B) = \lambda A + \lambda B,$$

$$(\lambda + \mu)A = \lambda A + \mu A.$$

A further consequence of the above definition is that

$$\frac{A}{|A|} = \frac{A}{A} = e$$

is a unit vector in the direction of **A**, where $A \neq 0$.

1.3 Scalar and vector products

Multiplication of two vectors can be achieved in two types of products, one which yields a scalar and the other a vector.

Definition 1.5   *The scalar product (dot product or inner product) of two vectors **A** and **B** is equal to the product of their absolute values and the cosine of the angle between their (positive) directions, i.e.*

$$A \cdot B = |A||B| \cos \theta = AB \cos \theta.$$

Observe that $|A| \cos \theta$ is the projection of vector **A** in the direction of **B**, while $|B| \cos \theta$ is the projection of the vector **B** in the direction of **A**. This leads to the conclusion that *the dot product of two vectors is equal to the length of either of them multiplied by the projection of the other upon it* (see Fig. 1.3). Several important cases are worth noting. Firstly, we have seen that A/A is a unit vector in the direction of **A**. From this, it follows that

$$\frac{A}{A} \cdot B$$

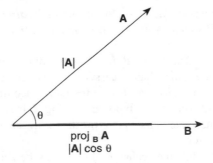

Figure 1.3   The geometrical interpretation of the scalar product.

is the *scalar projection* of **B** in the direction of **A** and that

$$\left(\frac{\mathbf{A}}{A} \cdot \mathbf{B}\right) \frac{\mathbf{A}}{A} = \left(\frac{\mathbf{A} \cdot \mathbf{B}}{A^2}\right) \mathbf{A}$$

is the *vector projection* of **B** in the direction of **A**. If **B** = **A**, then

$$\mathbf{A} \cdot \mathbf{A} = |\mathbf{A}|^2 = A^2. \tag{1.1}$$

Of the operation rules, the associative rule is valid for the scalar product only in the wider sense, whereas the commutative and distributive rules are always valid. The scalar product is commutative:

$$\mathbf{A} \cdot \mathbf{B} = \mathbf{B} \cdot \mathbf{A},$$

associative concerning a scalar factor:

$$\lambda(\mathbf{A} \cdot \mathbf{B}) = (\lambda\mathbf{A}) \cdot \mathbf{B} = \mathbf{A} \cdot (\lambda\mathbf{B})$$

and distributive:

$$(\mathbf{A} + \mathbf{B}) \cdot \mathbf{C} = \mathbf{A} \cdot \mathbf{C} + \mathbf{B} \cdot \mathbf{C}.$$

From the distributive property it follows that scalar multiplication of polynomial expressions is done in the same way as that of real numbers, i.e.

$$(\mathbf{A} + \mathbf{B}) \cdot (\mathbf{C} + \mathbf{D}) = \mathbf{A} \cdot \mathbf{C} + \mathbf{B} \cdot \mathbf{C} + \mathbf{A} \cdot \mathbf{D} + \mathbf{B} \cdot \mathbf{D}.$$

These statements follow directly from the definitions. The third type of product is the *vector* or *cross product* of two vectors, indicated by putting a cross ($\times$) between the vectors.

Definition 1.6   *The vector product of two vectors **A** and **B** is another vector. It is denoted by **A** $\times$ **B**. Its absolute value is:*

$$|\mathbf{A} \times \mathbf{B}| = |\mathbf{A}||\mathbf{B}|| \sin \phi|,$$

where ϕ is the angle between the vectors $\mathbf{A}$ and $\mathbf{B}$ and its direction is perpendicular to the vectors $\mathbf{A}$ and $\mathbf{B}$ so that $\mathbf{A}$, $\mathbf{B}$, $\mathbf{A} \times \mathbf{B}$ form a right-hand system.

In other words, if we measure $\mathbf{A}$, $\mathbf{B}$ and $\mathbf{A} \times \mathbf{B}$ from a common starting point, the rotation at an angle smaller than π carrying the vector $\mathbf{A}$ into the direction of the vector $\mathbf{B}$ is positive if we look at it from the half-space limited by the plane of vectors $\mathbf{A}$ and $\mathbf{B}$, towards which $\mathbf{A} \times \mathbf{B}$ points (see Fig. 1.4a and b). The geometrical meaning of the absolute value of the vector product is the area of the parallelogram spanned by the two vectors.

The vector product – like the scalar product – may be zero even if none of its vectors are zero vectors. Namely, if two vectors are parallel, the angle enclosed is 0 or π and so on account of $\sin \phi = 0$, the vector product is 0. And conversely, if $\mathbf{A} \times \mathbf{B} = \mathbf{0}$ and $\mathbf{A} \neq \mathbf{0}$, $\mathbf{B} \neq \mathbf{0}$, then the two vectors are parallel.

It follows then that a necessary and sufficient condition for two vectors to be parallel is that

$$\mathbf{A} \times \mathbf{B} = \mathbf{0}.$$

Among the operations neither the commutative nor the associative, but only the distributive rule is valid for the vector product. However, since the direction of $\mathbf{A} \times \mathbf{B}$ is determined by the right-hand rule, it is clear that interchanging $\mathbf{A}$ and $\mathbf{B}$ reverses the direction, or sign, of their product. Hence, *cross multiplication is not commutative*, and we have, in fact,

$$\mathbf{A} \times \mathbf{B} = -\mathbf{B} \times \mathbf{A}.$$

The multiplication in which products obey this rule is sometimes said to be *anti-commutative*. From the foregoing it is clear that we must be careful to preserve the proper order of vectors in any expression involving vector multiplication. Concerning the multiplication of scalars, the vector product is associative:

$$\alpha(\mathbf{A} \times \mathbf{B}) = (\alpha\mathbf{A}) \times \mathbf{B}$$

and distributive:

$$(\mathbf{A} + \mathbf{B}) \times \mathbf{C} = \mathbf{A} \times \mathbf{C} + \mathbf{B} \times \mathbf{C}.$$

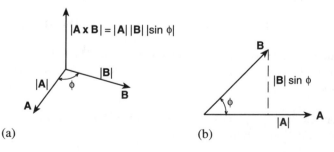

(a)   (b)

Figure 1.4   The geometrical interpretation of the vector product.

From distributivity, it follows that polynomial expressions are to be multiplied just as real numbers:

$$(\mathbf{A} + \mathbf{B}) \times (\mathbf{C} + \mathbf{D}) = \mathbf{A} \times \mathbf{C} + \mathbf{B} \times \mathbf{C} + \mathbf{A} \times \mathbf{D} + \mathbf{B} \times \mathbf{D}.$$

The scalar factor can be factored out of any factor of the scalar product:

$$\lambda \mathbf{A} \times \mu \mathbf{B} = \lambda \mu (\mathbf{A} \times \mathbf{B})$$

and

$$\lambda (\mathbf{A} \times \mathbf{B}) = (\lambda \mathbf{A} \times \mathbf{B}) = \mathbf{A} \times (\lambda \mathbf{B}).$$

1.4 Scalar and vector projections

We may use the definition of the inner product to obtain the *scalar projection* of one vector onto another. Recall that the projection of a line segment of a units in length onto a line is $a \cos \theta$, where θ is the angle between the two lines (see Fig. 1.5).

If we wish the scalar projection of $\mathbf{A}$ onto $\mathbf{B}$, symbolized $\text{proj}_{\mathbf{B}} \mathbf{A}$, we use the magnitude of $\mathbf{A}$ and the angle between $\mathbf{A}$ and $\mathbf{B}$ (Fig. 1.6). This leads us to the inner product of $\mathbf{A}$ and $\mathbf{B}$. Since

$$\mathbf{A} \cdot \mathbf{B} = |\mathbf{A}||\mathbf{B}| \cos \theta, \tag{1.2}$$

then

$$|\mathbf{A}| \cos \theta = \frac{\mathbf{A} \cdot \mathbf{B}}{|\mathbf{B}|}.$$

But

$$\text{proj}_{\mathbf{B}} \mathbf{A} = |\mathbf{A}| \cos \theta.$$

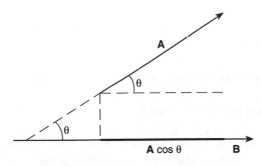

Figure 1.5   The scalar projection.

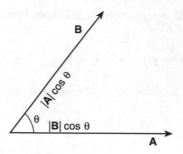

Figure 1.6   Geometrical interpretation of the scalar product.

Thus,

$$\text{proj}_{\mathbf{B}}\mathbf{A} = \frac{\mathbf{A} \cdot \mathbf{B}}{|\mathbf{B}|}.$$

Note: If the angle θ is such that $0 \leq \theta < \pi/2$, the projection is positive and if $\pi/2 < \theta < \pi$, the projection is negative. If $\theta = \pi/2$, the projection is zero and the vectors are perpendicular.

We may then wish to find the *vector projection* of **A** onto **B** in the direction of **B**. That is, we may wish to find a vector parallel to **B**, in the same direction as **B** and of a magnitude that is the absolute value of the scalar projection of **A** onto **B**. We symbolize the vector projection of **A** onto **B** by $\text{proj}_{\mathbf{B}}\mathbf{A}$. To find $\text{proj}_{\mathbf{B}}\mathbf{A}$, we find $|\text{proj}_{\mathbf{B}}\mathbf{A}|$, then multiply this scalar by a unit vector in the direction of **B**. Thus,

$$\begin{aligned}
\text{proj}_{\mathbf{B}}\mathbf{A} &= |\text{proj}_{\mathbf{B}}\mathbf{A}|\frac{\mathbf{B}}{|\mathbf{B}|} \\
&= \frac{|\mathbf{A} \cdot \mathbf{B}|}{|\mathbf{B}|}\frac{\mathbf{B}}{|\mathbf{B}|}. \\
&= \frac{|\mathbf{A} \cdot \mathbf{B}|}{|\mathbf{B}||\mathbf{B}|}\mathbf{B} \\
&= \frac{|\mathbf{A} \cdot \mathbf{B}|}{\mathbf{B} \cdot \mathbf{B}}\mathbf{B}.
\end{aligned}$$

1.5 Cartesian frame of reference

1.5.1 Definition of a position vector R

It is often convenient to be able to refer to vector expressions in a Cartesian frame of reference. To provide for this, we define **i**, **j** and **k** to be vectors of unit length, directed, respectively, along the positive x, y and z axes of a right-handed

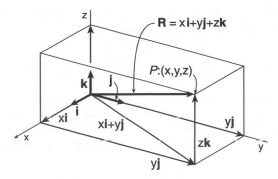

Figure 1.7   Vector representation using the unit vectors **i**, **j** and **k**.

rectangular coordinate system. Then $x\mathbf{i}$, $y\mathbf{j}$ and $z\mathbf{k}$ represent vectors of length $|x|$, $|y|$ and $|z|$ which run parallel to the respective axes. Thus, from the definition of vector addition it is evident that the vector joining the origin to a general point $P(x, y, z)$ (Fig. 1.7) can be written as

$$\mathbf{R} = x\mathbf{i} + y\mathbf{j} + z\mathbf{k},$$

which is usually known as the position vector. In the most general case, we introduce the following definition.

Definition 1.7   *Let* **i**, **j** *and* **k** *be unit vectors in the direction of the positive axes of the rectangular three-dimensional coordinate system. Any vector,* **V**, *of three-dimensional space can be represented in the form:*

$$\mathbf{V} = v_1\mathbf{i} + v_2\mathbf{j} + v_3\mathbf{k},$$

where v_1, v_2 and v_3 are called the components of the vector **V**. *The vectors*

$$v_1\mathbf{i} = v_x, \quad v_2\mathbf{j} = v_y \quad and \quad v_3\mathbf{k} = v_z$$

are called the projection vectors of the vector **V**.

If one of the components of the vector $\mathbf{V} = v_1\mathbf{i} + v_2\mathbf{j} + v_3\mathbf{k}$ is 0, then the vector lies in the plane formed by the other two coordinates axes (the so-called two-dimensional plane).

1.6 Vector algebra using coordinates

In order to solve geometric problems, it is often helpful to represent them algebraically. We now show that vectors can be represented algebraically. Suppose we denote by **AB** the vector from point A to point B. If A and B have coordinates (a_1, a_2, a_3) and (b_1, b_2, b_3) in the coordinate system shown, then the length of **AB** is

$$|\mathbf{AB}| = \sqrt{(b_1 - a_1)^2 + (b_2 - a_2)^2 + (b_3 - a_3)^2}. \tag{1.3}$$

Note that if we start at the point A, proceed $b_1 - a_1$ units in the x-direction, then $b_2 - a_2$ in the y-direction and finally $b_3 - a_3$ in the z-direction, we arrive at B. In other words, the three numbers $b_1 - a_1, b_2 - a_2$ and $b_3 - a_3$ characterize both the direction and the length of the vector joining A to B. This leads to the following definition.

Definition 1.8 *If the tail of a vector* **V** *is at* $A(a_1, a_2, a_3)$ *and its tip is at* $B(b_1, b_2, b_3)$, *then* **V** *shall be represented by the triple of numbers* $b_1 - a_1, b_2 - a_2$ *and* $b_3 - a_3$. *In such a case we enclose the numbers in parenthesis and write*

$$\mathbf{V} = (b_1 - a_1, b_2 - a_2, b_3 - a_3). \tag{1.4}$$

The equal sign in (1.4) means 'is represented by'. The number $b_1 - a_1$ is called the x-*component* of **V**, $b_2 - a_2$ the y-*component* of **V** and $b_3 - a_3$ the z-*component*. Vectors in the xy-plane have only an x- and y-component:

$$\mathbf{V} = (b_1 - a_1, b_2 - a_2), \tag{1.5}$$

where (a_1, a_2) and (b_1, b_2) are the coordinates of the tail and tip of **V**. We now have an algebraic representation for vectors. Each vector has associated with it a set of components that can be found by subtracting the coordinates of its tail from the coordinates of its tip. Conversely, given a set of real numbers (a, b, c), there is one and only one vector with these numbers as components. We now examine the basic operations on vectors using the algebraic representation of vectors.

1. *Addition of vectors with coordinates*: The coordinates of the sum of two vectors are equal to the sum of the respective coordinates. If

    $$\mathbf{A} = a_1\mathbf{i} + a_2\mathbf{j} + a_3\mathbf{k} \quad \text{and} \quad \mathbf{B} = b_1\mathbf{i} + b_2\mathbf{j} + b_3\mathbf{k},$$

    then the sum of the two vectors is given by:

    $$\mathbf{A} + \mathbf{B} = (a_1 + b_1)\mathbf{i} + (a_2 + b_2)\mathbf{j} + (a_3 + b_3)\mathbf{k}.$$

    For example, let $\mathbf{A} = 5\mathbf{i} + \mathbf{j} - 2\mathbf{k}$ and $\mathbf{B} = 3\mathbf{i} - 4\mathbf{j} + 6\mathbf{k}$, then the sum of the two vectors is

    $$\mathbf{A} + \mathbf{B} = 8\mathbf{i} - 3\mathbf{j} + 4\mathbf{k}.$$

2. *Subtraction of vectors with coordinates*: The coordinates of the difference of two vectors are equal to the difference of the respective coordinates. If $\mathbf{A} = a_1\mathbf{i} + a_2\mathbf{j} + a_3\mathbf{k}$ and $\mathbf{B} = b_1\mathbf{i} + b_2\mathbf{j} + b_3\mathbf{k}$, then the difference of the two vectors is:

    $$\mathbf{A} - \mathbf{B} = (a_1 - b_1)\mathbf{i} + (a_2 - b_2)\mathbf{j} + (a_3 - b_3)\mathbf{k}.$$

Using the two vectors from the previous example, the difference of the two is:

$$\mathbf{A} - \mathbf{B} = 2\mathbf{i} + 5\mathbf{j} - 8\mathbf{k}.$$

3. *Multiplication of vectors in coordinate form with scalars*: The coordinates of the product of a scalar, λ, and a vector are the λ-fold of the coordinates of the original vector. If $\mathbf{A} = a_1\mathbf{i} + a_1\mathbf{j} + a_3\mathbf{k}$, then the vector multiplied with a scalar λ is:

$$\lambda\mathbf{A} = \lambda(a_1\mathbf{i} + a_2\mathbf{j} + a_3\mathbf{k}) = \lambda a_1\mathbf{i} + \lambda a_2\mathbf{j} + \lambda a_3\mathbf{k}.$$

Let $\mathbf{A} = -7\mathbf{i} + 4\mathbf{j} + 5\mathbf{k}$, then

$$\frac{2}{3}\mathbf{A} = -\frac{14}{3}\mathbf{i} + \frac{8}{3}\mathbf{j} + \frac{10}{3}\mathbf{k}.$$

Division of vectors by scalars is defined as multiplication with the reciprocal value of the scalar, hence, the coordinates of the quotient of the scalar, λ, and the vector are equal to the $1/\lambda$ fold of the coordinates of the original vector.

4. *Coordinate form of the scalar product*: The scalar product of two vectors is equal to the sum of the products of the corresponding coordinates of the two vectors. If $\mathbf{A} = a_1\mathbf{i} + a_2\mathbf{j} + a_3\mathbf{k}$ and $\mathbf{B} = b_1\mathbf{i} + b_2\mathbf{j} + b_3\mathbf{k}$, then the scalar product of the two vectors is:

$$\mathbf{A} \cdot \mathbf{B} = (a_1\mathbf{i} + a_2\mathbf{j} + a_3\mathbf{k}) \cdot (b_1\mathbf{i} + b_2\mathbf{j} + b_3\mathbf{k}) = a_1b_1 + a_2b_2 + a_3b_3.$$

Recall that two vectors will be equal if and only if their respective components are equal. It follows that any vector equation implies three scalar equations. Then, since the dot product of perpendicular vectors (unit vectors in this case) is zero, it follows that

$$\mathbf{i} \cdot \mathbf{j} = \mathbf{j} \cdot \mathbf{k} = \mathbf{k} \cdot \mathbf{i} = 0.$$

Moreover, applying (1.1) to unit vectors, $\mathbf{i}$, $\mathbf{j}$ and $\mathbf{k}$, we have

$$\mathbf{i} \cdot \mathbf{i} = \mathbf{j} \cdot \mathbf{j} = \mathbf{k} \cdot \mathbf{k} = 1.$$

As an example, let $\mathbf{A} = 5\mathbf{i} + \mathbf{j} - 2\mathbf{k}$ and $\mathbf{B} = 3\mathbf{i} - 4\mathbf{j} + 6\mathbf{k}$. The scalar product of the two vectors is:

$$\mathbf{A} \cdot \mathbf{B} = 5 \cdot 3 + 1 \cdot (-4) + (-2) \cdot 6 = 15 - 4 - 12 = -1.$$

In the particular case that $\mathbf{A} = \mathbf{B}$, we have

$$\mathbf{A} \cdot \mathbf{A} = |\mathbf{A}|^2 = A^2 = a_1^2 + a_2^2 + a_3^2$$

or

$$|\mathbf{A}| = A = \sqrt{a_1^2 + a_2^2 + a_3^2}.$$

However, if the dot product is written as $\mathbf{A} \cdot \mathbf{B} = |\mathbf{A}||\mathbf{B}| \cos \theta$, then solved for $\cos \theta$, using the above expressions, we obtain the useful formula

$$\cos \theta = \frac{a_1 b_1 + a_2 b_2 + a_3 b_3}{\left(\sqrt{a_1^2 + a_2^2 + a_3^2} \right) \left(\sqrt{b_1^2 + b_2^2 + b_3^2} \right)},$$

which is the familiar analytic geometry result.

5. *Coordinate form of the vector product*: For the cross product of unit vectors, **i**, **j** and **k** we find that

$$\mathbf{i} \times \mathbf{i} = \mathbf{j} \times \mathbf{j} = \mathbf{k} \times \mathbf{k} = 0,$$

$$\mathbf{i} \times \mathbf{j} = -\mathbf{j} \times \mathbf{i} = \mathbf{k},$$

$$\mathbf{j} \times \mathbf{k} = -\mathbf{k} \times \mathbf{j} = \mathbf{i},$$

$$\mathbf{k} \times \mathbf{i} = -\mathbf{i} \times \mathbf{k} = \mathbf{j}.$$

Using these products along with the distributive rule, we get the vector product

$$\mathbf{A} \times \mathbf{B} = (a_1\mathbf{i} + a_2\mathbf{j} + a_3\mathbf{k}) \times (b_1\mathbf{i} + b_2\mathbf{j} + b_3\mathbf{k})$$

$$= (a_2 b_3 - a_3 b_2)\mathbf{i} + (a_3 b_1 - a_1 b_3)\mathbf{j} + (a_1 b_2 - a_2 b_1)\mathbf{k}. \quad (1.6)$$

This expression may be written in compact form as a determinant which is easier to remember:

$$\mathbf{A} \times \mathbf{B} = \begin{vmatrix} \mathbf{i} & \mathbf{j} & \mathbf{k} \\ a_1 & a_2 & a_3 \\ b_1 & b_2 & b_3 \end{vmatrix}. \quad (1.7)$$

Suppose we let $\mathbf{V}_1 = 5\mathbf{i} + \mathbf{j} - 2\mathbf{k}$ and $\mathbf{V}_2 = 3\mathbf{i} - 4\mathbf{j} + 6\mathbf{k}$. The vector product of these vectors is as follows:

$$\mathbf{V}_1 \times \mathbf{V}_2 = \begin{vmatrix} \mathbf{i} & \mathbf{j} & \mathbf{k} \\ 5 & 1 & -2 \\ 3 & -4 & 6 \end{vmatrix} = (6 - 8)\mathbf{i} - (30 + 6)\mathbf{j} + (-20 - 3)\mathbf{k}$$

$$= -2\mathbf{i} - 36\mathbf{j} - 23\mathbf{k}.$$

Note that the anti-commutative character of the vector multiplication corresponds to the fact that interchanging two rows of a determinant changes the sign of the determinant.

1.7 Mixed product in coordinate form

Products involving three, rather than two, vectors offer the following possibilities:

$$(\mathbf{A} \cdot \mathbf{B})\mathbf{C}, \quad \mathbf{A} \cdot (\mathbf{B} \times \mathbf{C}), \quad \mathbf{A} \times (\mathbf{B} \times \mathbf{C}).$$

1. $(\mathbf{A} \cdot \mathbf{B})\mathbf{C}$.

   This product can be defined as a vector, since $\mathbf{A} \cdot \mathbf{B}$ is a scalar and thus $(\mathbf{A} \cdot \mathbf{B})\mathbf{C}$ is a vector whose length is $|\mathbf{A} \cdot \mathbf{B}|$ times the length of $\mathbf{C}$ and whose direction is the same as that of $\mathbf{C}$ or opposite to it, according to whether $\mathbf{A} \cdot \mathbf{B}$ is positive or negative.

2. $\mathbf{A} \cdot (\mathbf{B} \times \mathbf{C})$.

   This product is known as a *scalar triple product* and we observe that the parentheses enclosing the vector $\mathbf{B} \times \mathbf{C}$ are superfluous. There is, in fact, only one alternative interpretation, namely $(\mathbf{A} \cdot \mathbf{B}) \times \mathbf{C}$, and this is meaningless since both factors in a cross product must be vectors, whereas $\mathbf{A} \cdot \mathbf{B}$ is a scalar. Therefore, no meaning except the one intended can be attached to the expression $\mathbf{A} \cdot \mathbf{B} \times \mathbf{C}$ and hence it is customary to omit the parentheses. If the vectors

$$\mathbf{A} = a_1\mathbf{i} + a_2\mathbf{j} + a_3\mathbf{k},$$
$$\mathbf{B} = b_1\mathbf{i} + b_2\mathbf{j} + b_3\mathbf{k},$$
$$\mathbf{C} = c_1\mathbf{i} + c_2\mathbf{j} + c_3\mathbf{k},$$

then the mixed product of the three vectors is

$$\mathbf{A} \cdot (\mathbf{B} \times \mathbf{C}) = (a_1\mathbf{i} + a_2\mathbf{j} + a_3\mathbf{k}) \cdot \begin{vmatrix} \mathbf{i} & \mathbf{j} & \mathbf{k} \\ b_1 & b_2 & b_3 \\ c_1 & c_2 & c_3 \end{vmatrix}.$$

From this, it can easily be shown that:

$$\mathbf{A} \cdot (\mathbf{B} \times \mathbf{C}) = \begin{vmatrix} a_1 & a_2 & a_3 \\ b_1 & b_2 & b_3 \\ c_1 & c_2 & c_3 \end{vmatrix}. \tag{1.8}$$

From (1.8) and the fact that interchanging any two rows in a determinant changes the sign of the determinant, it follows easily that $\mathbf{A} \cdot (\mathbf{B} \times \mathbf{C}) = (\mathbf{A} \times \mathbf{B}) \cdot \mathbf{C}$. In other words, in any scalar product, the dot and cross can be interchanged without altering the value of the product. For this reason, it is customary to omit these symbols and write a scalar triple product simply as $[\mathbf{ABC}]$. Again, either from (1.8) or directly from the anticommutative character of vector products, it is easy to see that $\mathbf{A} \cdot (\mathbf{B} \times \mathbf{C}) = -\mathbf{A} \cdot (\mathbf{C} \times \mathbf{B})$ and more generally,

$$[\mathbf{ABC}] = [\mathbf{BCA}] = [\mathbf{CAB}] = -[\mathbf{ACB}] = -[\mathbf{BAC}] = -[\mathbf{CBA}]. \tag{1.9}$$

Thus (1.9) asserts that any cyclic permutation of factors in a scalar triple product leaves the value of the product unchanged, whereas any permutation which reverses the original cyclic order changes the sign of the product. Geometrically, the absolute value of the triple product $\mathbf{A} \cdot (\mathbf{B} \times \mathbf{C})$ is the volume of the parallelepiped having vectors $\mathbf{A}$, $\mathbf{B}$ and $\mathbf{C}$ as concurrent edges. If we regard the parallelogram having $\mathbf{B}$ and $\mathbf{C}$ as adjacent sides as the base of this figure, then $\mathbf{B} \times \mathbf{C}$ is a vector perpendicular to the base and whose magnitude is the projection of $\mathbf{A}$ on $\mathbf{B} \times \mathbf{C}$ (see Fig. 1.8).

Hence, $\mathbf{A} \cdot (\mathbf{B} \times \mathbf{C})$, whose value is the magnitude of $\mathbf{B} \times \mathbf{C}$ multiplied by the projection of $\mathbf{A}$ on $\mathbf{B} \times \mathbf{C}$, is numerically equal to the volume of the parallelepiped. If the vectors, $\mathbf{A}$, $\mathbf{B}$ and $\mathbf{C}$ all lie in the same plane or are parallel to the same plane, they necessarily form a parallelepiped of zero volume and conversely. Hence, $[\mathbf{ABC}] = 0$ is a necessary and sufficient condition that the three vectors $\mathbf{A}$, $\mathbf{B}$ and $\mathbf{C}$ should be parallel to one and the same plane, i.e. coplanar. In particular, if two factors of a scalar triple product have the same direction, the product is zero.

The mixed product of the vectors $\mathbf{A} = 2\mathbf{i} - \mathbf{j} + 4\mathbf{k}$, $\mathbf{B} = -\mathbf{i} + 3\mathbf{j} + \mathbf{k}$, $\mathbf{C} = -5\mathbf{i} + 2\mathbf{j} - \mathbf{k}$ is as follows:

$$[\mathbf{ABC}] = \begin{vmatrix} 2 & -1 & 4 \\ -1 & 3 & 1 \\ -5 & 2 & -1 \end{vmatrix} = 2(-3 - 2) + 1(1 + 5) + 4(-2 + 15) = 48.$$

Definition 1.9　*If the line vectors* $\mathbf{A}$, $\mathbf{B}$ *and* $\mathbf{C}$ *are the concurrent edges of a parallelepiped, then the scalar triple product*

$$\mathbf{A} \cdot (\mathbf{B} \times \mathbf{C}) = \begin{vmatrix} a_1 & a_2 & a_3 \\ b_1 & b_2 & b_3 \\ c_1 & c_2 & c_3 \end{vmatrix} = [\mathbf{ABC}].$$

is defined as the volume of the parallelepiped.

3.　$\mathbf{A} \times (\mathbf{B} \times \mathbf{C})$.

An indicated product of this form is meaningful because, by definition, $\mathbf{B} \times \mathbf{C}$ is a vector and, hence, is an appropriate factor in a vector product with $\mathbf{A}$.

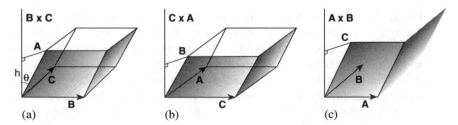

Figure 1.8　Geometrical interpretation of the scalar triple product.

To obtain a simpler, expanded form for a product such as $\mathbf{A} \times (\mathbf{B} \times \mathbf{C})$, we first recall the expansion for $\mathbf{B} \times \mathbf{C}$:

$$\mathbf{B} \times \mathbf{C} = \begin{vmatrix} \mathbf{i} & \mathbf{j} & \mathbf{k} \\ b_1 & b_2 & b_3 \\ c_1 & c_2 & c_3 \end{vmatrix}$$

$$= (b_2 c_3 - b_3 c_2)\mathbf{i} + (b_3 c_1 - b_1 c_3)\mathbf{j} + (b_1 c_2 - b_2 c_1)\mathbf{k},$$

then use (1.7) to express the vector product of $\mathbf{A}$ and $\mathbf{B} \times \mathbf{C}$:

$$\mathbf{A} \times (\mathbf{B} \times \mathbf{C}) = \begin{vmatrix} \mathbf{i} & \mathbf{j} & \mathbf{k} \\ a_1 & a_2 & a_3 \\ (b_2 c_3 - b_3 c_2) & (b_3 c_1 - b_1 c_3) & (b_1 c_2 - b_2 c_1) \end{vmatrix}.$$

If we expand this determinant in terms of the (vector) elements of the first row and group positive and negative terms in each cofactor, we obtain

$$
\begin{aligned}
\text{RHS} = {} & \mathbf{i}[a_2(b_1 c_2 - b_2 c_1) - a_3(b_3 c_1 - b_1 c_3)] \\
& + \mathbf{j}[a_3(b_2 c_3 - b_3 c_2) - a_1(b_1 c_2 - b_2 c_1)] \\
& + \mathbf{k}[a_1(b_3 c_1 - b_1 c_3) - a_2(a_2 c_3 - a_3 c_2)] \\
= {} & \mathbf{i}[b_1(a_2 c_2 + a_3 c_3) - c_1(a_2 b_2 + a_3 b_3)] \\
& + \mathbf{j}[b_2(a_3 c_3 + a_1 c_1) - c_2(a_3 b_3 + a_1 b_1)] \\
& + \mathbf{k}[b_3(a_1 c_1 + a_2 c_2) - c_3(a_1 b_1 + a_2 b_2)].
\end{aligned}
$$

If we now add and subtract the products $a_1 b_1 c_1$, $a_2 b_2 c_2$, and $a_3 b_3 c_3$ in the respective bracketed expressions and note the dot products we have thus formed, this becomes

$$
\begin{aligned}
\text{RHS} = {} & \mathbf{i}[b_1(\mathbf{A} \cdot \mathbf{C}) - c_1(\mathbf{A} \cdot \mathbf{B})] \\
& + \mathbf{j}[b_2(\mathbf{A} \cdot \mathbf{C}) - c_2(\mathbf{A} \cdot \mathbf{B})] \\
& + \mathbf{k}[b_3(\mathbf{A} \cdot \mathbf{C}) - c_3(\mathbf{A} \cdot \mathbf{B})] \\
= {} & (\mathbf{A} \cdot \mathbf{C})(b_1 \mathbf{i} + b_2 \mathbf{j} + b_3 \mathbf{k}) \\
& - (\mathbf{A} \cdot \mathbf{B})(c_1 \mathbf{i} + c_2 \mathbf{j} + c_3 \mathbf{k}) \\
= {} & (\mathbf{A} \cdot \mathbf{C})\mathbf{B} - (\mathbf{A} \cdot \mathbf{B})\mathbf{C}.
\end{aligned}
$$

Finally, by regrouping and factoring, this becomes the required formula:

$$\mathbf{A} \times (\mathbf{B} \times \mathbf{C}) = (\mathbf{A} \cdot \mathbf{C})\mathbf{B} - (\mathbf{A} \cdot \mathbf{B})\mathbf{C}. \qquad (1.10)$$

A straightforward application of (1.10) leads to

$$(\mathbf{A} \times \mathbf{B}) \times \mathbf{C} = -\mathbf{C} \times (\mathbf{A} \times \mathbf{B}) = -(\mathbf{C} \cdot \mathbf{B})\mathbf{A} + (\mathbf{C} \cdot \mathbf{A})\mathbf{B},$$

which is *not* equal to $\mathbf{A} \times (\mathbf{B} \times \mathbf{C})$. Hence, as mentioned before vector multiplication is not associative.

With a knowledge of scalar and vector triple products, products involving more than three vectors can be expanded without difficulty. For example,

$$(\mathbf{A} \times \mathbf{B}) \cdot (\mathbf{C} \times \mathbf{D})$$

can be regarded as a scalar triple product of the vectors, $\mathbf{A}$, $\mathbf{B}$ and $\mathbf{C} \times \mathbf{D}$. This allows us to write

$$(\mathbf{A} \times \mathbf{B}) \cdot (\mathbf{C} \times \mathbf{D}) = \mathbf{A} \cdot [\mathbf{B} \times (\mathbf{C} \times \mathbf{D})]$$
$$= \mathbf{A} \cdot [(\mathbf{B} \cdot \mathbf{D})\mathbf{C} - (\mathbf{B} \cdot \mathbf{C})\mathbf{D}]$$
$$= (\mathbf{A} \cdot \mathbf{C})(\mathbf{B} \cdot \mathbf{D}) - (\mathbf{A} \cdot \mathbf{D})(\mathbf{B} \cdot \mathbf{C}).$$

This result is referred to as *Lagrange's* identity. Similarly, $(\mathbf{A} \times \mathbf{B}) \times (\mathbf{C} \times \mathbf{D})$ can be thought of as the vector triple product of $\mathbf{A} \times \mathbf{B}$, $\mathbf{C}$ and $\mathbf{D}$ (Fig. 1.9). Then applying (1.10), we find

$$(\mathbf{A} \times \mathbf{B}) \times (\mathbf{C} \times \mathbf{D}) = (\mathbf{A} \times \mathbf{B} \cdot \mathbf{D})\mathbf{C} - (\mathbf{A} \times \mathbf{B} \cdot \mathbf{C})\mathbf{D}$$
$$= [\mathbf{ABD}]\mathbf{C} - [\mathbf{ABC}]\mathbf{D},$$

which is a vector in the plane of $\mathbf{C}$ and $\mathbf{D}$. From another point of view, we have

$$(\mathbf{A} \times \mathbf{B}) \times (\mathbf{C} \times \mathbf{D}) = -(\mathbf{C} \times \mathbf{D}) \times (\mathbf{A} \times \mathbf{B})$$
$$= -[(\mathbf{C} \times \mathbf{D} \cdot \mathbf{B})\mathbf{A} - (\mathbf{C} \times \mathbf{D} \cdot \mathbf{A})\mathbf{B}]$$
$$= -(\mathbf{C} \times \mathbf{D} \cdot \mathbf{B})\mathbf{A} + (\mathbf{C} \times \mathbf{D} \cdot \mathbf{A})\mathbf{B}$$
$$= -[\mathbf{CDB}]\mathbf{A} + [\mathbf{CDA}]\mathbf{B},$$

which is a vector in the plane of $\mathbf{A}$ and $\mathbf{B}$. These two results together show that $(\mathbf{A} \times \mathbf{B}) \times (\mathbf{C} \times \mathbf{D})$ is directed along the line of intersection of the plane of $\mathbf{A}$ and $\mathbf{B}$ and the plane of $\mathbf{C}$ and $\mathbf{D}$, provided these planes are distinct.

Example 1.1 If $\mathbf{A}$, $\mathbf{B}$ and $\mathbf{C}$ are any three vectors, prove that

$$\mathbf{A} \times (\mathbf{B} \times \mathbf{C}) + \mathbf{B} \times (\mathbf{C} \times \mathbf{A}) + \mathbf{C} \times (\mathbf{A} \times \mathbf{B}) = \mathbf{0}.$$

$\square$

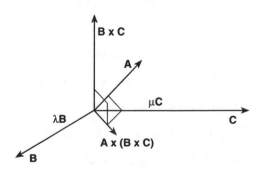

Figure 1.9 Geometric interpretation of the vector triple product.

Solution   Using Lagrange's identity, the left-hand side of the above can be expressed as

$$\begin{aligned}
\text{LHS} &= \mathbf{A} \times (\mathbf{B} \times \mathbf{C}) + \mathbf{B} \times (\mathbf{C} \times \mathbf{A}) + \mathbf{C} \times (\mathbf{A} \times \mathbf{B}) \\
&= [(\mathbf{A} \cdot \mathbf{C})\mathbf{B} - (\mathbf{A} \cdot \mathbf{B})\mathbf{C}] + [(\mathbf{B} \cdot \mathbf{A})\mathbf{C} - (\mathbf{B} \cdot \mathbf{C})\mathbf{A}] \\
&\quad + [(\mathbf{C} \cdot \mathbf{B})\mathbf{A} - (\mathbf{C} \cdot \mathbf{A})\mathbf{B}] \\
&= \mathbf{0} \\
&= \text{RHS}.
\end{aligned}$$

Hence the required proof.   ■

Example 1.2   Prove that

$$(\mathbf{A} \times \mathbf{B}) \cdot (\mathbf{C} \times \mathbf{D}) + (\mathbf{B} \times \mathbf{C}) \cdot (\mathbf{A} \times \mathbf{D}) + (\mathbf{C} \times \mathbf{A}) \cdot (\mathbf{B} \times \mathbf{D}) = 0.$$

☐

Solution   Again, we use Lagrange's identity as follows,

$$\begin{aligned}
\text{LHS} &= (\mathbf{A} \cdot \mathbf{C})(\mathbf{B} \cdot \mathbf{D}) - (\mathbf{A} \cdot \mathbf{D})(\mathbf{B} \cdot \mathbf{C}) \\
&\quad + (\mathbf{B} \cdot \mathbf{A})(\mathbf{C} \cdot \mathbf{D}) - (\mathbf{B} \cdot \mathbf{D})(\mathbf{C} \cdot \mathbf{A}) \\
&\quad + (\mathbf{C} \cdot \mathbf{B})(\mathbf{A} \cdot \mathbf{D}) - (\mathbf{C} \cdot \mathbf{D})(\mathbf{A} \cdot \mathbf{B}) \\
&= 0 \\
&= \text{RHS}.
\end{aligned}$$

Hence the required proof.   ■

Example 1.3   Prove that

$$(\mathbf{A} \times \mathbf{B}) \cdot (\mathbf{B} \times \mathbf{C}) \times (\mathbf{C} \times \mathbf{A}) = [\mathbf{ABC}]^2.$$

☐

Solution   Using Lagrange's identity, we can write the left-hand side as

$$\begin{aligned}
\text{LHS} &= (\mathbf{A} \times \mathbf{B}) \cdot (\mathbf{B} \times \mathbf{C}) \times (\mathbf{C} \times \mathbf{A}) \\
&= (\mathbf{A} \times \mathbf{B}) \cdot [(\mathbf{B} \times \mathbf{C} \cdot \mathbf{A})\mathbf{C} - (\mathbf{B} \times \mathbf{C} \cdot \mathbf{C})\mathbf{A}] \\
&= (\mathbf{A} \times \mathbf{B}) \cdot [(\mathbf{B} \times \mathbf{C} \cdot \mathbf{A})\mathbf{C} - 0] \\
&= [\mathbf{A} \times \mathbf{B} \cdot \mathbf{C}][\mathbf{B} \times \mathbf{C} \cdot \mathbf{A}] \\
&= [\mathbf{ABC}][\mathbf{ABC}] = [\mathbf{ABC}]^2 \\
&= \text{RHS}.
\end{aligned}$$

Hence the required proof.   ■

Example 1.4   Verify that if

$$A = ia_1 + ja_2 + ka_3,$$
$$B = ib_1 + jb_2 + kb_3,$$
$$C = ic_1 + jc_2 + kc_3$$

and

$$D = id_1 + jd_2 + kd_3,$$

then the system of equations

$$a_1x + b_1y + c_1z = d_1,$$
$$a_2x + b_2y + c_2z = d_2,$$
$$a_3x + b_3y + c_3z = d_3$$

is equivalent to the single vector equation

$$xA + yB + zC = D.$$

Assuming that $[ABC] \neq 0$, solve this vector equation for x, y and z. Show that the result is equivalent to that obtained from the algebraic form of the system by using Cramer's rule.   □

Solution   We can easily establish the linear system if we take the dot product of $xA + yB + zC = D$ with the unit vectors i, j and k.

$$i \cdot (xA + yB + zC) = i \cdot D,$$
$$a_1x + b_1y + c_1z = d_1.$$

Similarly, we can establish the other two equations. To find the solution x, y and z of the vector equation, we proceed as follows. First, we cross multiply by C to obtain

$$xA \times C + yB \times C + zC \times C = D \times C.$$

Since we know $C \times C = 0$,

$$xA \times C + yB \times C = D \times C.$$

Now, by scalar multiplication of the resulting equation by the vector B, we get

$$xA \times C \cdot B + yB \times C \cdot B = D \times C \cdot B,$$
$$xA \times C \cdot B = D \times C \cdot B,$$

since we know that $\mathbf{B} \times \mathbf{C} \cdot \mathbf{B} = 0$. Therefore, we get

$$x\mathbf{A} \times \mathbf{C} \cdot \mathbf{B} = \mathbf{D} \times \mathbf{C} \cdot \mathbf{B},$$

$$x\mathbf{A} \cdot \mathbf{B} \times \mathbf{C} = \mathbf{D} \cdot \mathbf{B} \times \mathbf{C}, \quad \text{using cyclic permutation property}$$

$$x[\mathbf{ABC}] = [\mathbf{DBC}]$$

$$x = \frac{[\mathbf{DBC}]}{[\mathbf{ABC}]}.$$

Similarly, we can obtain the values for y and z as

$$y = \frac{[\mathbf{ADC}]}{[\mathbf{ABC}]}, \quad z = \frac{[\mathbf{ABD}]}{[\mathbf{ABC}]}.$$

Solving the 3×3 system by Cramer's rule, we get

$$x = \frac{\begin{vmatrix} d_1 & b_1 & c_1 \\ d_2 & b_2 & c_2 \\ d_3 & b_3 & c_3 \end{vmatrix}}{\begin{vmatrix} a_1 & b_1 & c_1 \\ a_2 & b_2 & c_2 \\ a_3 & b_3 & c_3 \end{vmatrix}}.$$

From the vector method, we have

$$x = \frac{\begin{vmatrix} d_1 & d_2 & d_3 \\ b_1 & b_2 & b_3 \\ c_1 & c_2 & c_3 \end{vmatrix}}{\begin{vmatrix} a_1 & a_2 & a_3 \\ b_1 & b_2 & b_3 \\ c_1 & c_2 & c_3 \end{vmatrix}},$$

using the property of determinants in changing the columns to rows. The value of the determinant will not be changed if we transpose the rows and columns. ■

Example 1.5 Using vector methods, determine the law of cosines. □

Solution To obtain the law of cosines, let us assume that the sides of the given triangle are $\mathbf{A}$, $\mathbf{B}$ and $\mathbf{C}$ such that $\mathbf{C} = \mathbf{A} - \mathbf{B}$. Then by taking the dot product we have the following:

$$\mathbf{C} \cdot \mathbf{C} = (\mathbf{A} - \mathbf{B}) \cdot (\mathbf{A} - \mathbf{B})$$

$$= \mathbf{A} \cdot \mathbf{A} - 2\mathbf{A} \cdot \mathbf{B} + \mathbf{B} \cdot \mathbf{B}$$

$$= |A|^2 - 2|A||B|\cos\theta + |B||B|,$$

which yields

$$C^2 = A^2 - 2AB \cos\theta + B^2.$$

Here θ is the angle between **A** and **B**. This is the required law of cosines, which the reader is no doubt familiar with.   ■

Example 1.6   Find the unit vector, which is perpendicular to both $\mathbf{A} = \mathbf{i} - 2\mathbf{j} + \mathbf{k}$ and $\mathbf{B} = 3\mathbf{i} + \mathbf{j} - 2\mathbf{k}$.   □

Solution   We know that the cross product of two vectors is a vector which is perpendicular to both the original vectors. This means that we can write

$$\mathbf{A} \times \mathbf{B} = \begin{vmatrix} \mathbf{i} & \mathbf{j} & \mathbf{k} \\ 1 & -2 & 1 \\ 3 & 1 & -2 \end{vmatrix}$$

$$= 3\mathbf{i} + 5\mathbf{j} + 7\mathbf{k}$$

$$= \mathbf{C},$$

which is a vector perpendicular to both **A** and **B**. Then the unit vector of **C** is given by

$$\frac{\mathbf{C}}{|\mathbf{C}|} = \frac{3\mathbf{i} + 5\mathbf{j} + 7\mathbf{k}}{\sqrt{9 + 25 + 49}}$$

$$= \frac{3\mathbf{i} + 5\mathbf{j} + 7\mathbf{k}}{\sqrt{83}}$$

$$= \frac{3}{\sqrt{83}}\mathbf{i} + \frac{5}{\sqrt{83}}\mathbf{j} + \frac{7}{\sqrt{83}}\mathbf{k}.$$

■

Example 1.7   Find the unit vector which is parallel to the plane of $\mathbf{A} = \mathbf{i} + \mathbf{j} - 2\mathbf{k}$ and $\mathbf{B} = 3\mathbf{i} - 2\mathbf{j} + \mathbf{k}$ and perpendicular to $\mathbf{C} = 2\mathbf{i} + 2\mathbf{j} - \mathbf{k}$.   □

Solution   We know that the cross product of the vectors **A** and **B** is a vector perpendicular to both **A** and **B**, i.e. $\mathbf{A} \times \mathbf{B}$. We need to determine a vector which is perpendicular to both $\mathbf{A} \times \mathbf{B}$ and **C** such that the resulting vector will be parallel to **A** and **B** by the vector cross product. That is to say,

$$(\mathbf{A} \times \mathbf{B}) \times \mathbf{C} = -\mathbf{C} \times (\mathbf{A} \times \mathbf{B})$$

$$= -(\mathbf{C} \cdot \mathbf{B})\mathbf{A} + (\mathbf{C} \cdot \mathbf{A})\mathbf{B}$$

$$= -(6 - 4 - 1)(\mathbf{i} + \mathbf{j} - 2\mathbf{k}) + (2 + 2 + 2)(3\mathbf{i} - 2\mathbf{j} + \mathbf{k})$$

$$= -(\mathbf{i} + \mathbf{j} - 2\mathbf{k}) + 6(3\mathbf{i} - 2\mathbf{j} + \mathbf{k})$$

$$= 17\mathbf{i} - 13\mathbf{j} + 8\mathbf{k}.$$

Therefore, the required unit vector is given by

$$\frac{17\mathbf{i} - 13\mathbf{j} + 8\mathbf{k}}{\sqrt{17^2 + 13^2 + 8^2}} = \frac{17\mathbf{i} - 13\mathbf{j} + 8\mathbf{k}}{\sqrt{522}}.$$

■

1.8 Vector representation using coordinates

1. *Absolute value of a vector*: The absolute value of a vector – according to the definition – is the square root of the sum of the squares of the coordinates

$$|\mathbf{A}| = \sqrt{|\mathbf{A}|^2} = \sqrt{\mathbf{A} \cdot \mathbf{A}}$$
$$= \sqrt{(a_1\mathbf{i} + a_2\mathbf{j} + a_3\mathbf{k}) \cdot (a_1\mathbf{i} + a_2\mathbf{j} + a_3\mathbf{k})}$$
$$= \sqrt{a_1^2 + a_2^2 + a_3^2}.$$

The absolute value of the vector $\mathbf{A} = 4\mathbf{i} + 2\mathbf{j} - 3\mathbf{k}$ is

$$|\mathbf{A}| = \sqrt{16 + 4 + 9} = \sqrt{29}.$$

If $\mathbf{e}$ is a unit vector, i.e. $|\mathbf{e}| = 1$, then

$$\sqrt{e_1^2 + e_2^2 + e_3^2} = 1$$

and so $e_1^2 + e_2^2 + e_3^2 = 1$. This means that the sum of the squares of the coordinates of a unit vector is equal to one.

The vector $\mathbf{B} = \frac{1}{2}\mathbf{i} - \frac{5}{6}\mathbf{j} + \frac{\sqrt{2}}{6}\mathbf{k}$ is a unit vector because

$$e_1^2 + e_2^2 + e_3^2 = \frac{9}{36} + \frac{25}{36} + \frac{2}{36} = 1.$$

2. *Unit vectors represented by coordinates*: A unit vector pointing in the direction of an arbitrary vector $\mathbf{B} = 5\mathbf{i} - 3\mathbf{j} + \mathbf{k}$ is

$$\mathbf{e_B} = \frac{5\mathbf{i} - 3\mathbf{j} + \mathbf{k}}{\sqrt{25 + 9 + 1}} = \frac{5}{\sqrt{35}}\mathbf{i} - \frac{3}{\sqrt{35}}\mathbf{j} + \frac{1}{\sqrt{35}}\mathbf{k} = \frac{\mathbf{B}}{|\mathbf{B}|}.$$

3. *Direction cosines of a vector*: The cosines of the angles enclosed by a vector and the coordinate axes (respectively, the unit vectors $\mathbf{i}$, $\mathbf{j}$ and $\mathbf{k}$) are called the direction cosines of the vector (see Fig. 1.10).

Denoting the angles enclosed by the vector $\mathbf{A} = a_1\mathbf{i} + a_2\mathbf{j} + a_3\mathbf{k}$ and the unit vectors $\mathbf{i}$, $\mathbf{j}$ and $\mathbf{k}$ by α, β and γ, respectively, we can easily find the direction cosines of the line vector by using the dot products as follows

$$\mathbf{i} \cdot \mathbf{A} = |\mathbf{i}||\mathbf{A}| \cos \alpha$$

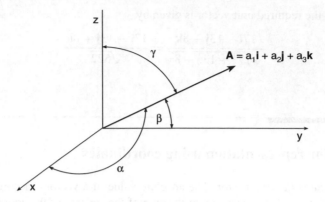

Figure 1.10   Direction cosines of a vector.

or

$$\cos \alpha = \frac{a_1}{|\mathbf{A}|} = \frac{a_1}{\sqrt{a_1^2 + a_2^2 + a_3^2}}.$$

Similarly, we can find that

$$\cos \beta = \frac{a_2}{|\mathbf{A}|} = \frac{a_2}{\sqrt{a_1^2 + a_2^2 + a_3^2}},$$

$$\cos \gamma = \frac{a_3}{|\mathbf{A}|} = \frac{a_3}{\sqrt{a_1^2 + a_2^2 + a_3^2}}.$$

Thus, for the vector $\mathbf{A} = a_1\mathbf{i} + a_2\mathbf{j} + a_3\mathbf{k}$, the direction cosines are given by

$$\cos \alpha = \frac{a_1}{\sqrt{a_1^2 + a_2^2 + a_3^2}},$$

$$\cos \beta = \frac{a_2}{\sqrt{a_1^2 + a_2^2 + a_3^2}},$$

and

$$\cos \gamma = \frac{a_3}{\sqrt{a_1^2 + a_2^2 + a_3^2}}.$$

Hence, the direction cosines of $\mathbf{A}$ are proportional to a_1, a_2 and a_3, respectively.

The direction cosines of an arbitrary vector are identical with the direction cosines of the unit vector pointing in its direction. As the sum of the squares of the coordinates of a unit vector is one, the sum of the squares of the direction cosines of any direction is also one:

$$\cos^2 \alpha + \cos^2 \beta + \cos^2 \gamma = 1.$$

The direction cosines of the vector $\mathbf{A} = 2\mathbf{i} - \mathbf{j} + \mathbf{k}$ are the following:

$$\cos \alpha = \frac{2}{\sqrt{4+1+1}} = \frac{2}{\sqrt{6}},$$

$$\cos \beta = \frac{-1}{\sqrt{4+1+1}} = \frac{-1}{\sqrt{6}},$$

$$\cos \gamma = \frac{1}{\sqrt{4+1+1}} = \frac{1}{\sqrt{6}}.$$

1.9 Lines and planes using vector algebra

1.9.1 Lines

A straight line can be determined by one of its points and its direction (i.e. the direction vector that is parallel to the straight line) or by two of its points.

1. *Equation of a straight line determined by a point and a direction*: The vector equation of a straight line passing through the point $P_0(x_0, y_0, z_0)$ and parallel to the vector $\mathbf{C} = a\mathbf{i} + b\mathbf{j} + c\mathbf{k}$ can be determined as follows:

Definition 1.10   *If we denote a moving point of the straight line by $P(x, y, z)$, then for the points of the straight line passing through the point P_0 in the direction $\mathbf{C}$ (and only for these points) the following equality holds:*

$$\mathbf{P_0P} \times \mathbf{C} = \mathbf{0}.$$

If we denote the radius of an arbitrary point of the straight line by $\mathbf{r}$, and the radius vector leading up to P_0 by $\mathbf{r}_0$, then

$$\mathbf{P_0P} = \mathbf{r} - \mathbf{r}_0.$$

Thus, the vector equation of the straight line is given by:

$$(\mathbf{r} - \mathbf{r}_0) \times \mathbf{C} = \mathbf{0}.$$

Expressed by coordinates:

$$\mathbf{r} - \mathbf{r}_0 = (x - x_0)\mathbf{i} + (y - y_0)\mathbf{j} + (z - z_0)\mathbf{k}$$

and

$$C = a\mathbf{i} + b\mathbf{j} + c\mathbf{k},$$

thus,

$$(\mathbf{r} - \mathbf{r}_0) \times \mathbf{C} = \begin{vmatrix} \mathbf{i} & \mathbf{j} & \mathbf{k} \\ (x - x_0) & (y - y_0) & (z - z_0) \\ a & b & c \end{vmatrix}$$

$$= [(y - y_0)c - (z - z_0)b]\mathbf{i} - [(x - x_0)c - (z - z_0)a]\mathbf{j}$$
$$+ [(x - x_0)b - (y - y_0)a]\mathbf{k} = \mathbf{0}.$$

Hence, the vector equation of the straight line expressed by coordinates is:

$$[(y - y_0)c - (z - z_0)b]\mathbf{i} + [(z - z_0)a - (x - x_0)c]\mathbf{j}$$
$$+ [(x - x_0)b - (y - y_0)a]\mathbf{k} = \mathbf{0}.$$

Recall that a vector is zero if and only if all its coordinates are zero, i.e.

$$(y - y_0)c - (z - z_0)b = 0,$$
$$(z - z_0)a - (x - x_0)c = 0,$$
$$(x - x_0)b - (y - y_0)a = 0,$$

provided none of a, b or c is zero. After reduction, these become:

$$\frac{x - x_0}{a} = \frac{y - y_0}{b} = \frac{z - z_0}{c}. \qquad (1.11)$$

This is the scalar system for equations of the straight line. Among the three equations of this system, only two are independent. If any two of them hold, the third one follows.

Example 1.8   Find the equation or system of equations of the straight line passing through the point $P_0(-2, 3, 5)$ and running in the direction $\mathbf{C} = 4\mathbf{i} - \mathbf{j} + 6\mathbf{k}$. □

Solution   The vector equation of the straight line is:

$$(\mathbf{r} - \mathbf{r}_0) \times \mathbf{C} = \begin{vmatrix} \mathbf{i} & \mathbf{j} & \mathbf{k} \\ x + 2 & y - 3 & z - 5 \\ 4 & -1 & 6 \end{vmatrix} = \mathbf{0}. \qquad (1.12)$$

The scalar system of equations of the straight line is

$$\frac{x + 2}{4} = \frac{y - 3}{-1} = \frac{z - 5}{6}.$$

■

2. *Equation of the straight line determined by two points*: The point $P(x, y, z)$ is on the straight line passing through the points $P_1(x_1, y_1, z_1)$ and $P_2(x_2, y_2, z_2)$ if and only if the vector $\mathbf{P_1P}$ is parallel to the vector $\mathbf{P_1P_2}$, determining the direction of the straight line. As we can determine the direction of the straight line from two points, the equation of the straight line determined by two points can be reduced to the equation of the straight line given by a point and a direction. The vector equation of the line passing through the points $P_1(x_1, y_1, z_1)$ and $P_2(x_2, y_2, z_2)$ is given by:

$$(\mathbf{r} - \mathbf{r}_i) \times \mathbf{P_1P_2} = \mathbf{0},$$

where $\mathbf{r}$ is the radius vector of an arbitrary point on the straight line and $\mathbf{r}_i$ is the radius vector of the point $P_i, i = 1$ or 2. Expressed in coordinate form, this is

$$\begin{vmatrix} \mathbf{i} & \mathbf{j} & \mathbf{k} \\ x - x_1 & y - y_1 & z - z_1 \\ x_2 - x_1 & y_2 - y_1 & z_2 - z_1 \end{vmatrix} = \mathbf{0}. \tag{1.13}$$

Thus, the equation of the straight line passing through two points is given by

$$\frac{x - x_1}{x_2 - x_1} = \frac{y - y_1}{y_2 - y_1} = \frac{z - z_1}{z_2 - z_1}. \tag{1.14}$$

1.10 The equation of a plane

A plane in space can be characterized in various ways:

1. by two intersecting lines,
2. by a line and a point on the line,
3. by three non-collinear points.

For our present purpose, we use the fact that, given a point $P(x, y, z)$ and a vector (a, b, c), there is one and only one plane through P that is perpendicular to (a, b, c).

To find the equation of the plane, we consider a point $Q(x, y, z)$ which is any other point on the plane. Then the vector $\mathbf{PQ} = (x - x_1, y - y_1, z - z_1)$ lies in the plane. But $\mathbf{PQ}$ must be perpendicular to the given vector (a, b, c) and, hence, by the vector dot product we have

$$(a, b, c) \cdot (x - x_1, y - y_1, z - z_1) = 0,$$

which is equivalent to

$$a(x - x_1) + b(y - y_1) + c(z - z_1) = 0.$$

This is the required equation of a plane passing through a point and perpendicular (or normal) to a given vector.

Example 1.9   Find the equation of the plane through the point $(4, -3, 5)$ and normal to the vector $(4, -8, 3)$. □

Solution   The equation of the plane is

$$4(x - 4) - 8(y + 3) + 3(z - 5) = 0$$

or $4x - 8y + 3z = 55$. ∎

1.11 Summary

- Physical quantities that have only magnitude associated with them are called *scalars*. A *scalar field* is a function $\phi(x, y, z)$ that defines a scalar at each point of space.
- Quantities that have both magnitude and direction associated with them are called *vectors*. A *vector field* has the form $\mathbf{i}A_1(x, y, z) + \mathbf{j}A_2(x, y, z) + \mathbf{k}A_3(x, y, z) = \mathbf{A}$ and defines a vector at each point of space.
- The scalar product of two vectors $\mathbf{A}$ and $\mathbf{B}$ is defined as

$$\mathbf{A} \cdot \mathbf{B} = |\mathbf{A}||\mathbf{B}| \cos \theta,$$

where θ is the angle included by these two vectors. Another definition of the scalar product of two vectors is given by

$$\mathbf{A} \cdot \mathbf{B} = a_1 b_1 + a_2 b_2 + a_3 b_3.$$

- The *scalar projection* of $\mathbf{B}$ in the direction of $\mathbf{A}$ is

$$\frac{\mathbf{A}}{A} \cdot \mathbf{B}.$$

- The *vector projection* of $\mathbf{B}$ in the direction of $\mathbf{A}$ is

$$\left(\frac{\mathbf{A}}{A} \cdot \mathbf{B}\right)\frac{\mathbf{A}}{A} = \left(\frac{\mathbf{A} \cdot \mathbf{B}}{A^2}\right)\mathbf{A}.$$

- The cross product or vector product of two vectors $\mathbf{A}$ and $\mathbf{B}$ is defined as

$$\mathbf{A} \times \mathbf{B} = |\mathbf{A}||\mathbf{B}| \sin \theta \mathbf{n},$$

where θ is the included angle and $\mathbf{n}$ is the unit normal vector to both these vectors. Another definition of cross product is the following:

$$\mathbf{A} \times \mathbf{B} = \begin{vmatrix} \mathbf{i} & \mathbf{j} & \mathbf{k} \\ a_1 & a_2 & a_3 \\ b_1 & b_2 & b_3 \end{vmatrix}.$$

- A *scalar triple product* of three vectors is

$$\mathbf{A} \cdot (\mathbf{B} \times \mathbf{C}) = \begin{vmatrix} a_1 & a_2 & a_3 \\ b_1 & b_2 & b_3 \\ c_1 & c_2 & c_3 \end{vmatrix}.$$

The physical implication of this formula is the volume of a parallelepiped having the congruent sides as $\mathbf{A}$, $\mathbf{B}$ and $\mathbf{C}$. This scalar triple product can be written as

$$[\mathbf{ABC}] = [\mathbf{BCA}] = [\mathbf{CAB}] = -[\mathbf{ACB}] = -[\mathbf{BAC}] = -[\mathbf{CBA}].$$

- A *vector triple product* of three vectors is

$$\mathbf{A} \times (\mathbf{B} \times \mathbf{C}) = (\mathbf{A} \cdot \mathbf{C})\mathbf{B} - (\mathbf{A} \cdot \mathbf{B})\mathbf{C}.$$

- Lagrange's identity:

$$\begin{aligned} (\mathbf{A} \times \mathbf{B}) \cdot (\mathbf{C} \times \mathbf{D}) &= \mathbf{A} \cdot [\mathbf{B} \times (\mathbf{C} \times \mathbf{D})] \\ &= \mathbf{A} \cdot [(\mathbf{B} \cdot \mathbf{D})\mathbf{C} - (\mathbf{B} \cdot \mathbf{C})\mathbf{D}] \\ &= (\mathbf{A} \cdot \mathbf{C})(\mathbf{B} \cdot \mathbf{D}) - (\mathbf{A} \cdot \mathbf{D})(\mathbf{B} \cdot \mathbf{C}). \end{aligned}$$

This result is referred to as *Lagrange's* identity.
- Similarly, $(\mathbf{A} \times \mathbf{B}) \times (\mathbf{C} \times \mathbf{D})$ can be thought of as the vector triple product of $\mathbf{A} \times \mathbf{B}$, $\mathbf{C}$ and $\mathbf{D}$. Then applying (1.10), we find

$$\begin{aligned} (\mathbf{A} \times \mathbf{B}) \times (\mathbf{C} \times \mathbf{D}) &= (\mathbf{A} \times \mathbf{B} \cdot \mathbf{D})\mathbf{C} - (\mathbf{A} \times \mathbf{B} \cdot \mathbf{C})\mathbf{D} \\ &= [\mathbf{ABD}]\mathbf{C} - [\mathbf{ABC}]\mathbf{D}, \end{aligned}$$

which is a vector in the plane of $\mathbf{C}$ and $\mathbf{D}$.
- From another point of view, we have

$$\begin{aligned} (\mathbf{A} \times \mathbf{B}) \times (\mathbf{C} \times \mathbf{D}) &= -(\mathbf{C} \times \mathbf{D}) \times (\mathbf{A} \times \mathbf{B}) \\ &= -[(\mathbf{C} \times \mathbf{D} \cdot \mathbf{B})\mathbf{A} - (\mathbf{C} \times \mathbf{D} \cdot \mathbf{A})\mathbf{B}] \\ &= -(\mathbf{C} \times \mathbf{D} \cdot \mathbf{B})\mathbf{A} + (\mathbf{C} \times \mathbf{D} \cdot \mathbf{A})\mathbf{B} \\ &= -[\mathbf{CDB}]\mathbf{A} + [\mathbf{CDA}]\mathbf{B}, \end{aligned}$$

which is a vector in the plane of $\mathbf{A}$ and $\mathbf{B}$.
- Direction cosine of a vector: The direction cosine of a vector $\mathbf{A} = a_1\mathbf{i} + a_2\mathbf{j} + a_3\mathbf{k}$ is defined as

$$\cos \alpha = \frac{a_1}{\sqrt{a_1^2 + a_2^2 + a_3^2}},$$

$$\cos \beta = \frac{a_2}{\sqrt{a_1^2 + a_2^2 + a_3^2}}$$

and

$$\cos \gamma = \frac{a_3}{\sqrt{a_1^2 + a_2^2 + a_3^2}}.$$

Hence, the direction cosines of A are proportional to a_1, a_2 and a_3, respectively.
- The direction cosines of an arbitrary vector are identical with the direction cosines of the unit vector pointing in its direction. As the sum of the squares of the coordinates of a unit vector is one, the sum of the squares of the direction cosines of any direction is also one:

$$\cos^2 \alpha + \cos^2 \beta + \cos^2 \gamma = 1.$$

Exercises

1. Show by two methods that the vectors $A = 2i - 3j - k$, $B = -6i + 9j + 3k$ are parallel.
2. Find a unit vector perpendicular to the vectors

$$A = i - j + k \quad \text{and} \quad B = i + j - k.$$

3. Find the cosine of the angle between the two vectors $A = 2i - 3j + k$ and $B = 3i - j - 2k$.
4. Find the unit vector which is parallel to the plane of $i + j - 2k$ and $3i - 2j + k$ and perpendicular to $2i + 2j - k$.
5. Show that $(A \times B) \cdot (B \times C) \times (C \times A) = [ABC]^2$.
6. Determine whether the given vectors are orthogonal, parallel or neither
   (a) $A = 2i - 4j$, $B = -i + 2j$.
   (b) $A = 2i + j$, $B = 4i + 2j$.
   (c) $A = 2i + 8j$, $B = -i + 2j$.
7. Find a unit vector that is orthogonal to $i + 3j$.
8. Find the scalar and vector projections of B onto A.
   (a) $A = 2i + 3j$, $B = 4i + j$.
   (b) $A = 3i - j$, $B = 2i + 3j$.
   (c) $A = 4i + 2j$, $B = i + j$.
   (d) $A = -i - 2j$, $B = 3i + 3j$.
   (e) $A = i$, $B = i - j$.
   (f) $A = 2i - 3j$, $B = i + 6j$.
9. (a) Show that

$$A \times (B \times C) + B \times (C \times A) + C \times (A \times B) = 0.$$

   (b) Show that if four vectors A, B, C and D are coplanar, then $(A \times B) \times (C \times D) = 0$.
10. (a) Find the distance from the point $(3,7)$ to the line $y = 2x$.
    (b) Find the distance from the point $(-2, 3)$ to the line $3x + 4y + 5 = 0$.
11. Find the distance between the parallel lines $y = -4x + 3$ and $y = -4x + 12$.

12. Show that the volume of the parallelepiped having $\mathbf{A} + \mathbf{B}$, $\mathbf{B} + \mathbf{C}$ and $\mathbf{C} + \mathbf{A}$ as concurrent edges is twice the volume of the parallelepiped having $\mathbf{A}$, $\mathbf{B}$ and $\mathbf{C}$ as concurrent edges.

13. (a) Use a scalar projection to show that the distance from a point $P_1(x_1, y_1)$ to the line $ax + by + c = 0$ is

$$\frac{|ax_1 + by_1 + c|}{\sqrt{a^2 + b^2}}.$$

   (b) Use this formula to find the distance from the point $(-2, 3)$ to the line $3x - 4y + 5 = 0$.

14. Find the area of a parallelogram if the vertices are $(1, 1)$, $(4, -2)$, $(9, 3)$ and $(12, 0)$.

15. Find the area of a parallelogram if the vertices are $(1, 1, 1)$, $(4, 4, 4)$, $(8, -3, 14)$ and $(11, 0, 17)$.

16. Find a normal vector to the plane through the points $(1, 3, 0)$, $(2, 0, 8)$ and $(0, 2, 2)$ and from it find an equation for the plane.

17. Find the volume of the parallelepiped if the edge vectors are:
   (a) $\mathbf{i} + \mathbf{j}$, $-2\mathbf{i} + 2\mathbf{k}$, $-2\mathbf{i} - 3\mathbf{k}$.
   (b) $(4, 9, -1)$, $(2, 0, 0)$, $(5, -4, 2)$.

18. Find the length of each of the following vectors and the angle it makes with the positive x-axis:
   (a) $\mathbf{i} + \mathbf{j}$.
   (b) $-\mathbf{i} + \mathbf{j}$.
   (c) $\mathbf{i} + \sqrt{3}\mathbf{j}$.
   (d) $\mathbf{i} - \sqrt{3}\mathbf{j}$.

19. Write the equation of the plane
   (a) passing through $P_0(1, 2, 3)$ and parallel to $3x - 2y + 4z - 5 = 0$.
   (b) passing through $P_0(1, 2, 3)$ and $P_1(3, -2, 1)$, and perpendicular to the plane $3x - 2y + 4z - 5 = 0$.
   (c) through $P_0(1, 2, 3)$, $P_1(3, -2, 1)$ and $P_2(5, 0, -4)$. Let $P(x, y, z)$ be a general point in the required plane.

20. Prove that

$$(\mathbf{A} \times \mathbf{B}) \cdot (\mathbf{C} \times \mathbf{D}) = (\mathbf{A} \cdot \mathbf{C})(\mathbf{B} \cdot \mathbf{D}) - (\mathbf{A} \cdot \mathbf{D})(\mathbf{B} \cdot \mathbf{C}).$$

21. If $\mathbf{A}$, $\mathbf{B}$ and $\mathbf{C}$ are any three independent vectors, the vectors

$$\mathbf{U} = \frac{\mathbf{B} \times \mathbf{C}}{[\mathbf{ABC}]},$$

$$\mathbf{V} = \frac{\mathbf{C} \times \mathbf{A}}{[\mathbf{ABC}]},$$

$$\mathbf{W} = \frac{\mathbf{A} \times \mathbf{B}}{[\mathbf{ABC}]},$$

are said to form a set *reciprocal* to the set of **A, B** and **C**, respectively. Show that

$$\mathbf{A} \cdot \mathbf{U} = \mathbf{B} \cdot \mathbf{V} = \mathbf{C} \cdot \mathbf{W} = 1$$

and

$$[\mathbf{UVW}] = \frac{1}{[\mathbf{ABC}]}.$$

If $\mathbf{A} = \mathbf{i} + 2\mathbf{j} - 2\mathbf{k}, \mathbf{B} = \mathbf{i} + 8\mathbf{j} + 4\mathbf{k}$ and $\mathbf{C} = 12\mathbf{i} - 4\mathbf{j} + 3\mathbf{k}$, express the vector $\mathbf{D} = \mathbf{i} + 2\mathbf{j} + 3\mathbf{k}$ as a linear combination of **A, B** and **C** and also as a linear combination of the vectors **U, V** and **W** of the set reciprocal to **A, B** and **C**.

22.   In mechanics, the *moment* **M** of a *force* **F** about a point O is defined by the magnitude of **F** times the perpendicular distance from the point O to the line of action of **F**. If the *vector moment* **M** is defined as the vector whose magnitude is M and whose direction is perpendicular to the plane of O and **F**, show that $\mathbf{M} = \mathbf{R} \times \mathbf{F}$, where **R** is the position vector from O to any point on the line of action of **F**. Would $\mathbf{M} = \mathbf{F} \times \mathbf{R}$ be an equally acceptable definition? Explain.

If the force $\mathbf{F} = 3\mathbf{i} - 2\mathbf{j} + 8\mathbf{k}$ is applied at the point $(1, 2, -1)$ about the point $(3, 1, 4)$, determine the moment of this force.

Chapter 2

Vector functions of one variable

2.1 Introduction

When solving problems in physics and mathematics, we must often examine vector quantities depending on scalar or vector variables, or scalar quantities depending on vectors. For example, the speed of a moving point depends on the time, i.e. on a scalar variable; the intensity of electric field, which is a scalar quantity, depends on the radius vector of the point, that is to say on a vector variable. The branch of mathematics that deals with variable vector quantities, vector functions, is called vector analysis. There are three types of functions in vector analysis:

1. *Vector–scalar* functions: the independent variable is a scalar and the dependent variable is a vector.
2. *Scalar–vector* functions: the independent variable is a vector and the dependent variable is a scalar.
3. *Vector–vector* functions: both the independent and the dependent variables are vectors.

If t is a scalar variable, and if to each value of t in some interval there corresponds a value of a vector $\mathbf{V}$, we say that $\mathbf{V}$ is a *vector function* of t. We use the letter t to denote the independent variable because it represents time in most applications of vector functions. Since the component of a vector in any direction is known whenever the vector itself is known, it follows that if $\mathbf{V}$ is a function of t, so too are its components in the directions of the unit vectors $\mathbf{i}$, $\mathbf{j}$ and $\mathbf{k}$. We denote the vector function with one variable by $\mathbf{V}(t)$, where the independent variable t is a scalar quantity and the dependent variable $\mathbf{V}$ is a vector quantity. Any n-dimensional vector function $\mathbf{V}(t)$ can be given by means of n scalar functions. The three-dimensional case is dealt with in this analysis.

Let the coordinates of the vector $\mathbf{V}$ be V_1, V_2 and V_3. Some value of the variable t determines the vector $\mathbf{V}(t)$ and thus determines its coordinates, V_1, V_2 and V_3. If we denote the function relations that determine the dependence of the coordinates V_1, V_2 and V_3 upon t by $V_1(t)$, $V_2(t)$ and $V_3(t)$, then we can write down the vector function $\mathbf{V}(t)$ in the following form:

$$\mathbf{V}(t) = V_1(t)\mathbf{i} + V_2(t)\mathbf{j} + V_3(t)\mathbf{k}. \tag{2.1}$$

In particular, we say that $\mathbf{V}(t)$ is continuous if and only if the three scalar functions $V_1(t)$, $V_2(t)$ and $V_3(t)$ are continuous. The scalar variable of the vector function is called the *parameter* of the space curve. Thus, the vector function

$$\mathbf{V}(t) = V_1(t)\mathbf{i} + V_2(t)\mathbf{j} + V_3(t)\mathbf{k}$$

written by means of coordinates:

$$\mathbf{V}(t) = \begin{cases} V_1 = V_1(t) \\ V_2 = V_2(t) \\ V_3 = V_3(t) \end{cases}$$

is at the same time a parametric system of equations of the space curve. Note that if one of the coordinates of the vector functions $\mathbf{V} = V_1(t)\mathbf{i} + V_2(t)\mathbf{j} + V_3(t)\mathbf{k}$ is a constant, the curve is a plane lying – according to the constant coordinate – parallel to one of the coordinate planes.

2.2 Vector differentiation

Suppose the independent variable t of a vector function $\mathbf{V}(t)$ changes by an amount Δt, then the function will change both in magnitude and in direction. Corresponding to the scalar increment Δt, we have the vector increment

$$\begin{aligned} \Delta \mathbf{V} &= \mathbf{V}(t + \Delta t) - \mathbf{V}(t) \\ &= [V_1(t + \Delta t)\mathbf{i} + V_2(t + \Delta t)\mathbf{j} + V_3(t + \Delta t)\mathbf{k}] \\ &\quad - [V_1(t)\mathbf{i} + V_2(t)\mathbf{j} + V_3(t)\mathbf{k}] \\ &= \Delta V_1(t)\mathbf{i} + \Delta V_2(t)\mathbf{j} + \Delta V_3(t)\mathbf{k}. \end{aligned} \tag{2.2}$$

If we now divide $\Delta \mathbf{V}$ by Δt and take the limit as $\Delta t \to 0$, we obtain the new vector known as the derivative of a vector function $\mathbf{V}(t)$ with respect to t

$$\frac{d\mathbf{V}}{dt} = \lim_{\Delta t \to 0} \frac{\mathbf{V}(t + \Delta t) - \mathbf{V}(t)}{\Delta t} = \lim_{\Delta t \to 0} \frac{\Delta \mathbf{V}}{\Delta t}. \tag{2.3}$$

Using (2.2), this becomes

$$\frac{d\mathbf{V}}{dt} = \lim_{\Delta \to 0} \frac{\Delta V_1(t)}{\Delta t}\mathbf{i} + \lim_{\Delta t \to 0} \frac{\Delta V_2(t)}{\Delta t}\mathbf{j} + \lim_{\Delta t \to 0} \frac{\Delta V_3(t)}{\Delta t}\mathbf{k}$$

$$= \frac{dV_1(t)}{dt}\mathbf{i} + \frac{dV_2(t)}{dt}\mathbf{j} + \frac{dV_3(t)}{dt}\mathbf{k}. \tag{2.4}$$

From (2.4), we can define the *differential of a vector function* $\mathbf{V}(t)$ to be

$$d\mathbf{V} = dV_1\mathbf{i} + dV_2\mathbf{j} + dV_3\mathbf{k}. \tag{2.5}$$

The notation $\mathbf{V}'$ is often used to signify the derivative when the variable of differentiation is clear from the context. It is also common to place a dot above the vector to indicate total differentiation with respect to time t, i.e.

$$\dot{\mathbf{V}} = \frac{d\mathbf{V}}{dt}.$$

The above expression implies that in terms of components with respect to a given Cartesian coordinate system, $\mathbf{V}(t)$ is differentiable at a point t if and only if its three components are differentiable at t. The derivative $\dot{\mathbf{V}}(t)$ is obtained by differentiating each component separately. For the position vector

$$\mathbf{R} = x\mathbf{i} + y\mathbf{j} + z\mathbf{k}$$

drawn from the origin to the point (x, y, z), we have

$$d\mathbf{R} = dx\mathbf{i} + dy\mathbf{j} + dz\mathbf{k} \quad \text{or} \quad \frac{d\mathbf{R}}{dt} = \frac{dx}{dt}\mathbf{i} + \frac{dy}{dt}\mathbf{j} + \frac{dz}{dt}\mathbf{k}.$$

As can be seen from the above expression, on a practical level, the derivative of a vector function is obtained by simply differentiating its component functions. When the components of a vector function $\mathbf{R}$ have continuous first derivatives defined as $d\mathbf{R}/dt = \mathbf{R}'$ and $\mathbf{R}' \neq 0$ for all t in the open interval (a, b), then $\mathbf{R}$ is said to be a *smooth function* and the curve C traced by $\mathbf{R}$ is called a smooth curve. We shall denote the absolute value of $\mathbf{R}$ by r, i.e. $|\mathbf{R}| = r$.

Remark 2.1   In the next section, we shall see that $d\mathbf{R}/dt$ is a tangent vector. Since the components of a tangent vector to the curve C do not vanish simultaneously, this definition implies that along a smooth curve, small changes in t produce small changes in the direction of the tangent vector that is known as curvature. In other words, the tangent vector turns gradually or 'smoothly'. ■

2.3 Geometric interpretation of $\mathbf{R}'$

If the vector $\mathbf{R}'$ is not $\mathbf{0}$ at a point P, then it may be drawn at a tangent to the curve at P. As seen in Fig. 2.1a and b, the vectors

$$\Delta\mathbf{R} = \mathbf{R}(t + \Delta t) - \mathbf{R}(t) \quad \text{and} \quad \frac{\Delta\mathbf{R}}{\Delta t} = \frac{1}{\Delta t}[\mathbf{R}(t + \Delta t) - \mathbf{R}(t)]$$

are parallel. If we assume $\lim_{\Delta t \to 0} \Delta\mathbf{R}/\Delta t$ exists, it is reasonable to conclude that as $\Delta\mathbf{R} \to 0$, $\mathbf{R}(t)$ and $\mathbf{R}(t + \Delta t)$ become close and, as a consequence, the limiting position of the vector $\Delta\mathbf{R}/\Delta t$ is the tangent line at P. In fact, the tangent line at P is defined as that line through P parallel to $\mathbf{R}'(t)$.

From the definition of the derivative of a vector function of one variable, it follows that sums, differences and products of vectors can be differentiated by formulae similar to those of ordinary calculus, provided the proper order of factors is maintained wherever the order is significant. Specifically, we have

$$\frac{d(\mathbf{U} \pm \mathbf{V})}{dt} = \frac{d\mathbf{U}}{dt} \pm \frac{d\mathbf{V}}{dt}, \tag{2.6}$$

$$\frac{d(\phi\mathbf{V})}{dt} = \frac{d\phi}{dt}\mathbf{V} + \phi\frac{d\mathbf{V}}{dt}, \tag{2.7}$$

$$\frac{d(\mathbf{U} \cdot \mathbf{V})}{dt} = \frac{d\mathbf{U}}{dt} \cdot \mathbf{V} + \mathbf{U} \cdot \frac{d\mathbf{V}}{dt}, \tag{2.8}$$

$$\frac{d(\mathbf{U} \times \mathbf{V})}{dt} = \frac{d\mathbf{U}}{dt} \times \mathbf{V} + \mathbf{U} \times \frac{d\mathbf{V}}{dt}, \tag{2.9}$$

$$\frac{d[\mathbf{U} \times (\mathbf{V} \times \mathbf{W})]}{dt} = \frac{d\mathbf{U}}{dt} \times (\mathbf{V} \times \mathbf{W}) + \mathbf{U} \times \left\{\frac{d\mathbf{V}}{dt} \times \mathbf{W}\right\}$$

$$+ \mathbf{U} \times \left\{\mathbf{V} \times \frac{d\mathbf{W}}{dt}\right\}. \tag{2.10}$$

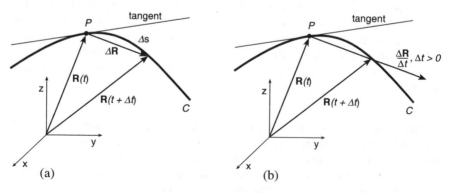

Figure 2.1 Geometrical interpretation of vector function.

Since every vector equation implies three simultaneous scalar equations, it follows from the definition of vector differentiation that the integral of a vector function of one variable is found by integrating each component, provided only that the constant of integration is an arbitrary constant vector. Therefore, if

$$\mathbf{V}(t) = V_1(t)\mathbf{i} + V_2(t)\mathbf{j} + V_3(t)\mathbf{k},$$

then

$$\int \mathbf{V}(t)dt = \mathbf{i}\int V_1(t)dt + \mathbf{j}\int V_2(t)dt + \mathbf{k}\int V_3(t)dt + \mathbf{C},$$

where the constant vector $\mathbf{C} = C_1\mathbf{i} + C_2\mathbf{j} + C_3\mathbf{k}$. When $\Delta\mathbf{R}$ tends to zero, the elementary arc length is given by

$$\Delta s = |\Delta\mathbf{R}|.$$

If s and $\mathbf{R}$ are functions of the independent time variable t, and as $\Delta t \to 0$, then

$$\frac{ds}{dt} = \left|\frac{d\mathbf{R}}{dt}\right| = \sqrt{\frac{d\mathbf{R}}{dt} \cdot \frac{d\mathbf{R}}{dt}}.$$

Therefore, the total arc length can be obtained from

$$s = \int_{a(t)}^{b(t)} \sqrt{\left(\frac{dx}{dt}\right)^2 + \left(\frac{dy}{dt}\right)^2 + \left(\frac{dz}{dt}\right)^2}\, dt.$$

2.4 Higher-order derivatives

Higher-order derivatives of a vector function are also obtained by differentiating its components. In the case of the second derivative, we have

$$\mathbf{R}'' = f''(t)\mathbf{i} + g''(t)\mathbf{j} + h''(t)\mathbf{k},$$

where $f(t)$, $g(t)$, and $g(t)$ are the components of $\mathbf{R}(t)$ with first and second derivatives.

As an example, consider $\mathbf{R}(t) = (t^3 - 2t^2)\mathbf{i} + 4t\mathbf{j} + e^{-t}\mathbf{k}$. It follows that

$$\mathbf{R}'(t) = (3t^2 - 4t)\mathbf{i} + 4\mathbf{j} - e^{-t}\mathbf{k}$$

$$\text{and} \quad \mathbf{R}'' = (6t - 4)\mathbf{i} + e^{-t}\mathbf{k}.$$

Also, as an example, if $\mathbf{R}(t) = 6t^2\mathbf{i} + 4e^{-2t}\mathbf{j} + 8\cos 4t\mathbf{k}$, then

$$\int \mathbf{R}(t)dt = \left(\int 6t^2 dt\right)\mathbf{i} + \left(\int 4e^{-2t}dt\right)\mathbf{j} + \left(\int 8\cos 4t dt\right)\mathbf{k}$$

$$= (2t^3 + C_1)\mathbf{i} + (-2e^{-2t} + C_2)\mathbf{j} + (2\sin 4t + C_3)\mathbf{k}$$

$$= 2t^3\mathbf{i} - 2e^{-2t}\mathbf{j} + 2\sin 4t\mathbf{k} + \mathbf{C},$$

where the constant vector $\mathbf{C} = C_1\mathbf{i} + C_2\mathbf{j} + C_3\mathbf{k}$.

2.5 Curves, length and arc length

Curves in space are important in calculus and physics (i.e. as paths of moving bodies), and we show in the following sections that they constitute a major field of application in vector calculus. Their theory together with that of surfaces in space is called *differential geometry*. It plays a significant role in engineering design, geodesy, geography, space travel and relativity theory.

A curve C in space can be represented by a vector function,

$$\mathbf{R}(t) = (x(t), y(t), z(t)) = x(t)\mathbf{i} + y(t)\mathbf{j} + z(t)\mathbf{k}, \tag{2.11}$$

where x, y and z are cartesian coordinates. This is known as a *parametric representation* of the curve, with t being the parameter of the representation (Fig. 2.2). To each value t_0 there corresponds a point C with position vector $\mathbf{R}(t_0)$, i.e. with coordinates $x(t_0)$, $y(t_0)$ and $z(t_0)$. The parameter t may represent time or some other physical quantity. In equation (2.11), the coordinates x, y and z play the same role – all three are dependent variables. We shall see that very often this is a great practical advantage and the key property of a parametric representation.

We will now define the length l of a curve. This length will be the limit of the lengths of broken lines of n chords (see Fig. 2.3) with larger and larger n. To represent C, let us take $\mathbf{R}(t)$, $a \le t \le b$. For each $n = 1, 2, \ldots$, we subdivide (or partition) the interval $a \le t \le b$ by points

$$t_0(= a), t_1, \ldots, t_{n-1}, t_n(= b),$$

where $t_0 < t_1 < \cdots < t_n$. This gives a broken line of chords with endpoints $\mathbf{R}(t_0), \ldots, \mathbf{R}(t_n)$. This is done arbitrarily but such that the greatest $|\Delta t_m| = |t_m - t_{m-1}|$ approaches zero as $n \to \infty$. The lengths $l_1, l_2, \ldots$ of these lines of chords can be obtained from the Pythagorean theorem.

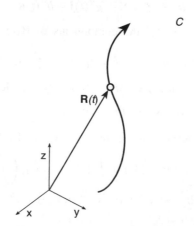

Figure 2.2   Parametric representation of a curve.

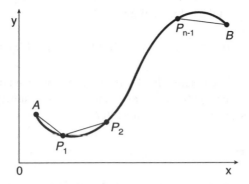

Figure 2.3   Length of a curve.

If $\mathbf{R}(t)$ has a continuous derivative $\mathbf{R}'(t)$, it can be shown that the sequence $l_1, l_2, \ldots$ has a limit, which is independent of the particular choice of representation of C and of the choice of subdivisions. This limit is given by the integral

$$s = \int_a^b \sqrt{\mathbf{R}' \cdot \mathbf{R}'} \, dt, \tag{2.12}$$

where s is called the *length* of C and C is called *rectifiable*. The practical evaluation of the integral (2.12) will be difficult, in general.

The length s of a curve is a constant, a positive number. In expression (2.12), if we replace the fixed upper limit with the variable t, then the integral becomes a function of t. This function of t is commonly denoted by $s(t)$ and called the *arc length* of C. If $\mathbf{R}(t) = f(t)\mathbf{i} + g(t)\mathbf{j} + h(t)\mathbf{k}$ is a smooth function, then it can be shown that the length of the smooth curve traced by $\mathbf{R}$ is given by

$$s = \int_a^b \sqrt{(f'(t))^2 + (g'(t))^2 + (h'(t))^2} dt$$
$$= \int_a^b |\mathbf{R}'(t)| dt.$$

A curve in the plane or in space can be parameterized in terms of the arc length s. Consider the helix $\mathbf{R}(t) = 2\cos t\mathbf{i} + 2\sin t\mathbf{j} + t\mathbf{k}$, $t \geq 0$. Since $|\mathbf{R}'(t)| = \sqrt{5}$, it follows that the length of the curve from $\mathbf{R}(0)$ to an arbitrary point $\mathbf{R}(t)$ is

$$s = \int_0^t \sqrt{5} du = \sqrt{5}t,$$

where we have used u as a dummy variable of integration. Using $t = s/\sqrt{5}$, we obtain a vector equation of the helix as a function of arc length:

$$\mathbf{R}(s) = 2\cos(s/\sqrt{5})\mathbf{i} + 2\sin(s/\sqrt{5})\mathbf{j} + (s/\sqrt{5})\mathbf{k}.$$

Parametric equations of the helix are then

$$f(s) = 2\cos(s/\sqrt{5}), \quad g(s) = 2\sin(s/\sqrt{5}), \quad h(s) = \frac{s}{\sqrt{5}}.$$

The derivative of a vector function $\mathbf{R}(t)$ with respect to the parameter t is a tangent vector to the curve traced by $\mathbf{R}$. However, if the curve is parameterized in terms of arc length s, then $\mathbf{R}'(s)$ is a unit tangent vector.

2.6 Motion on a curve, velocity and acceleration vectors

Suppose a body or a particle moves along a curve C so that its position at time t is given by

$$\mathbf{R}(t) = f(t)\mathbf{i} + g(t)\mathbf{j} + h(t)\mathbf{k}.$$

If f, g and h have second derivatives, then the vectors

$$\mathbf{V}(t) = \mathbf{R}'(t) = f'(t)\mathbf{i} + g'(t)\mathbf{j} + h'(t)\mathbf{k}, \tag{2.13}$$

$$\mathbf{a}(t) = \mathbf{R}''(t) = f''(t)\mathbf{i} + g''(t)\mathbf{j} + h''(t)\mathbf{k} \tag{2.14}$$

are called *velocity* and *acceleration* of the particle, respectively. The scalar function $|\mathbf{V}(t)|$ is the *speed* of the particle. Since the position vector $\mathbf{R}$ is usually defined by $\mathbf{R} = x\mathbf{i} + y\mathbf{j} + z\mathbf{k}$, we can write

$$|\mathbf{V}(t)| = \left|\frac{d\mathbf{R}}{dt}\right| = \sqrt{\left(\frac{dx}{dt}\right)^2 + \left(\frac{dy}{dt}\right)^2 + \left(\frac{dz}{dt}\right)^2}.$$

Speed is related to the arc length s by $s'(t) = |\mathbf{V}(t)|$. In other words, the arc length is given by

$$s = \int_{t_0}^{t_1} |\mathbf{V}(t)|\,dt.$$

Example 2.1   The position of a moving particle is given by $\mathbf{R}(t) = t^2\mathbf{i} + t\mathbf{j} + \frac{5}{2}t\mathbf{k}$. Represent graphically the curve defined by $\mathbf{R}(t)$ and the vectors $\mathbf{V}(2)$ and $\mathbf{a}(2)$. $\square$

Solution   Since $x = t^2$, $y = t$ and $z = \frac{5}{2}t$, the path of the particle is above the parabola $x = y^2$. When $t = 2$, the position vector $\mathbf{R}(2) = 4\mathbf{i} + 2\mathbf{j} + 5\mathbf{k}$ indicates that the particle is at the point $P(4, 2, 5)$. Now,

$$\mathbf{V}(t) = \mathbf{R}'(t) = 2t\mathbf{i} + \mathbf{j} + \frac{5}{2}\mathbf{k} \quad \text{and} \quad \mathbf{a}(t) = \mathbf{R}''(t) = 2\mathbf{i},$$

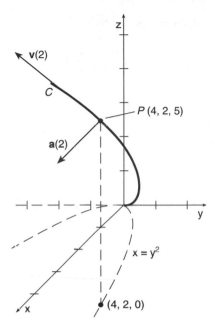

Figure 2.4   Curves defined by $\mathbf{R}(t)$, $\mathbf{V}(2)$ and $\mathbf{a}(2)$.

so that

$$\mathbf{V}(2) = 4\mathbf{i} + \mathbf{j} + \frac{5}{2}\mathbf{k} \quad \text{and} \quad \mathbf{a}(2) = 2\mathbf{i}.$$

These vectors are shown in Fig. 2.4.   ■

2.7 Curvature, components of acceleration

Let C be a smooth curve in 2- or 3-space traced out by a vector function $\mathbf{R}(t)$. In this section, we will discuss in greater detail the acceleration vector $\mathbf{a}(t) = \mathbf{R}''(t)$ introduced in the last section. Before doing this, we need to examine a scalar quantity called the *curvature* of a curve. It is known that $\mathbf{R}'(t)$ is a tangent vector to the curve C and consequently

$$\mathbf{T} = \frac{\mathbf{R}'(t)}{|\mathbf{R}'(t)|} \tag{2.15}$$

is a *unit tangent* vector. However, if C is parameterized by arc length s, then a unit tangent to the curve is also given by $d\mathbf{R}/ds$. The quantity $|\mathbf{R}'(t)|$ in (2.15) is related to arc length s by $ds/dt = |\mathbf{R}'(t)|$. Since the curve C is smooth, it is known that

$ds/dt > 0$. Hence, by the chain rule, we have:

$$\frac{d\mathbf{R}}{dt} = \frac{d\mathbf{R}}{ds}\left(\frac{ds}{dt}\right).$$

We know that

$$\frac{ds}{dt} = V = \text{Speed of the particle along } C$$

and

$$\frac{d\mathbf{R}}{ds} = \mathbf{T} = \text{Unit vector tangent to C.}$$

Therefore,

$$\frac{d\mathbf{R}}{dt} = v\mathbf{T}.$$

Now, $\mathbf{T}$ can be written as

$$\mathbf{T} = \frac{d\mathbf{R}}{ds} = \frac{d\mathbf{R}/dt}{ds/dt} = \frac{\mathbf{R}'(t)}{|\mathbf{R}'(t)|}.$$

Suppose C is as shown in Fig. 2.5, As s increases, $\mathbf{T}$ moves along C, changing direction but not length (it is always of unit length). Along the portion of the curve between P_1 and P_2, the vector $\mathbf{T}$ varies little in direction; along the curve between P_2 and P_3, where C obviously bends more sharply, the change in the direction of the tangent $\mathbf{T}$ is more pronounced. We use the *rate* at which the unit vector $\mathbf{T}$ changes direction with respect to arc length as an indicator of a smooth curve C. This leads to the following definition.

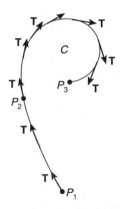

Figure 2.5   The tangent $\mathbf{T}$ as it moves along the arc.

Definition 2.1   *Let* $\mathbf{R}(t)$ *be a vector function defining a smooth curve* C. *If* s *is the arc length parameter and* $\mathbf{T} = d\mathbf{R}/ds$ *is the unit tangent vector, then the* curvature *of* C *at a point is defined as*

$$\kappa = \left| \frac{d\mathbf{T}}{ds} \right|. \tag{2.16}$$

The symbol κ in (2.16) is the Greek letter *kappa*. Now, since curves are generally not parameterized by arc length, it is convenient to express (2.16) in terms of a general parameter t. We can write $d\mathbf{T}'/ds$ as

$$\frac{d\mathbf{T}'}{ds} = \frac{d\mathbf{T}'/dt}{ds/dt}.$$

So,

$$\left| \frac{d\mathbf{T}}{ds} \right| = \frac{|\mathbf{T}'(t)|}{|\mathbf{R}'(t)|}.$$

In other words, the curvature is given by

$$\kappa = \frac{|\mathbf{T}'(t)|}{|\mathbf{R}'(t)|}. \tag{2.17}$$

Example 2.2   Find the curvature of a circle of radius a. $\square$

Solution   A circle can be described by the vector function $\mathbf{R}(t) = a\cos t\mathbf{i} + a\sin t\mathbf{j}$. Now, from $\mathbf{R}'(t) = -a\sin t\mathbf{i} + a\cos t\mathbf{j}$ and $|\mathbf{R}'(t)| = a$, we get

$$\mathbf{T}(t) = \frac{\mathbf{R}'(t)}{|\mathbf{R}'(t)|} = -\sin t\mathbf{i} + \cos t\mathbf{j} \quad \text{and} \quad \mathbf{T}'(t) = -\cos t\mathbf{i} - \sin t\mathbf{j}.$$

Hence, the curvature is given by

$$\kappa = \frac{|\mathbf{T}'(t)|}{|\mathbf{R}'(t)|} = \frac{\sqrt{\cos^2 t + \sin^2 t}}{a} = \frac{1}{a}.$$

■

2.8 Curvature, tangential and normal components of acceleration

Suppose a particle moves in 2- or 3-space on a smooth curve C described by the vector function $\mathbf{R}(t)$. Then the velocity of the particle on C is $\mathbf{V}(t) = \mathbf{R}'(t)$, whereas its speed is $ds/dt = V = |\mathbf{V}(t)|$. Thus, we have

$$\mathbf{T} = \frac{\mathbf{V}(t)}{|\mathbf{V}(t)|} = \frac{\mathbf{V}(t)}{V},$$

so that $\mathbf{V}(t) = V\mathbf{T}$. Differentiating this last expression with respect to t gives acceleration:

$$\mathbf{a}(t) = V\frac{d\mathbf{T}}{dt} + \frac{dV}{dt}\mathbf{T}. \tag{2.18}$$

Furthermore, it follows from the differentiation of $\mathbf{T} \cdot \mathbf{T} = 1$ that $\mathbf{T} \cdot d\mathbf{T}/dt = 0$. Hence, at a point P on C, the vectors $\mathbf{T}$ and $d\mathbf{T}/dt$ are orthogonal. If $|d\mathbf{T}/dt| \neq 0$, the vector

$$\mathbf{N} = \frac{d\mathbf{T}/dt}{|d\mathbf{T}/dt|} \tag{2.19}$$

is a unit normal to the curve C at P with direction given by $d\mathbf{T}/dt$. The vector $\mathbf{N}$ is also called the *principal normal*. But since curvature is $\kappa = |d\mathbf{T}/dt|/V$, it follows from (2.19) that $d\mathbf{T}/dt = \kappa v\mathbf{N}$. Thus, (2.18) becomes

$$\mathbf{a}(t) = \kappa V^2\mathbf{N} + \frac{dV}{dt}\mathbf{T}. \tag{2.20}$$

Thus, we have

$$\frac{d\mathbf{T}}{ds} = \frac{d\mathbf{T}/dt}{ds/dt} = \frac{\kappa V}{V}\mathbf{N} = \kappa\mathbf{N}. \tag{2.21}$$

A third unit vector defined by

$$\mathbf{B} = \mathbf{T} \times \mathbf{N}$$

is called the *binormal*. The three unit vectors, $\mathbf{T}$, $\mathbf{N}$ and $\mathbf{B}$ form a right-handed set of mutually orthogonal vectors called the *moving trihedral*. The plane of $\mathbf{T}$ and $\mathbf{N}$ is called the *osculating plane* (literally, this means the 'kissing' plane), the plane of $\mathbf{N}$ and $\mathbf{B}$ is said to be the *normal plane*, and the plane of $\mathbf{T}$ and $\mathbf{B}$ is the *rectifying plane* (see Fig. 2.6) If we take the derivative of $\mathbf{B}$ with respect to s (arc length) we obtain the following

$$\frac{d\mathbf{B}}{ds} = \frac{d\mathbf{T}}{ds} \times \mathbf{N} + \mathbf{T} \times \frac{d\mathbf{N}}{ds}$$

$$= \kappa\mathbf{N} \times \mathbf{N} + \mathbf{T} \times \frac{d\mathbf{N}}{ds}$$

$$= \mathbf{T} \times \frac{d\mathbf{N}}{ds} = -\tau\mathbf{N}. \tag{2.22}$$

The quantity $|d\mathbf{B}/ds|$ is called the *torsion* of the curve at P. It is a measure of the arc rate of change of $\mathbf{B}$. This leads to the definition

$$\tau = \left|\frac{d\mathbf{B}}{ds}\right|. \tag{2.23}$$

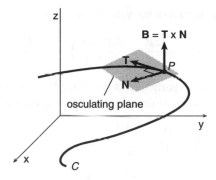

Figure 2.6   Geometric view of an osculating plane.

The larger the value of τ, the 'quicker' the curve moves away from the osculating plane at P.

In fact, it can be shown that the condition for a curve to be planar is that it has zero torsion. Now since

$$\mathbf{N} = \mathbf{B} \times \mathbf{T}$$

and

$$
\begin{aligned}
\frac{d\mathbf{N}}{ds} &= \frac{d\mathbf{B}}{ds} \times \mathbf{T} + \mathbf{B} \times \frac{d\mathbf{T}}{ds} \\
&= -\tau(\mathbf{N} \times \mathbf{T}) + \mathbf{B} \times (\kappa\mathbf{N}) \\
&= -\tau(-\mathbf{B}) - \kappa\mathbf{T} \\
&= \tau\mathbf{B} - \kappa\mathbf{T}.
\end{aligned}
$$

The three formulae:

$$\frac{d\mathbf{T}}{ds} = \kappa\mathbf{N},$$

$$\frac{d\mathbf{B}}{ds} = -\tau\mathbf{N}$$

and

$$\frac{d\mathbf{N}}{ds} = \tau\mathbf{B} - \kappa\mathbf{T}$$

are called the *Serret–Frenet* equations and are fundamental to the study of the differential geometry of curves.

Remark 2.2   If the position vector $\mathbf{R} = x\mathbf{i} + y\mathbf{j} + z\mathbf{k}$ of a particle moving along a curve in space is known, then its velocity and acceleration vectors can easily be determined as follows. We know the velocity vector is

$$\mathbf{V}(t) = \frac{d\mathbf{R}}{dt} = \frac{d\mathbf{R}}{ds}\left(\frac{ds}{dt}\right) = V\frac{d\mathbf{R}}{ds}.$$

where V is the speed of the particle and is defined by $V = ds/dt$. Now the acceleration vector $\mathbf{a}(t)$ can be determined in the following manner:

$$
\begin{aligned}
\mathbf{a}(t) &= \frac{d\mathbf{V}}{dt} = \frac{d^2\mathbf{R}}{dt^2} \\
&= \mathbf{T}\frac{dV}{dt} + V\frac{d\mathbf{T}}{dt} \\
&= \mathbf{T}\frac{dV}{dt} + V\left(\frac{d\mathbf{T}}{ds}\right)\left(\frac{ds}{dt}\right) \\
&= \mathbf{T}\frac{dV}{dt} + V^2\left(\frac{d\mathbf{T}}{ds}\right) \\
&= \mathbf{T}\frac{dV}{dt} + \kappa V^2\mathbf{N}.
\end{aligned}
$$

Since $\mathbf{T} \cdot \mathbf{T} = 1$, it follows that $\mathbf{T} \cdot d\mathbf{T}/ds = 0$ which implies that $d\mathbf{T}/ds$ is a vector normal to $\mathbf{T}$ and consequently $d\mathbf{T}/ds \propto \mathbf{N}$, where $\mathbf{N}$ is a unit normal vector to $\mathbf{T}$. Thus, $d\mathbf{T}/ds = \kappa\mathbf{N}$, where κ is a constant of proportionality, usually known as the curvature of the curve C. Hence, we have that

$$\left|\frac{d\mathbf{T}}{ds}\right| = \kappa.$$

Let $a_T = dV/dt$ and $a_N = \kappa V^2$. Then the acceleration vector will be given by

$$
\begin{aligned}
\mathbf{a}(t) &= a_T\mathbf{T} + a_N\mathbf{N} \\
|\mathbf{a}(t)| &= a_T^2 + a_N^2.
\end{aligned}
$$

To find the formula for the tangential and normal acceleration components, we start by taking the dot product

$$
\begin{aligned}
\mathbf{V} \cdot \mathbf{a} &= \mathbf{V} \cdot (a_T\mathbf{T} + a_N\mathbf{N}) \\
&= V\mathbf{T} \cdot (a_T\mathbf{T} + a_N\mathbf{N}) \\
&= Va_T,
\end{aligned}
$$

so that finally we have

$$a_T = \frac{\mathbf{V} \cdot \mathbf{a}}{V} = \frac{\dot{\mathbf{R}} \cdot \ddot{\mathbf{R}}}{|\dot{\mathbf{R}}|}$$

$$= \text{Tangential acceleration.}$$

Alternatively, if we take the cross product, we will obtain

$$\mathbf{V} \times \mathbf{a} = V\mathbf{T} \times (a_\mathrm{T}\mathbf{T} + a_\mathrm{N}\mathbf{N})$$
$$= Va_\mathrm{N}(\mathbf{T} \times \mathbf{N})$$
$$= Va_\mathrm{N}\mathbf{B},$$

from which we get

$$a_\mathrm{N} = \frac{|\mathbf{V} \times \mathbf{a}|}{V} = \frac{|\dot{\mathbf{R}} \times \ddot{\mathbf{R}}|}{|\dot{\mathbf{R}}|}$$
$$= \text{Normal acceleration.}$$

To find the curvature, κ, in terms of the position vector, observe the following

$$\kappa = \left|\frac{d\mathbf{T}}{ds}\right| = \frac{|d\mathbf{T}/dt|}{|ds/dt|} = \frac{|d\mathbf{T}/dt|}{|\dot{\mathbf{R}}|}.$$

We know that $\mathbf{V} = d\mathbf{R}/dt = V\mathbf{T}$ and that

$$\mathbf{a} = \mathbf{T}\frac{dV}{dt} + V\left|\frac{d\mathbf{T}}{dt}\right|\mathbf{N}.$$

We have seen that the vector product $\mathbf{V} \times \mathbf{a}$ can be written as

$$|\mathbf{V} \times \mathbf{a}| = V^2\left|\frac{d\mathbf{T}}{dt}\right|$$

from which we have

$$\left|\frac{d\mathbf{T}}{dt}\right| = \frac{|\mathbf{V} \times \mathbf{a}|}{V^2},$$

and since

$$\kappa = \frac{|d\mathbf{T}/dt|}{V},$$

then we have

$$\kappa = \frac{|\dot{\mathbf{R}} \times \ddot{\mathbf{R}}|}{|\dot{\mathbf{R}}|^3}$$

in terms of the position vector.

The last line follows from the fact that $\mathbf{V} = \dot{\mathbf{R}}$, $\mathbf{a} = \ddot{\mathbf{R}}$ and $V = |\dot{\mathbf{R}}|$.

$$\frac{d}{ds}(\mathbf{T} \cdot \mathbf{B}) = \frac{d\mathbf{T}}{ds} \cdot \mathbf{B} + \mathbf{T} \cdot \frac{d\mathbf{B}}{ds}$$
$$= \kappa\mathbf{N} \cdot \mathbf{B} + \mathbf{T} \cdot \frac{d\mathbf{B}}{ds}$$
$$= \mathbf{T} \cdot \frac{d\mathbf{B}}{ds}$$
$$\frac{d}{ds}(\mathbf{T} \cdot \mathbf{B}) = 0 = \mathbf{T} \cdot \frac{d\mathbf{B}}{ds}.$$

Thus, $\mathbf{T} \cdot d\mathbf{B}/ds = 0$, which implies that $d\mathbf{B}/ds$ is perpendicular to $\mathbf{T}$. Also, since $\mathbf{B}$ is a unit vector, $d\mathbf{B}/ds$ is perpendicular to $\mathbf{B}$. Therefore, $d\mathbf{B}/ds$ must be parallel to $\mathbf{N}$. Therefore, we can define

$$\frac{d\mathbf{B}}{ds} = -\tau \mathbf{N},$$

$$\left| \frac{d\mathbf{B}}{ds} \right| = \tau,$$

where τ is the torsion of the curve C, which measures the rate at which the osculating plane turns as we move along C. ∎

Example 2.3   Determine the unit vectors $\mathbf{T}$, $\mathbf{N}$ and $\mathbf{B}$ and the constants $\kappa = $ curvature and $\tau = $ torsion for the helix

$$\mathbf{R} = \mathbf{R}(t) = \mathbf{i} \cos t + \mathbf{j} \sin t + \mathbf{k}t.$$

□

Solution

$$\mathbf{T} = \frac{d\mathbf{R}}{ds} = \frac{d\mathbf{R}}{dt}\left(\frac{dt}{ds}\right) = \frac{d\mathbf{R}/dt}{(ds/dt)} = \frac{d\mathbf{R}/dt}{|d\mathbf{R}/dt|},$$

so that we have the following

$$\frac{d\mathbf{R}}{dt} = -\mathbf{i} \sin t + \mathbf{j} \cos t + \mathbf{k},$$

$$\left| \frac{d\mathbf{R}}{dt} \right| = \sqrt{\sin^2 t + \cos^2 t + 1} = \sqrt{2} = \left(\frac{ds}{dt} \right).$$

Thus, $\mathbf{T} = (-\mathbf{i} \sin t + \mathbf{j} \cos t + \mathbf{k})/\sqrt{2}$. Now we have

$$\frac{d\mathbf{T}}{dt} = \frac{1}{\sqrt{2}}(-\mathbf{i} \cos t - \mathbf{j} \sin t),$$

$$\frac{d\mathbf{T}}{ds} = \frac{d\mathbf{T}}{dt}\left(\frac{dt}{ds}\right) = \frac{d\mathbf{T}/dt}{(ds/dt)}$$

$$= \frac{1}{\sqrt{2}}\left(\frac{d\mathbf{T}}{dt}\right) = \frac{1}{2}(-\mathbf{i} \cos t - \mathbf{j} \sin t),$$

$$\frac{d\mathbf{T}}{ds} = \kappa \mathbf{N}, \quad \kappa = \left| \frac{d\mathbf{T}}{ds} \right| = \frac{1}{2}$$

and

$$N = \frac{dT/ds}{\kappa} = \frac{\frac{1}{2}(-i\cos t - j\sin t)}{\frac{1}{2}} = -i\cos t - j\sin t.$$

Using this quantity, we can then find

$$B = T \times N = \frac{1}{\sqrt{2}}\begin{vmatrix} i & j & k \\ -\sin t & \cos t & 1 \\ -\cos t & -\sin t & 0 \end{vmatrix}$$

$$= \frac{1}{\sqrt{2}}(i\sin t - j\cos t + k).$$

Therefore,

$$\frac{dB}{ds} = \frac{dB/dt}{(ds/dt)} = \frac{\frac{1}{\sqrt{2}}(i\cos t + j\sin t)}{\sqrt{2}}$$

$$= \frac{1}{2}(i\cos t + j\sin t),$$

$$\left|\frac{dB}{ds}\right| = \frac{1}{2}(1) = \frac{1}{2} = \tau = \text{Torsion.}$$

■

Example 2.4   Show that for a smooth curve $y = y(x)$ in the xy-plane

$$\kappa(x) = \frac{|y''|}{[1 + (y')^2]^{3/2}}.$$

□

Solution   The parametric equation to the curve is given by $x = x$ and $y = y(x)$. Then our position, velocity and acceleration vectors, respectively, are given by

$$R = xi + yj + 0k,$$
$$\dot{R} = i + y'j + 0k,$$
$$\ddot{R} = 0i + y''j + 0k.$$

Now we know that the curvature, κ, is given by

$$\kappa = \frac{|\dot{R} \times \ddot{R}|}{|\dot{R}|^3},$$

which then becomes

$$\kappa = \frac{|y''|}{(1 + (y')^2)^{3/2}}.$$

The last line follows since the quantity in the numerator is

$$\dot{\mathbf{R}} \times \ddot{\mathbf{R}} = \begin{vmatrix} \mathbf{i} & \mathbf{j} & \mathbf{k} \\ 1 & y' & 0 \\ 0 & y'' & 0 \end{vmatrix} = \mathbf{k} y''.$$

∎

Remark 2.3   We note here that both the curvature and torsion are constant for every point on the curve. It can be shown that the circular helix is the only curve for which this is true.   ∎

2.9 Equation of an osculating plane

Although a brief mention has been made of the osculating plane, we have not yet discussed how such a plane can be obtained. A close examination of Fig. 2.6 will reveal that the vectors associated with the plane all appear to be in the same plane. If we let $P(x, y, z)$ be a general point in the osculating plane and $P_1(x_1, y_1, z_1)$ be another vector in the plane, then we can impose the condition that the vector joining P to P_1 be coplanar with the vectors $\mathbf{T}$ and $d\mathbf{T}/ds$ at P_1. It follows that we can apply the scalar triple product to these three vectors and equate it to zero to obtain the equation of the required plane. In other words, we have

$$\mathbf{P_1 P} \cdot \left(\mathbf{T} \times \frac{d\mathbf{T}}{ds} \right) = 0.$$

Example 2.5   Find the equation of the osculating plane to the curve $x = \cos t$, $y = \sin t$ and $z = t$ at the point $P_1(1, 0, 0)$ when $t = 0$.   □

Solution   We know that $\mathbf{R} = \mathbf{i} \cos t + \mathbf{j} \sin t + \mathbf{k} t$, $|d\mathbf{R}/dt| = \sqrt{2}$ and

$$\mathbf{T} = \frac{1}{2}(-\mathbf{i} \sin t + \mathbf{j} \cos t + \mathbf{k}),$$

$$\frac{d\mathbf{T}}{ds} = \frac{1}{2}(-\mathbf{i} \cos t - \mathbf{j} \sin t),$$

$$\mathbf{T}(0) = \frac{1}{\sqrt{2}}(-\mathbf{i}(0) + \mathbf{j}(1) + \mathbf{k}),$$

$$\frac{d\mathbf{T}}{ds}(0) = -\frac{1}{2}(\mathbf{i}(1) + \mathbf{j}(0)),$$

$$\mathbf{P_1 P} = \mathbf{i}(x - 1) + \mathbf{j}(y - 0) + \mathbf{k}(z - 0).$$

Thus, the equation of the osculating plane is given by

$$P_1P \cdot \left(T \times \frac{dT}{ds} \right) = 0,$$

$$\begin{vmatrix} x & y & z \\ 0 & \frac{1}{\sqrt{2}} & \frac{1}{\sqrt{2}} \\ -\frac{1}{2} & 0 & 0 \end{vmatrix} = 0,$$

i.e.

$$y\left(-\frac{1}{2\sqrt{2}} \right) + z\left(\frac{1}{2\sqrt{2}} \right) = 0, \quad y = z.$$

∎

There is an easier alternate formula to determine the osculating plane. Here, we define T in the following manner

$$T = \frac{dR}{ds} = \frac{dR}{dt}\left(\frac{dt}{ds} \right),$$

$$\frac{dT}{ds} = \frac{d^2R}{dt^2}\left(\frac{dt}{ds} \right)^2 + \frac{dR}{dt}\left(\frac{d^2t}{ds^2} \right).$$

We can now define the scalar triple product as follows

$$P_1P \cdot \left(T \times \frac{dT}{ds} \right) = 0,$$

$$P_1P \cdot \left(\frac{dR}{dt} \right)\left(\frac{dt}{ds} \right) \times \left[\frac{d^2R}{dt^2}\left(\frac{dt}{ds} \right)^2 + \frac{dR}{dt}\left(\frac{d^2t}{ds^2} \right) \right] = 0,$$

$$P_1P \cdot \left[\frac{dR}{dt} \times \frac{d^2R}{dt^2} \right]\left(\frac{dt}{ds} \right)^3 = 0,$$

$$P_1P \cdot \left\{ \frac{dR}{dt} \times \frac{d^2R}{dt^2} \right\}_{t_0} = 0.$$

This final equation can be rewritten with the notation

$$P_1P \cdot (\dot{R} \times \ddot{R}) = 0. \tag{2.24}$$

The following example illustrates the use of this simple formula.

Example 2.6 What is the equation of the osculating plane to the space curve $x = t^4$, $y = t^2$ and $z = t^3$ at the point $P_1(1, 1, 1)$ when $t = 1$?   □

Solution   Let our position vector $\mathbf{R}$ be given by $\mathbf{R}(t) = t^4\mathbf{i} + t^2\mathbf{j} + t^3\mathbf{k}$. Then our corresponding derivatives are given by

$$\dot{\mathbf{R}} = (4t^3, 2t, 3t^2),$$
$$\ddot{\mathbf{R}} = (12t^2, 2, 6t)$$

and our vector is given by

$$\mathbf{P_1P} = (x - 1, y - 1, z - 1).$$

At $t = 1$, the derivatives above become $\dot{\mathbf{R}} = (4, 2, 3)$ and $\ddot{\mathbf{R}} = (12, 2, 6)$. Then forming the triple product, we have

$$\mathbf{P_1P} \cdot (\dot{\mathbf{R}} \times \ddot{\mathbf{R}}) = \begin{vmatrix} x - 1 & y - 1 & z - 1 \\ 4 & 2 & 3 \\ 12 & 2 & 6 \end{vmatrix} = 0,$$

$$3(x - 1) + 6(y - 1) - 8(z - 1) = 0,$$
$$3x + 6y - 8z = 1,$$

which is the equation of the osculating plane.   ∎

2.10 Summary

- Ordinary derivatives of sums, differences and products of vector functions of one independent variable:

$$\frac{d(\mathbf{U} \pm \mathbf{V})}{dt} = \frac{d\mathbf{U}}{dt} \pm \frac{d\mathbf{V}}{dt}$$

$$\frac{d(\phi\mathbf{V})}{dt} = \frac{d\phi}{dt}\mathbf{V} + \phi\frac{d\mathbf{V}}{dt}$$

$$\frac{d(\mathbf{U} \cdot \mathbf{V})}{dt} = \frac{d\mathbf{U}}{dt} \cdot \mathbf{V} + \mathbf{U} \cdot \frac{d\mathbf{V}}{dt}$$

$$\frac{d(\mathbf{U} \times \mathbf{V})}{dt} = \frac{d\mathbf{U}}{dt} \times \mathbf{V} + \mathbf{U} \times \frac{d\mathbf{V}}{dt}$$

$$\frac{d[\mathbf{U} \times (\mathbf{V} \times \mathbf{W})]}{dt} = \frac{d\mathbf{U}}{dt} \times (\mathbf{V} \times \mathbf{W}) + \mathbf{U} \times \left\{ \frac{d\mathbf{V}}{dt} \times \mathbf{W} \right\}$$

$$+ \mathbf{U} \times \left\{ \mathbf{V} \times \frac{d\mathbf{W}}{dt} \right\}.$$

- Integrals of vector functions of one independent variable:
  If
$$\mathbf{V}(t) = \mathbf{V}_1(t)\mathbf{i} + \mathbf{V}_2(t)\mathbf{j} + \mathbf{V}_3(t)\mathbf{k},$$

  then

$$\int \mathbf{V}(t)dt = \mathbf{i} \int \mathbf{V}_1(t)dt + \mathbf{j} \int \mathbf{V}_2(t)dt + \mathbf{k} \int \mathbf{V}_3(t)dt + \mathbf{C},$$

  where the constant vector is $\mathbf{C} = C_1\mathbf{i} + C_2\mathbf{j} + C_3\mathbf{k}$. When $\Delta\mathbf{R}$ tends to zero, the elementary arc length is given by

$$\Delta s = |\Delta\mathbf{R}|.$$

  If s and $\mathbf{R}$ are functions of the independent time variable t, and as $\Delta t \to 0$, then

$$\frac{ds}{dt} = \left|\frac{d\mathbf{R}}{dt}\right| = \sqrt{\frac{d\mathbf{R}}{dt} \cdot \frac{d\mathbf{R}}{dt}}.$$

  Therefore, the total arc length can be obtained from

$$s = \int_{a(t)}^{b(t)} \sqrt{\left(\frac{dx}{dt}\right)^2 + \left(\frac{dy}{dt}\right)^2 + \left(\frac{dz}{dt}\right)^2}\, dt.$$

- The following are important formulae from this chapter. Suppose $\mathbf{R} = \mathbf{R}(t)$ is a position vector as discussed in this chapter, then we have:

  1. $\mathbf{R} = x\mathbf{i} + y\mathbf{j} + z\mathbf{k}$.
  2. $\mathbf{T} = \frac{d\mathbf{R}/dt}{|d\mathbf{R}/dt|} = d\mathbf{R}/ds$.
  3. $\mathbf{N} = \frac{d\mathbf{T}/dt}{|d\mathbf{T}/dt|} = $ *Unit normal.*
  4. $d\mathbf{T}/ds = \kappa\mathbf{N}$.
  5. $\kappa = |d\mathbf{T}/ds| = \frac{|d\mathbf{T}/dt|}{(ds/dt)} = $ *Curvature.*
  6. $\mathbf{B} = \mathbf{T} \times \mathbf{N} = $ *Unit binormal.*
  7. $d\mathbf{B}/ds = \mathbf{T} \times d\mathbf{N}/ds = -\tau\mathbf{N}$.
  8. $\tau = |d\mathbf{B}/ds| = \frac{|d\mathbf{B}/dt|}{(ds/dt)} = $ *Torsion.*
  9. $d\mathbf{N}/ds = \tau\mathbf{B} - \kappa\mathbf{T}$.
  10. *Serret–Frenet formulae:*
      (a) $d\mathbf{T}/ds = \kappa\mathbf{N}$,
      (b) $d\mathbf{B}/ds = -\tau\mathbf{N}$,
      (c) $d\mathbf{N}/ds = \tau\mathbf{B} - \kappa\mathbf{T}$.
  11. *The equation of an osculating plane at $t = t_0$:*

$$\mathbf{P_1P} \cdot \left\{\frac{d\mathbf{R}}{dt} \times \frac{d^2\mathbf{R}}{dt^2}\right\}_{t_0} = 0.$$

  12. *Tangential acceleration* $= a_T = (\dot{\mathbf{R}} \cdot \ddot{\mathbf{R}})/|\dot{\mathbf{R}}|$; *Normal acceleration* $= a_N = |\dot{\mathbf{R}} \times \ddot{\mathbf{R}}|/|\dot{\mathbf{R}}|$; *Curvature* $= \kappa = |\dot{\mathbf{R}} \times \ddot{\mathbf{R}}|/|\dot{\mathbf{R}}|^3$.

Exercises

1. Let $\mathbf{R}(t) = \sin t\,\mathbf{i} + \cos t\,\mathbf{j} + \mathbf{k}$.
   (a) Find $\mathbf{R}'(t)$.
   (b) Show that $\mathbf{R}'(t)$ is always parallel to the xy-plane.
   (c) For what values of t is $\mathbf{R}'(t)$ parallel to the xz-plane?
   (d) Does $\mathbf{R}(t)$ have constant magnitude?
   (e) Does $\mathbf{R}'(t)$ have constant magnitude?
   (f) Compute $\mathbf{R}''(t)$.

2. Find $\mathbf{R}'(t)$ in each of the following cases.
   (a) $\mathbf{R}(t) = \sin t\,\mathbf{i} + e^{-t}\mathbf{j} + 3\mathbf{k}$.
   (b) $\mathbf{R}(t) = (e^t\mathbf{i} + \mathbf{j} + t^2\mathbf{k}) \times (t^3\mathbf{i} + \mathbf{j} - \mathbf{k})$.
   (c) $\mathbf{R}(t) = (\sin t + t^3)(\mathbf{i} + \mathbf{j} + 2\mathbf{k})$.

3. Show that

$$\frac{d}{dt}\left(\mathbf{R} \times \frac{d\mathbf{R}}{dt}\right) = \mathbf{R} \times \frac{d^2\mathbf{R}}{dt^2}.$$

4. For the curve $x = \sin t - t\cos t$, $y = \cos t + t\sin t$, $z = t^2$, find
   (a) the arc length between $(0, 1, 0)$ and $(-2\pi, 1, 4\pi^2)$.
   (b) $\mathbf{T}(t)$.
   (c) $\mathbf{T}(\pi)$.

In the following four problems, the coordinates of a moving particle are given as a function of time t. Find: (a) the speed; (b) the tangential and normal components of acceleration; (c) the unit tangent vector $\mathbf{T}$; and (d) the curvature of the curve, as functions of time.

5. $x = e^t\cos t$, $y = e^t\sin t$, $z = 0$.
6. $x = 3t\cos t$, $y = 3t\sin t$, $z = 4t$.
7. $x = e^t\cos t$, $y = e^t\sin t$, $z = e^t$.
8. $x = 5\sin 4t$, $y = 5\cos 4t$, $z = 10t$.

9. The position vector of a moving particle is

$$\mathbf{R}(t) = \cos t\,(\mathbf{i} - \mathbf{j}) + \sin t\,(\mathbf{i} + \mathbf{j}) + \frac{1}{2}t\mathbf{k}.$$

   (a) Determine the velocity and speed of the particle.
   (b) Determine the acceleration of the particle.
   (c) Find a unit tangent to the path of the particle in the direction of motion.
   (d) Show that the curve traversed by the particle has constant curvature κ, and find its value.

For the following problems, find: (a) the unit tangent and unit normal vectors $\mathbf{T}(t)$ and $\mathbf{N}(t)$; (b) the curvature.

10. $\mathbf{R}(t) = (\sin 4t,\, 3t,\, \cos 4t)$.
11. $\mathbf{R}(t) = (6t,\, 3\sqrt{2}t^2,\, 2t^3)$.

12. $\mathbf{R}(t) = (\sqrt{2}\cos t, \sin t, \sin t)$.
13. $\mathbf{R}(t) = (t^2, 2t, \ln t)$.

14. Find the vectors $\mathbf{T}$, $\mathbf{N}$ and $\mathbf{B}$ at the given point
    (a) $\mathbf{R}(t) = (t^2, \frac{2}{3}t^3, t)$, $(1, \frac{2}{3}, 1)$.
    (b) $\mathbf{R}(t) = (e^t, e^t \sin t, e^t \cos t)$, $(1, 0, 1)$.
15. Find the equations of the normal plane and osculating plane of the curve at the given point:
    (a) $x = 2\sin 3t$, $y = t$, $z = 2\cos 3t$, $(0, \pi, -2)$.
    (b) $x = t$, $y = t^2$, $z = t^3$, $(1, 1, 1)$.
16. The position function of a particle is given by $\mathbf{R}(t) = (t^2, 5t, t^2 - 16t)$. When is the speed a minimum?
17. A force with magnitude 20 N acts directly upward from the xy-plane on an object with mass 4 kg. The object starts at the origin with initial velocity $\mathbf{V}(0) = \mathbf{i} - \mathbf{j}$. Find its position function and its speed at time t.
18. The position function of a spaceship is

$$\mathbf{R}(t) = (3 + t)\mathbf{i} + (2 + \ln t)\mathbf{j} + \left(z - \frac{4}{t^2 + 1}\right)\mathbf{k}$$

and the coordinates of a space station are $(6, 4, 9)$. If the captain wants the spaceship to coast into the space station, when should the engines be turned off?
19. Let C be the curve with equations $x = 2 - t^3$, $y = 2t - 1$, $z = \ln t$. Find:
    (a) the point where C intersects the xz-plane;
    (b) parametric equations of the tangent line $(1, 1, 0)$;
    (c) an equation of the normal plane to C at $(1, 1, 0)$.
20. (a) Find an equation of the osculating plane of the curve $x = \sin 2t$, $y = t$, $z = \cos 2t$ at the point $(0, \pi, 1)$.
    (b) A particle starts at the origin with initial velocity $\mathbf{i} + 2\mathbf{j} + \mathbf{k}$. Its acceleration is $\mathbf{a}(t) = t\mathbf{i} + \mathbf{j} + t^2\mathbf{k}$. Find its position function.
    (c) Find the tangential and normal components of the acceleration vector of a particle with position function $\mathbf{R}(t) = t\mathbf{i} + 2t\mathbf{j} + t^2\mathbf{k}$.
21. If $\mathbf{P} = \mathbf{A}\cos \omega t + \mathbf{B}\sin \omega t$, where $\mathbf{A}$ and $\mathbf{B}$ are arbitrary constant vectors, show that $\mathbf{P} \times d\mathbf{P}/dt$ is a constant and that $d^2\mathbf{P}/dt^2 + \omega^2\mathbf{P} = \mathbf{0}$.
22. If $\mathbf{R} = t^2\mathbf{i} - t^3\mathbf{j} + t^4\mathbf{k}$ is the vector from the origin to a moving particle, find the velocity of the particle when $t = 1$. What is the component of this velocity in the direction of the vector $8\mathbf{i} - \mathbf{j} + 4\mathbf{k}$? What is the vector acceleration of the particle? What are the tangential and normal components of this acceleration?
23. If a particle starts to move from rest at the point $(0, 1, 2)$ with component acceleration $a_x = 1 + t$, $a_y = t^3$, $a_z = 2t - t^2$, find the vector from the origin to the instantaneous position of the particle.
24. At $t = 0$, a body starts to move from the point $(-1, 0, 1)$ with velocity $v_0 = \mathbf{i} + 2\mathbf{j} - 3\mathbf{k}$. If the acceleration of the body is $a(t) = 3t^2\mathbf{i} + \mathbf{j} + (\cos 2t)\mathbf{k}$, find the position of the body at any subsequent time. At what velocity is the body moving at $t = 1$?

Chapter 3

Partial derivatives of functions of several variables

3.1 Introduction

In the previous chapter, we have seen how the analysis of the motion of a body depends only on one parameter, i.e. time, and the mathematical investigation leads to an ordinary differential equation that consists of ordinary derivatives. However, in our real world problems, the dependent variable does depend on more than one variable, specially the space variables. In that situation, the problem has to be formulated in such a manner that the dependent variable depends on more than one independent variable, and the mathematical equation must be formulated with the partial derivatives. That leads to the partial differential equation, the solution of which will eventually contain several independent variables. In this chapter, we shall discuss such equations as they commonly arise in applied mathematics. We shall begin by first defining the partial derivatives of a function with two independent variables and then extend it to three or more variables. Some practical examples are illustrated.

3.2 Partial derivatives

Before proceeding to the next chapter, it is necessary to briefly discuss and introduce the concept of partial derivatives. From elementary calculus, we are familiar with functions of a single variable. In real life, physical quantities usually depend on two or more variables. Fortunately, we merely need to extend the ideas of single-variable calculus to functions of several variables.

The temperature T at a point on the surface of the earth at any given time depends on the longitude x and latitude y of the point. We can think of T as being a function of the two variables x and y, or as a function of the pair (x, y). This functional dependence is indicated by writing $T = f(x, y)$. The volume V of a circular cylinder depends on its radius r and height h. We know from elementary calculus that $V = \pi r^2 h$. Here, V is said to be a function of r and h, and we write $V(r, h) = \pi r^2 h$. As we have observed in the two examples, we often write $z = f(x, y)$ to make explicit the value taken on by f at the general point (x, y). The variables x and y are *independent variables* and z is the *dependent variable*. One can compare this with the single variable notation $y = f(x)$. Functions of two or more variables are also known as multivariable functions.

Suppose that f is a function of two variables x and y. If we let only x vary while keeping y fixed, say $y = b$, then we are only considering a function of a single variable x, namely $z = f(x, b)$. If z has a derivative at a, then we call it the partial derivative of f with respect to x at (a, b). Similarly, the partial derivative of f with respect to y at (a, b) is obtained by keeping x fixed $(x = a)$ and finding the ordinary derivative at b of the function $z = f(a, y)$.

The same limiting notation for single variable functions is applicable to multivariable functions. This leads us to the following definition for partial differentiation.

Definition 3.1   *The partial derivative of a function $f(x, y)$ with respect to x is*

$$\frac{\partial f}{\partial x} = \lim_{\Delta x \to 0} \frac{f(x + \Delta x, y) - f(x, y)}{\Delta x}, \tag{3.1}$$

and the partial derivative with respect to y is

$$\frac{\partial f}{\partial y} = \lim_{\Delta y \to 0} \frac{f(x, y + \Delta y) - f(x, y)}{\Delta y}. \tag{3.2}$$

It can be clearly seen from (3.1) that the partial derivative of $f(x, y)$ with respect to x is simply the ordinary derivative of $f(x, y)$ with respect to x, where y is considered a constant. Similarly, $\partial f / \partial y$ is the ordinary derivative of $f(x, y)$ with respect to y, holding x constant. For partial derivatives of functions of more than two independent variables, we again permit only one variable to vary, while holding all the others constant. As an example, the partial derivative of $f(x, y, z, t, \ldots)$ with respect to z is

$$\frac{\partial f}{\partial z} = \lim_{\Delta z \to 0} \frac{f(x, y, z + \Delta z, t, \ldots) - f(x, y, z, t, \ldots)}{\Delta z}.$$

Therefore, we differentiate with respect to z, while treating $x, y, t, \ldots$ as constants. There are other commonly used notations for the partial derivative. If $z = f(x, y)$, we write

$$f_x(x, y) = f_x = \frac{\partial f}{\partial x} = \frac{\partial}{\partial x} f(x, y) = \frac{\partial z}{\partial x} = D_x f,$$

$$f_y(x, y) = f_y = \frac{\partial f}{\partial y} = \frac{\partial}{\partial y} f(x, y) = \frac{\partial z}{\partial y} = D_y f.$$

Note that here we cannot interpret $\partial f / \partial x$ as a ratio of differentials.

Example 3.1   If $f(x, y) = x^3 + x^2 y^3 - 2y^2$, find $f_x(2, 1)$ and $f_y(2, 1)$.   □

Solution   Holding y constant and differentiating partially with respect to x, we get

$$f_x(x, y) = 3x^2 + 2xy^3$$

and so

$$f_x(2, 1) = 3 \cdot 2^2 + 2 \cdot 2 \cdot 1^3 = 16.$$

Holding x constant and differentiating partially with respect to y, we get

$$f_y(x, y) = 3x^2 y^2 - 4y,$$
$$f_y(2, 1) = 3 \cdot 2^2 \cdot 1^2 - 4 \cdot 1 = 8.$$

∎

Example 3.2   Find f_x, f_y and f_z if $f(x, y, z) = e^{xy} \ln z$.   □

Solution   Holding y and z constant and differentiating partially with respect to x, we have

$$f_x = ye^{xy} \ln z.$$

Similarly,

$$f_y = xe^{xy} \ln z \quad \text{and} \quad f_z = \frac{e^{xy}}{z}.$$

∎

Example 3.3   Find $\partial z / \partial x$ and $\partial z / \partial y$ if $z = f(x, y) = x^2 y^3 + e^{xy}$.   □

Solution   For this function,

$$\frac{\partial z}{\partial x} = 2xy^3 + ye^{xy} \quad \text{and} \quad \frac{\partial z}{\partial y} = 3x^2 y^2 + xe^{xy}.$$

∎

3.3 Iterated partial derivatives

If f is a function of two variables, then its partial derivatives f_x and f_y are also functions of two variables. We can, therefore, consider their partial derivatives to be $(f_x)_x$, $(f_x)_y$, $(f_y)_x$ and $(f_y)_y$, which are called the *second partial derivatives* of f. If $z = f(x, y)$, we use the following notation for the second partial derivatives:

$$(f_x)_x = f_{xx} = \frac{\partial}{\partial x}\left(\frac{\partial f}{\partial x}\right) = \frac{\partial^2 f}{\partial x^2} = \frac{\partial^2 z}{\partial x^2},$$

$$(f_x)_y = f_{xy} = \frac{\partial}{\partial y}\left(\frac{\partial f}{\partial x}\right) = \frac{\partial^2 f}{\partial y \partial x} = \frac{\partial^2 z}{\partial y \partial x},$$

$$(f_y)_x = f_{yx} = \frac{\partial}{\partial x}\left(\frac{\partial f}{\partial y}\right) = \frac{\partial^2 f}{\partial x \partial y} = \frac{\partial^2 z}{\partial x \partial y},$$

$$(f_y)_y = f_{yy} = \frac{\partial}{\partial y}\left(\frac{\partial f}{\partial y}\right) = \frac{\partial^2 f}{\partial y^2} = \frac{\partial^2 z}{\partial y^2}.$$

The notation f_{xy} (or $\partial^2 f / \partial x \partial y$) means that we first differentiate with respect to x and then with respect to y, as opposed to f_{yx} where the order is reversed.

Example 3.4   Find the second partial derivatives of

$$f(x, y) = x^3 + x^2 y^3 - 2y^2.$$

$\square$

Solution   In this example, we find that the first derivatives are

$$f_x(x, y) = 3x^2 + 2xy^3 \quad \text{and} \quad f_y(x, y) = 3x^2 y^2 - 4y.$$

Therefore, we obtain

$$f_{xx} = \frac{\partial}{\partial x}(3x^2 + 2xy^3) = 6x + 2y^3,$$

$$f_{xy} = \frac{\partial}{\partial y}(3x^2 + 2xy^3) = 6xy^2,$$

$$f_{yx} = \frac{\partial}{\partial x}(3x^2 y^2 - 4y) = 6xy^2,$$

$$f_{yy} = \frac{\partial}{\partial y}(3x^2 y^2 - 4y) = 6x^2 y - 4.$$

From this example, we can note that $f_{xy} = f_{yx}$. This is not merely coincidence. It has been proven that the mixed partial derivatives f_{xy} and f_{yx} are equal for most functions that one meets in practice. Most texts on advanced calculus cite the French mathematician Alexis Clairaut's (1713–1765) theorem giving the conditions under which we can assert that $f_{xy} = f_{yx}$.

Example 3.5   The general form of the wave equation is

$$u_{tt} = c^2 u_{xx}.$$

Verify that $u = f(x - ct) + g(x + ct)$ is a solution of the above equation, where f and g are arbitrary progressive and regressive wave functions, respectively.   □

Solution   Given that $u = f(x - ct) + g(x + ct)$ is a solution, we need to find the partial derivatives u_{tt} and u_{xx} first. Therefore, we have

$$u_t = -cf'(x - ct) + cg'(x + ct),$$

so that

$$u_{tt} = c^2 f''(x - ct) + c^2 g''(x + ct),$$

and also,

$$u_x = f'(x - ct) + g'(x + ct),$$

which leads to

$$u_{xx} = f''(x - ct) + g''(x + ct)$$

and one can see that

$$u_{tt} = c^2 (f''(x - ct) + g''(x + ct))$$
$$= c^2 u_{xx}.$$

■

3.4 Solitary wave solution

The theory of shallow-water waves, and subsequently the solitary wave, is an important area of scientific investigation because of its enormous practical applications. The KdV equation, which leads to the theory of solitons, has been the object of intensive scientific research in different branches of physics. In the following example, we shall demonstrate how the partial derivatives play important roles in the solution process of this important problem.

Example 3.6   The partial differential equation $u_t + uu_x + u_{xxx} = 0$, called the *Korteweg–de Vries equation* (KdV equation), describes the motion of water waves in a shallow channel (see Rahman 1995).

(a)   Show that for any positive constant c, the function

$$u(x, t) = 3c \, \text{sech}^2 \left[\frac{\sqrt{c}}{2} (x - ct) \right]$$

is a solution of the KdV equation. (This solution represents a travelling *hump* of water in the channel and is called a *soliton*, which was first observed by J. Scott Russell around 1840 in barge canals near Edinburgh.)

(b)   How do the shape and speed of the soliton depend on c?   ☐

Solution   (a) We compute u_t, u_x, u_{xx} and u_{xxx} using the chain rule and the differentiation formula $(d/dx)\text{sech}\, x = -\text{sech}\, x \tanh x$ from single-variable calculus. Then by letting $\eta = (x - ct)\sqrt{c}/2$, we have

$$u = 3c \, \text{sech}^2 \eta,$$

$$u_t = 6c \, \text{sech}\, \eta \frac{\partial}{\partial t} \text{sech}\, \eta$$

$$= -6c \, \text{sech}^2 \eta \tanh \eta \frac{\partial \eta}{\partial t}$$

$$= 3c^{5/2} \, \text{sech}^2 \eta \tanh \eta$$

$$= c^{3/2} u \tanh \eta.$$

Also,

$$u_x = -6c \, \text{sech}^2 \eta \tanh \eta \frac{\partial \eta}{\partial x}$$

$$= -3c^{3/2} \, \text{sech}^2 \eta \tanh \eta$$

$$= -\sqrt{c} u \tanh \eta,$$

$$cu_x = -3c^{5/2} \text{sech}^2 \eta \tanh \eta$$

$$= -u_t,$$

so that $u_t = -cu_x$ and this leads to

$$u_{xx} = -\sqrt{c}(\tanh \eta)u_x - \sqrt{c} u \, \text{sech}^2 \eta \frac{\sqrt{c}}{2}$$

$$= -\sqrt{c} u_x \tanh \eta - \frac{c}{2} u \, \text{sech}^2 \eta$$

$$= -\sqrt{c} u_x \tanh \eta - \frac{u^2}{6}.$$

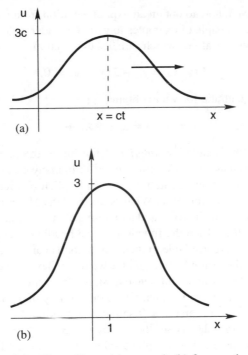

Figure 3.1 The soliton: (a) general; (b) for $c = 1, t = 1$.

Now using the value of u_x and also the formula $\tanh^2 \eta = 1 - \operatorname{sech}^2 \eta$, u_{xx} can be reduced to a simple form as

$$u_{xx} = cu - \frac{u^2}{2}.$$

And, hence, differentiating partially with respect to x once, we have

$$u_{xxx} = cu_x - uu_x.$$

Therefore, using the relation $u_t = -cu_x$ and transposing the result, we have

$$u_t + uu_x + u_{xxx} = 0.$$

(b) The speed of the soliton is c, since the graph in Fig. 3.1a moves c units in x in one unit of t. The shape at time $t = 1$ is shown in Fig. 3.1b for $c = 1$.   ■

3.5 Implicit functions and the concept of Jacobians

This section deals with the implicit functions and their differentiations with the relevant concept of Jacobians. It is important to know that if an algebraic equation is implicitly involved, how to obtain an explicit solution from its implicit relation. We shall start with a couple of examples to give the reader a clear-cut view.

If y is a function of x alone, which satisfies the relation

$$F(x, y) = y^2 - 2xy - x^3 = 0, \tag{3.3}$$

we can solve the quadratic for y and obtain

$$y = f(x) = x \pm x\sqrt{1 + x}. \tag{3.4}$$

The first formulation is called an *implicit* definition of the function $y(x)$, but in the second y is given as an *explicit* function of x. In many cases, it is not possible to give a simple explicit representation from an implicit relation $F(x, y)$, yet we want to treat y as a function of x. What is needed, therefore, is some assurance that y does exist as a function of x and a precise statement of the conditions under which this will be true. Then the function can be handled with confidence and if no explicit formulae are available, numerical methods can be resorted to. Care is needed because in the example given above we really have two explicit functions, corresponding to the positive and negative signs, from the one implicit relation. Accordingly, we have to restrict attention to the vicinity of a particular point. For example, the values $x = 3$ and $y = 9$ satisfy the relation $F(x, y) = 0$, but only one of the explicit formulae. Also, the values $x = 3$ and $y = -3$ will satisfy the given relation. So, there are two explicit values of y obtained from one value of x. We can assert, however, that there exists a function $y = f(x)$, which satisfies $F(x, y) = F(x, f(x)) = 0$ and for which $f(3) = 9, -3$, even though we cannot obtain $f(x) = x \pm x\sqrt{1 + x}$. This is useful if we want to treat x as a function of y, say $g(y)$, for here no explicit formula exists. At the point $x = -1, y = -1$ (or $x = y = 0$) we run into trouble because here both the formulae in (3.4) are valid. We shall obtain criteria for such points as these.

We shall state some general theorems, without proof, that are required in the text and give some specific applications.

Theorem 3.1   Let $x = x_0$ and $y = y_0$ be the pair of values satisfying $F(x, y) = 0$ and let F and its first derivatives be continuous in the neighbourhood of this point. Then, if $\partial F / \partial y$ does not vanish at $x = x_0$ and $y = y_0$, there exists one and only one continuous function $f(x)$ such that $F(x, f(x)) = 0$ and $f(x_0) = y_0$.

A simple illustration will clarify this theorem. Let us consider the above example (3.3) where $F(x, y) = y^2 - 2xy - x^3$. Implicit differentiation with respect to x gives

$$2yy' - 2y - 2xy' - 3x^2 = 0$$

after solving for y', we obtain

$$\frac{dy}{dx} = \frac{2y + 3x^2}{2(y - x)}.$$

However, the same result can be obtained if we use the chain rule to differentiate the function $F(x, y) = 0$. We can differentiate it with respect to x and obtain

$$\frac{dF}{dx} = \frac{\partial F}{\partial x} + \frac{\partial F}{\partial y}\frac{dy}{dx}.$$

We equate the derivatives of the right-hand side of the equation and we find

$$\frac{\partial F}{\partial x} + \frac{\partial F}{\partial y}\frac{dy}{dx} = 0,$$

which reduces to

$$\frac{dy}{dx} = -\frac{\partial F/\partial x}{\partial F/\partial y} = \frac{2y + 3x^2}{2(y - x)}.$$

It can be easily seen that if $y = x$, that means $\partial F/\partial y = 0$, the gradient $dy/dx \to \infty$, which in turn implies that $\partial F/\partial y = 0$, confirming that there exists no solution. Hence, to have a solution, the condition is that $\partial F/\partial y$ must not vanish anywhere in the given domain. This confirms the theorem without any formal proof. This derivative $\partial F/\partial y$ in essence can be regarded as the one-dimensional Jacobian. The formal definition of *Jacobian* will be given in the following theorem with a general set of functions with several variables.

Theorem 3.2   *Between the $m + n$ variables $x_1, \ldots, x_m$; $y_1, \ldots, y_n$ there is a system of n relations*

$$F_j(x_1, \ldots, x_m; y_1, \ldots, y_n) = 0, \quad j = 1, \ldots, n.$$

If these equations are satisfied for $x_i = x_i^0$, $y_j = y_j^0$ and F_j are continuous with continuous first partial derivatives in the neighbourhood of these values and

$$J = \frac{\partial(F_1, \ldots, F_n)}{\partial(y_1, \ldots, y_n)} = \begin{vmatrix} \frac{\partial F_1}{\partial y_1} & \cdots & \frac{\partial F_1}{\partial y_n} \\ \cdots & \cdots & \cdots \\ \frac{\partial F_n}{\partial y_1} & \cdots & \frac{\partial F_n}{\partial y_n} \end{vmatrix} \quad (3.5)$$

does not vanish there, then there is one and only one set of solutions

$$y_j = f_j(x_1, \ldots, x_m),$$

which are continuous, satisfy relation $F_j = 0$ and for which $y_j^0 = f_j(x_1^0, \ldots, x_m^0)$.

Relation (3.5) is usually known as the *Jacobian* for n variables. This is the key to changing variables in the multiple integral. A little life sketch of Carl Gustav Jacob Jacobi (1804–1851) is worth noting. Born into a rich German family, the young Carl Gustav excelled in many areas of study, but his ability and love for intricate algebraic calculations led him to the life of a working and impoverished mathematician and teacher. His PhD dissertation was related to a topic now known to every student of calculus: partial fractions. Jacobi's greatest contributions to mathematics were in the fields of elliptic functions and number theory. He also made major contributions to the theory of determinants and to the simplification of that theory. He was principally a *pure* mathematician, but every student of dynamics and quantum mechanics will recognize Jacobi's contribution to these areas through the famous Hamilton–Jacobi equations.

The particular form in which this theorem is most needed is the inversion of a functional transformation. If $m = n$ and F_j has the form

$$g_j(y_1, \ldots, y_n) - x_j = 0,$$

where the g_j are continuous with continuous derivatives, then the theorem takes the following form.

Theorem 3.3   *If $x_j = g_j(y_1, \ldots, y_n)$, $j = 1, \ldots, n$, are n continuous functions of the variables $y_1, \ldots, y_n$ with continuous first partial derivatives, and if the Jacobian*

$$J = \frac{\partial(x_1, \ldots, x_n)}{\partial(y_1, \ldots, y_n)}$$

does not vanish in a given domain, then the transformation from y to x can be uniquely inverted to give

$$y_i = f_i(x_1, \ldots, x_n).$$

Example 3.7   Carry out the proof of Theorem 3.2 in detail to show that

$$F(x, y, z; u, v) = 0,$$
$$G(x, y, z; u, v) = 0,$$

give $u = f(x, y, z)$ and $v = g(x, y, z)$ if the Jacobian

$$\frac{\partial(F, G)}{\partial(u, v)} \neq 0.$$

$\square$

Proof   We pose the problem of finding the six first-order partial derivatives of u and v with respect to x, y and z, supposing that it is undesirable or even impossible to obtain the explicit form of the functions. To do this, we note that by substituting the values of u and v into the functional equations we have

$$F(x, y, z; f(x, y, z), g(x, y, z)) = 0,$$
$$G(x, y, z; f(x, y, z), g(x, y, z)) = 0,$$

which would be identities in x, y and z. As a result, we can differentiate each equation with respect to x partially and obtain using the chain rule

$$\frac{\partial F}{\partial x} + \frac{\partial F}{\partial u}\frac{\partial u}{\partial x} + \frac{\partial F}{\partial v}\frac{\partial v}{\partial x} = 0,$$

$$\frac{\partial G}{\partial x} + \frac{\partial G}{\partial u}\frac{\partial u}{\partial x} + \frac{\partial G}{\partial v}\frac{\partial v}{\partial x} = 0. \tag{3.6}$$

Equation (3.6) can be written as

$$\frac{\partial F}{\partial u}\frac{\partial u}{\partial x} + \frac{\partial F}{\partial v}\frac{\partial v}{\partial x} = -\frac{\partial F}{\partial x},$$

$$\frac{\partial G}{\partial u}\frac{\partial u}{\partial x} + \frac{\partial G}{\partial v}\frac{\partial v}{\partial x} = -\frac{\partial G}{\partial x}. \tag{3.7}$$

There are two unknowns $\partial u/\partial x$ and $\partial v/\partial x$ with two equations, and because the equations are linear in unknowns, solutions can be obtained by Cramer's rule. In particular,

$$\frac{\partial u}{\partial x} = \frac{\begin{vmatrix} -F_x & F_v \\ -G_x & G_v \end{vmatrix}}{\begin{vmatrix} F_u & F_v \\ G_u & G_v \end{vmatrix}}$$

$$= -\frac{\begin{vmatrix} F_x & F_v \\ G_x & G_v \end{vmatrix}}{\begin{vmatrix} F_u & F_v \\ G_u & G_v \end{vmatrix}}.$$

The solution will exist if the Jacobian

$$J = \begin{vmatrix} F_u & F_v \\ G_u & G_v \end{vmatrix}$$

does not vanish.   ■

In a similar manner, we could determine the values of $\partial v/\partial x$ and also other values.

Example 3.8   If $x^2 y^2 u^3 + ux \sin y = 5$ defines u as a function of x and y, determine $\partial u/\partial x$.   □

Solution   If we set $F(x, y, , u) = x^2 y^2 u^3 + ux \sin y - 5$, then

$$\frac{\partial u}{\partial x} = -\frac{F_x}{F_u} = -\frac{2xy^2u^3 + u \sin y}{3x^2 y^2 u^2 + x \sin y}.$$

■

Example 3.9 The equation of state for a thermodynamic system is $F(P, V, T) = 0$, where P, V and T are pressure, volume and temperature, respectively. If the equation defines V as a function of P and T, and also defines T as a function of P and V, show that

$$\frac{\partial V}{\partial T} = -\frac{\partial F/\partial T}{\partial F/\partial V} = \frac{1}{\partial T/\partial V}.$$

$\square$

Proof Given that $F(P, V, T) = 0$. Then differentiating F partially with respect to V and T in turn, we obtain

$$\frac{\partial F}{\partial V} + \frac{\partial F}{\partial T}\frac{\partial T}{\partial V} = 0,$$

$$\frac{\partial F}{\partial T} + \frac{\partial F}{\partial V}\frac{\partial V}{\partial T} = 0.$$

Thus solving for $\partial V/\partial T$ and $\partial T/\partial V$, we obtain

$$\frac{\partial V}{\partial T} = -\frac{\partial F/\partial T}{\partial F/\partial V} = \frac{1}{\partial T/\partial V}.$$

This is the desired proof.   ■

Example 3.10 If the equations $F(x, y, u, v) = 0$ and $G(x, y, u, v) = 0$ define u and v as functions of x and y, and also define x and y as functions of u and v, show that

$$\frac{\partial(u, v)}{\partial(x, y)} = \frac{1}{\partial(u, v)/\partial(x, y)}.$$

$\square$

Solution From the given statement, we can write the two equations in a simple form

$$u = u(x, y),$$
$$v = v(x, y).$$

And also if the inversion exists, then

$$x = x(u, v),$$
$$y = y(u, v).$$

Now, the total derivatives of these functions by the chain rule are given by

$$du = \frac{\partial u}{\partial x}dx + \frac{\partial u}{\partial y}dy,$$

$$dv = \frac{\partial v}{\partial x}dx + \frac{\partial v}{\partial y}dy,$$

$$dx = \frac{\partial x}{\partial u}du + \frac{\partial x}{\partial v}dv,$$

$$dy = \frac{\partial y}{\partial u}du + \frac{\partial y}{\partial v}dv.$$

From the first two equations we can obtain the value of dx and dy by using Cramer's rule to give

$$dx = \frac{\begin{vmatrix} du & u_y \\ dv & v_y \end{vmatrix}}{\begin{vmatrix} u_x & u_y \\ v_x & v_y \end{vmatrix}} = \frac{1}{J}\{v_y du - u_y dv\},$$

$$dy = \frac{\begin{vmatrix} u_x & du \\ v_x & dv \end{vmatrix}}{\begin{vmatrix} u_x & u_y \\ v_x & v_y \end{vmatrix}} = \frac{1}{J}\{u_x dv - v_x du\},$$

where $J = \partial(u, v)/\partial(x, y)$.

But using the differential value of dx and dy, we obtain

$$x_u du + x_v dv = \frac{1}{J}\{v_y du - u_y dv\},$$

$$y_u du + y_v dv = \frac{1}{J}\{u_x dv - v_x du\}.$$

Comparing the coefficients of du and dv yields

$$\frac{\partial x}{\partial u} = \frac{1}{J}\frac{\partial v}{\partial y},$$

$$\frac{\partial x}{\partial v} = -\frac{1}{J}\frac{\partial u}{\partial y},$$

$$\frac{\partial y}{\partial u} = -\frac{1}{J}\frac{\partial v}{\partial x},$$

$$\frac{\partial y}{\partial v} = \frac{1}{J}\frac{\partial u}{\partial x}.$$

It is an easy matter to determine the solutions for $\partial x/\partial u$ and $\partial u/\partial x$ from the above information, and they are given by

$$\frac{\partial u}{\partial x} = \frac{y_v}{\begin{vmatrix} x_u & x_v \\ y_u & y_v \end{vmatrix}},$$

$$\frac{\partial x}{\partial u} = \frac{v_y}{\begin{vmatrix} u_x & v_x \\ u_y & v_y \end{vmatrix}}.$$

Hence, upon multiplication it is evident that

$$\begin{vmatrix} u_x & u_y \\ v_x & v_y \end{vmatrix} = \frac{1}{\begin{vmatrix} x_u & x_v \\ y_u & y_v \end{vmatrix}}.$$

In our familiar notation it can be written as

$$\frac{\partial(u, v)}{\partial(x, y)} = \frac{1}{\partial(x, y)/\partial(u, v)}.$$

■

3.5.1 Properties of Jacobians

Jacobians satisfy a chain rule. If x and y are functions of u and v, while u and v are functions of λ and μ, then

$$\frac{\partial(x, y)}{\partial(u, v)} \frac{\partial(u, v)}{\partial(\lambda, \mu)} = \frac{\partial(x, y)}{\partial(\lambda, \mu)}.$$

By choosing $\lambda = x$ and $\mu = y$, we immediately deduce the important result that

$$\frac{\partial(u, v)}{\partial(x, y)} = \frac{1}{\frac{\partial(x,y)}{\partial(u,v)}}.$$

The reader should note that this result does not contradict the general rule that $\partial u/\partial x \neq 1/(\partial x/\partial u)$.

3.5.2 Plane polar coordinates

When the region of integration is bounded by circular arcs, or when the integrand is simplified by using plane polar coordinates r and θ, where $x = r \cos\theta$ and $y = r \sin\theta$, it is usually advisable to express the double integral in terms of r and θ. The relevant relation is:

$$\iint_S f(x, y)\,dxdy = \iint_{S'} f(r\cos\theta, r\sin\theta) \left\{ \frac{\partial(x, y)}{\partial(r, \theta)} \right\} drd\theta,$$

$$= \iint_{S'} f(r\cos\theta, r\sin\theta)\, r\,drd\theta,$$

where S' is the region of integration in the r, θ plane. The above equation provides a simple example of a change of variables, the Jacobian of transformation from Cartesian to plane polar coordinates being

$$\frac{\partial(x, y)}{\partial(r, \theta)} = \begin{vmatrix} x_r & x_\theta \\ y_r & y_\theta \end{vmatrix} = \begin{vmatrix} \cos\theta & -r\sin\theta \\ \sin\theta & r\cos\theta \end{vmatrix} = r.$$

Here r is called the scale factor. Thus, $dxdy = rd\theta dr$. Alternatively, we can deduce from first principles that the elementary rectangle with area $dS = dxdy$ at a typical point in Cartesian coordinates has sides of length dr in the radial direction and of length $rd\theta$ in the transverse direction in cylindrical polar coordinates, giving $dS = rd\theta dr$.

Theorem 3.4 *To change variables in a triple integral, let $x = f(u, v, w)$, $y = g(u, v, w)$, $z = h(u, v, w)$ be a one-to-one transformation from a region E in uvw-space to a region D in xyz. If F is continuous on D, then*

$$\iiint_D F(x, y, z)dxdydz = \iiint_E F(f(u, v, w), g(u, v, w), h(u, v, w))$$

$$\left\{ \frac{\partial(x, y, z)}{\partial(u, v, w)} \right\} dudvdw,$$

where

$$\frac{\partial(x, y, z)}{\partial(u, v, w)} = \begin{vmatrix} x_u & x_v & x_w \\ y_u & y_v & y_w \\ z_u & z_v & z_w \end{vmatrix}.$$

We leave it as an exercise for the student to show that if T is the transformation from spherical to rectangular coordinates defined by $x = r\sin\theta\cos\phi$, $y = r\sin\theta\sin\phi$ and $z = r\cos\theta$, then

$$\frac{\partial(x, y, z)}{\partial(r, \theta, \phi)} = r^2\sin\theta.$$

3.6 Summary

- The ordinary derivative of a function $f(x)$ with respect to x is

$$\frac{df}{dx} = \lim_{\Delta x \to 0} \frac{f(x + \Delta x) - f(x)}{\Delta x}.$$

- The partial derivative of a function $f(x, y, z)$ with respect to x is

$$\frac{\partial f}{\partial x} = \lim_{\Delta x \to 0} \frac{f(x + \Delta x, y, z) - f(x, y, z)}{\Delta x}.$$

- The partial derivative with respect to y is

$$\frac{\partial f}{\partial y} = \lim_{\Delta y \to 0} \frac{f(x, y + \Delta y, z) - f(x, y, z)}{\Delta y}.$$

- The partial derivative with respect to z is

$$\frac{\partial f}{\partial z} = \lim_{\Delta y \to 0} \frac{f(x, y, z + \Delta z) - f(x, y, z)}{\Delta z}.$$

- Iterative partial derivatives are

$$(f_x)_x = f_{xx} = \frac{\partial}{\partial x}\left(\frac{\partial f}{\partial x}\right) = \frac{\partial^2 f}{\partial x^2} = \frac{\partial^2 z}{\partial x^2},$$

$$(f_x)_y = f_{xy} = \frac{\partial}{\partial y}\left(\frac{\partial f}{\partial x}\right) = \frac{\partial^2 f}{\partial y \partial x} = \frac{\partial^2 z}{\partial y \partial x},$$

$$(f_y)_x = f_{yx} = \frac{\partial}{\partial x}\left(\frac{\partial f}{\partial y}\right) = \frac{\partial^2 f}{\partial x \partial y} = \frac{\partial^2 z}{\partial x \partial y},$$

$$(f_y)_y = f_{yy} = \frac{\partial}{\partial y}\left(\frac{\partial f}{\partial y}\right) = \frac{\partial^2 f}{\partial y^2} = \frac{\partial^2 z}{\partial y^2}.$$

The notation f_{xy} (or $\partial^2 f / \partial x \partial y$) means that we first differentiate with respect to x and then with respect to y, as opposed to f_{yx} where the order is reversed.

- **Theorem 3.3** *If $x_j = g_j(y_1, \ldots, y_n)$, $j = 1, \ldots, n$ are n continuous functions of the variables $y_1, \ldots, y_n$ with continuous first partial derivatives, and if the Jacobian*

$$J = \frac{\partial(x_1, \ldots, x_n)}{\partial(y_1, \ldots, y_n)}$$

does not vanish in a given domain, then the transformation from y to x can be uniquely inverted to give

$$y_i = f_i(x_1, \ldots, x_n).$$

Exercises

1. Evaluate ϕ_y at $(1, 1, 0)$ if $\phi(x, y, z) = xy(x^2 + y^2 + z^2)^{\frac{1}{3}}$.
2. If $\phi = x + y + ze^{y/x}$, show that $x^2\phi_{xx} + y^2\phi_{yy} + z\phi_{zz} + 2xy\phi_{xy} + 2yz\phi_{yz} + 2xz\phi_{xz} = 0$.
3. If $\phi(x, y) = x^3 y/(x - y)$, show that $x\phi_x + y\phi_y = 3\phi(x, y)$.

4. If $\phi(x, y, z) = (x^4 + y^4 + z^4)/xyz$, show that $x\phi_x + y\phi_y + z\phi_z = \phi(x, y, z)$.

5. If $\phi(x, y, z) = (x^2 + y^2) \cos[(y + z)/x]$, show that $x\phi_x + y\phi_y + z\phi_z = 2\phi(x, y, z)$.

6. The equation of continuity for three-dimensional unsteady flow of a compressible fluid is given by $\rho_t + \nabla \cdot (\rho \mathbf{V}) = 0$, where $\rho(x, y, z, t)$ is the density of the fluid, $\mathbf{V} = u\mathbf{i} + v\mathbf{j} + w\mathbf{k}$ is the velocity of the fluid at the position (x, y, z) and time t, and $\nabla = \mathbf{i}(\partial/\partial x) + \mathbf{j}(\partial/\partial y) + \mathbf{k}(\partial/\partial z)$. Determine whether the continuity equation is satisfied if: $\rho = xy + zt$, $u = x^2 y + t$, $v = y^2 z - 2t^2$ and $w = 5x + 2z$.

7. In complex variable theory, two functions $u(x, y)$ and $v(x, y)$ are said to be *harmonic conjugates* in a region R if in R they satisfy the Cauchy–Riemann equations

$$\frac{\partial u}{\partial x} = \frac{\partial v}{\partial y}, \quad \frac{\partial v}{\partial x} = -\frac{\partial u}{\partial y}.$$

Show that the following pairs of functions are harmonic conjugates:
(a) $u(x, y) = (x^2 + x + y^2)/(x^2 + y^2)$, $v(x, y) = -y/(x^2 + y^2)$.
(b) $u(x, y) = e^x(x \cos y - y \sin y)$, $v(x, y) = e^x(x \sin y + y \cos y)$.

8. If r and θ are polar coordinates, then the Cauchy–Riemann equations from the previous question for functions $u(r, \theta)$ and $v(r, \theta)$ take the form

$$\frac{\partial u}{\partial r} = \frac{1}{r}\frac{\partial v}{\partial \theta}, \quad \frac{1}{r}\frac{\partial u}{\partial \theta} = -\frac{\partial v}{\partial r}, \quad r \neq 0.$$

Show that the following pairs of functions satisfy these equations:
(a) $u(r, \theta) = (r^2 + r \cos \theta)/(1 + r^2 + 2r \cos \theta)$, $v(r, \theta) = r \sin \theta/(1 + r^2 + 2r \cos \theta)$.
(b) $u(r, \theta) = \sqrt{r} \cos(\theta/2)$, $v(r, \theta) = \sqrt{r} \sin(\theta/2)$.
(c) $u(r, \theta) = \ln r$, $v(r, \theta) = \theta$.

9. Verify that $u = e^{-\alpha t} \sin x$, where α is a constant thermal diffusivity, is a solution of the heat equation

$$u_t = \alpha u_{xx}.$$

10. Verify that $u = \ln r$, where $r = \sqrt{(x - x_0)^2 + (y - y_0)^2}$ is a solution of the two-dimensional Laplace equation

$$u_{xx} + u_{yy} = 0.$$

11. Verify that $u = 1/r$, where

$$r = \sqrt{(x - x_0)^2 + (y - y_0)^2 + (z - z_0)^2}$$

is a solution of the three-dimensional Laplace equation

$$u_{xx} + u_{yy} + u_{zz} = 0.$$

12. In the theory of thermodynamics, pressure P, temperature T, volume V and internal energy U are related by two equations of state:

$$F(P, T, V, U) = 0$$

and

$$G(P, T, V, U) = 0.$$

The second law of thermodynamics implies that if U and P are regarded as functions of T and V, then the functions $U(T, V)$ and $P(T, V)$ must satisfy the equation

$$\frac{\partial U}{\partial V} - T\frac{\partial P}{\partial T} + P = 0.$$

Show that if U and V are chosen as independent variables rather than T and V, then the second law takes the form

$$\frac{\partial T}{\partial V} + T\frac{\partial P}{\partial U} - P\frac{\partial T}{\partial U} = 0.$$

13. Find the Jacobian of transformation from the uv-plane to the xy-plane in the following problems: (a) $u = y/x^2$; $v = y^2/x$. (b) $u = 2x/(x^2 + y^2)$; $v = -2y/(x^2 + y^2)$.

14. The molecular concentration $C(x, t)$ of a chemical species is given by $C(x, t) = (1/\sqrt{t})e^{-x^2/\alpha t}$. Verify that this function satisfies the *diffusion equation*

$$\frac{\partial C}{\partial t} = \frac{\alpha}{4}\frac{\partial^2 C}{\partial x^2}.$$

15. The pressure P exerted by an enclosed ideal gas is given by $P = \kappa(T/V)$, where κ is a constant, T is the temperature and V is the volume. Find: (a) the rate of change of P with respect to V; (b) the rate of change of V with respect to T; and (c) the rate of change of T with respect to P.

Chapter 4

The Del operator

4.1 Introduction

Let $\phi(x, y, z)$ be a scalar function of a position possessing first partial derivatives with respect to x, y and z throughout some region of space, and let $\mathbf{R} = x\mathbf{i} + y\mathbf{j} + z\mathbf{k}$ be the vector drawn from the origin to a general point $P: (x, y, z)$. If we move from P to a neighbouring point $Q: (x + \Delta x, y + \Delta y, z + \Delta z)$ (Fig. 4.1), the function ϕ will change by an amount $\Delta\phi$ whose exact value, as derived in calculus, is

$$\Delta\phi = \phi(x + \Delta x, y + \Delta y, z + \Delta z) - \phi(x, y, z).$$

By Taylor series expansion, the first term on the right-hand side can be written as

$$\phi(x + \Delta x, y + \Delta y, z + \Delta z) = \phi(x, y, z) + \frac{\partial\phi}{\partial x}\Delta x + \frac{\partial\phi}{\partial y}\Delta y + \frac{\partial\phi}{\partial z}\Delta z$$
$$+ \mathbf{O}\left(\Delta x^2, \Delta y^2, \Delta z^2\right),$$

where $\mathbf{O}$ stands for order of expansion and contains all the remaining terms of the Taylor series expansion. Therefore,

$$\Delta\phi = \frac{\partial\phi}{\partial x}\Delta x + \frac{\partial\phi}{\partial y}\Delta y + \frac{\partial\phi}{\partial z}\Delta z + \mathbf{O}(\Delta x^2, \Delta y^2, \Delta z^2). \tag{4.1}$$

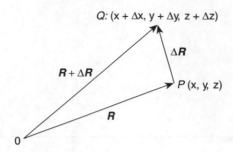

Figure 4.1 Coordinate vectors of two neighbouring points.

Now, if we divide the change $\Delta\phi$ by the distance $\Delta s = |\Delta \mathbf{R}|$ between P and Q, we obtain a measure of the rate at which ϕ changes when we move from P to Q:

$$\frac{\Delta\phi}{\Delta s} = \frac{\partial\phi}{\partial x}\frac{\Delta x}{\Delta s} + \frac{\partial\phi}{\partial y}\frac{\Delta y}{\Delta s} + \frac{\partial\phi}{\partial z}\frac{\Delta z}{\Delta s}$$

$$+ \mathbf{O}\left(\left(\frac{\partial x}{\partial s}\right)\Delta x, \left(\frac{\partial y}{\partial s}\right)\Delta y, \left(\frac{\partial z}{\partial s}\right)\Delta z\right), \qquad (4.2)$$

where the limiting value of the last term on the right-hand side tends to zero as Δx, Δy, and Δz tend to zero.

For example, if $\phi(x, y, z)$ is the temperature at the general point $P(x, y, z)$, then $\Delta\phi/\Delta s$ is the average rate of change of temperature in the direction in which Δs is measured. The limiting value of $\Delta\phi/\Delta s$ as Q approaches P along the segment PQ is called the *derivative* of Q *in the direction of* PQ or simply the *directional derivative* of Q. Clearly, in the limit, the last three terms in (4.2) become zero and we have explicitly

$$\frac{d\phi}{ds} = \frac{\partial\phi}{\partial x}\frac{dx}{ds} + \frac{\partial\phi}{\partial y}\frac{dy}{ds} + \frac{\partial\phi}{\partial z}\frac{dz}{ds}. \qquad (4.3)$$

4.2 Gradient of a scalar function

The first factor in each product on the right in (4.3) depends only on ϕ and the coordinates of the point at which the derivatives of ϕ are evaluated. The second factor in each product is independent of ϕ and only depends on the direction in which the derivative is being computed. This observation suggests that $d\phi/ds$ can be thought of as the dot product of two vectors, one depending only on ϕ and the coordinates of P and the other depending only on the direction of ds. In fact, we can write

$$\frac{d\phi}{ds} = \left(\frac{\partial\phi}{\partial y}\mathbf{i} + \frac{\partial\phi}{\partial y}\mathbf{j} + \frac{\partial\phi}{\partial z}\mathbf{k}\right) \cdot \left(\frac{dx}{ds}\mathbf{i} + \frac{dy}{ds}\mathbf{j} + \frac{dz}{ds}\mathbf{k}\right)$$

$$= \left(\frac{\partial\phi}{\partial x}\mathbf{i} + \frac{\partial\phi}{\partial y}\mathbf{j} + \frac{\partial\phi}{\partial z}\mathbf{k}\right) \cdot \frac{d\mathbf{R}}{ds}. \tag{4.4}$$

We have thus introduced the vector function

$$\frac{\partial\phi}{\partial x}\mathbf{i} + \frac{\partial\phi}{\partial y}\mathbf{j} + \frac{\partial\phi}{\partial z}\mathbf{k},$$

which is known as the *gradient of* ϕ or simply grad ϕ, and in this notation equation (4.4) can be rewritten as

$$\frac{d\phi}{ds} = (\text{grad }\phi) \cdot \frac{d\mathbf{R}}{ds}. \tag{4.5}$$

To determine the significance of grad ϕ, we observe first that since Δs is by definition just the length of $\Delta\mathbf{R}$, it follows that $d\mathbf{R}/ds$ is a unit vector. Hence, the dot product $(\text{grad }\phi) \cdot d\mathbf{R}/ds$ is just the projection of grad ϕ in the direction of $d\mathbf{R}/ds$. Furthermore, grad ϕ has the property that its projection in any direction is equal to the derivative of ϕ in that direction (see Fig. 4.2).

Since the maximum projection of a vector is the vector itself, it is clear that grad ϕ extends in the direction of the greatest rate of change of ϕ and has that rate of change for its length. If we set $\phi(x, y, z) = c$, we obtain, as c takes on different values, a family of surfaces known as *level surfaces* of ϕ, and on the assumption that ϕ is single-valued, one and only one level surface which passes through P. Fixing our attention on neighbouring points Q, which lie on this surface, we have

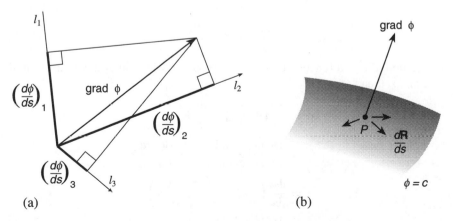

(a)   (b)

Figure 4.2   Geometric interpretation of the gradient.

$d\phi/ds = 0$, since $\Delta\phi = 0$, because by definition ϕ has the same value at all points of a level surface. Therefore, by (4.5)

$$(\text{grad } \phi) \cdot \frac{d\mathbf{R}}{ds} = 0 \qquad (4.6)$$

for any vector $d\mathbf{R}/ds$, which has the limiting direction of a secant PQ of the level surface. Clearly, such vectors are all tangents to $\phi = c$ at the point P. Hence, from the vanishing of the dot product in (4.6), it follows that grad ϕ is perpendicular to every tangent to the level surface at P. In other words, the gradient of ϕ at any point is perpendicular to the level surface of ϕ which passes through that point (see Fig. 4.2b).

Evidently, grad ϕ is related to the level surfaces of ϕ in a way which is independent of the particular coordinate system used to describe ϕ. In other words, grad ϕ depends only on the intrinsic properties of ϕ. Therefore, in the expression

$$\text{grad } \phi = \frac{\partial\phi}{\partial x}\mathbf{i} + \frac{\partial\phi}{\partial y}\mathbf{j} + \frac{\partial\phi}{\partial z}\mathbf{k}, \qquad (4.7)$$

$\mathbf{i}$, $\mathbf{j}$ and $\mathbf{k}$ can be replaced by any other set of mutually perpendicular unit vectors provided that $\partial\phi/\partial x$, $\partial\phi/\partial y$, $\partial\phi/\partial z$ are replaced by the directional derivatives of ϕ along the new axes. The gradient of a function is frequently written in its operational form as

$$\text{grad } \phi = \left(\mathbf{i}\frac{\partial}{\partial x} + \mathbf{j}\frac{\partial}{\partial y} + \mathbf{k}\frac{\partial}{\partial z}\right)\phi.$$

The operational *vector* or differential operator is denoted by the symbol ∇ (read *Del*), i.e.

$$\nabla \equiv \mathbf{i}\frac{\partial}{\partial x} + \mathbf{j}\frac{\partial}{\partial y} + \mathbf{k}\frac{\partial}{\partial z},$$

which leads to the notation

$$\text{grad } \phi = \nabla\phi$$

$$\frac{d\phi}{ds} = \nabla\phi \cdot \frac{d\mathbf{R}}{ds}.$$

In addition, if ϕ is a function of a single variable u, which in turn is a function of x, y and z, then

$$\begin{aligned}
\nabla\phi &= \frac{\partial\phi}{\partial x}\mathbf{i} + \frac{\partial\phi}{\partial y}\mathbf{j} + \frac{\partial\phi}{\partial z}\mathbf{k} \\
&= \frac{d\phi}{du}\frac{\partial u}{\partial x}\mathbf{i} + \frac{d\phi}{du}\frac{\partial u}{\partial y}\mathbf{j} + \frac{d\phi}{du}\frac{\partial u}{\partial z}\mathbf{k} \\
&= \frac{d\phi}{du}\left(\frac{\partial u}{\partial x}\mathbf{i} + \frac{\partial u}{\partial y}\mathbf{j} + \frac{\partial u}{\partial z}\mathbf{k}\right) \\
&= \frac{d\phi}{du} \cdot \nabla u.
\end{aligned} \qquad (4.8)$$

Example 4.1   What is the directional derivative of the function $\phi(x, y, z) = xy^2 + yz^3$ at the point $(2, -1, 1)$ in the direction of the vector $\mathbf{i} + 2\mathbf{j} + 2\mathbf{k}$?   □

Solution   Our first step must be to find the gradient of ϕ at the point $(2, -1, 1)$. This is

$$\nabla\phi = \frac{\partial(xy^2 + yz^3)}{\partial x}\mathbf{i} + \frac{\partial(xy^2 + yz^3)}{\partial y}\mathbf{j} + \frac{\partial(xy^2 + yz^3)}{\partial z}\mathbf{k} \quad \text{at } (2, -1, 1)$$

$$= y^2\mathbf{i} + (2xy + z^3)\mathbf{j} + 3yz^2\mathbf{k} \quad \text{at} \quad (2, -1, 1)$$

$$= \mathbf{i} - 3\mathbf{j} - 3\mathbf{k}.$$

The projection of this in the direction of the given vector will be the required directional derivative. Since this projection can be found at once as the dot product of $\nabla\phi$ and a unit vector in the given direction, we next reduce $\mathbf{i} + 2\mathbf{j} + 2\mathbf{k}$ to a unit vector by dividing its magnitude as follows:

$$\frac{\mathbf{i} + 2\mathbf{j} + 2\mathbf{k}}{\sqrt{1 + 4 + 4}} = \frac{1}{3}\mathbf{i} + \frac{2}{3}\mathbf{j} + \frac{2}{3}\mathbf{k}.$$

The answer to our problem is now given by

$$\nabla\phi \cdot \left(\frac{1}{3}\mathbf{i} + \frac{2}{3}\mathbf{j} + \frac{2}{3}\mathbf{k}\right) = (\mathbf{i} - 3\mathbf{j} - 3\mathbf{k}) \cdot \left(\frac{1}{3}\mathbf{i} + \frac{2}{3}\mathbf{j} + \frac{2}{3}\mathbf{k}\right)$$

$$= -\frac{11}{3}.$$

The negative sign, of course, indicates that ϕ decreases in the given direction.   ■

Definition 4.1   *If $\phi = \phi(x, y, z)$ is a differentiable function of x, y and z and $\mathbf{U}$ is any unit vector, then*

$$D_{\mathbf{U}}\phi = \nabla\phi \cdot \mathbf{U}$$

is defined as the directional derivative of $\phi = \phi(x, y, z)$ in the direction of the unit vector $\mathbf{U}$.

Note that it is similar to the formula $d\phi/ds = \nabla\phi \cdot d\mathbf{R}/ds$, provided we substitute $D_{\mathbf{U}}\phi$ for $d\phi/ds$ and $\mathbf{U}$ for $d\mathbf{R}/ds$. Thus, $D_{\mathbf{U}}\phi = \nabla\phi \cdot \mathbf{U}$ shows that the directional derivative of ϕ in the direction of $\mathbf{U}$ is just the scalar projection of $\nabla\phi$ along the direction of $\mathbf{U}$.

Example 4.2   What is the unit normal to the surface $xy^3z^2 = 4$ at the point $(-1, -1, 2)$?   □

Solution   Let us regard the given surface as a particular level surface of the function at the point $(-1, -1, 2)$ that will be perpendicular to the level surface through

$(-1, -1, 2)$, which is the given surface. When this gradient has been found, the unit normal can be obtained at once by dividing the gradient by its magnitude:

$$\nabla\phi = \frac{\partial(xy^3z^2)}{\partial x}\mathbf{i} + \frac{\partial(xy^3z^2)}{\partial y}\mathbf{j} + \frac{\partial(xy^3z^2)}{\partial z}\mathbf{k} \quad \text{at } (-1, -1, 2)$$

$$= -4\mathbf{i} - 12\mathbf{j} + 4\mathbf{k},$$

$$|\nabla\phi| = \sqrt{16 + 144 + 16} = 4\sqrt{11},$$

$$\frac{\nabla\phi}{|\nabla\phi|} = \frac{-4\mathbf{i} - 12\mathbf{j} + 4\mathbf{k}}{4\sqrt{11}} = -\frac{1}{\sqrt{11}}\mathbf{i} - \frac{3}{\sqrt{11}}\mathbf{j} + \frac{1}{\sqrt{11}}\mathbf{k}.$$

It may be necessary to reverse the direction of this result by multiplying it by -1, depending on which side of the surface we wish the normal to extend. ∎

Example 4.3 Find the unit normal $\mathbf{n}$ to the surface $x^2 + y^2 - z = 0$ at the point $(1, 1, 2)$. ☐

Solution Let us define our function to be $f(x, y, z) = x^2 + y^2 - z = 0$, so that the surface is $f = 0$. It follows that $\nabla f = (2x, 2y, -1)$ (using vector notation). At the point $(1, 1, 2)$, $\nabla f = (2, 2, -1)$. This is a vector normal to the surface. To find the unit normal we must divide by the magnitude, which in this case is $(2^2 + 2^2 + 1^2)^{1/2} = 3$. We thus have $\nabla f/|\nabla f| = (2/3, 2/3, -1/3)$. We note here that the unit normal is not uniquely defined as mentioned at the end of the previous example. The vector $-\mathbf{n} = (-2/3, -2/3, 1/3)$ is also a unit normal to the surface. ∎

4.3 Gradient characterizes maximum increase

We will first show that the length and direction of grad f are independent of the particular choice of Cartesian coordinates. By the definition of a scalar function, the value of f at a point P depends on P but is independent of the coordinates, and s, the arc length of that ray C, is also independent of the choice of coordinates. Hence, the directional derivative is also independent of the particular choice of coordinates. Now we have the directional derivative defined as

$$D_\mathbf{U}\phi = \nabla\phi \cdot \mathbf{U} = |\nabla\phi||\mathbf{U}|\cos\gamma = |\nabla\phi|\cos\gamma \quad (|\mathbf{U}| = 1),$$

where γ is the angle between $\nabla\phi$ and $\mathbf{U}$. Because $0 \leq \gamma \leq \pi$, we have $-1 \leq \cos\gamma \leq 1$ and, consequently,

$$-|\nabla\phi| \leq D_\mathbf{U}\phi \leq |\nabla\phi|.$$

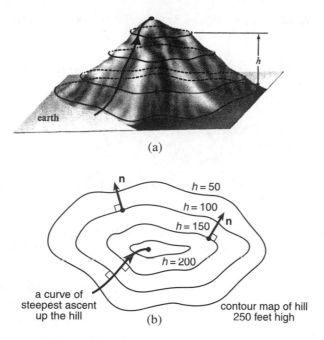

Figure 4.3 Orthogonal contours of a hill.

In other words, the maximum value of the directional derivative is $|\nabla\phi|$ and it occurs when $\mathbf{U}$ has the same direction as $\nabla\phi$ (when $\cos\gamma = 1$). The minimum value of the directional derivative is $-|\nabla f|$ and it occurs when $\mathbf{U}$ and $\nabla\phi$ have opposite directions(when $\cos\gamma = -1$).

The gradient $\nabla\phi$ is often spoken of as a *gradient vector field*.It must be noted that $\nabla\phi$ assigns a vector to each point in the domain of ϕ. This gradient vector field has an important geometric significance. It shows the direction in which ϕ is increasing the fastest and the direction that is orthogonal to the level surfaces (or curves in the plane) of ϕ.

In order to understand this, imagine a hill such as the one shown in Fig. 4.3. We let h be the height function, clearly a function of two variables. If the level curves of h are drawn, we see that they are the contours of the hill. One could imagine them as level paths on the hill (see Fig. 4.3b).

From the figure, it is obvious that to get to the top of the hill as fast as possible, one must walk perpendicular to the level contours. This is consistent with the theory in the preceding paragraphs. Thus, we can state that the direction of fastest increase (the gradient) is orthogonal to the level curves, which is the direction $\nabla\phi$ in Fig. 4.3a.

Example 4.4 Find the directional derivative of $\phi(x, y, z) = xy^2 - 4x^2y + z^2$ at $(1, -1, 2)$ in the direction of $6\mathbf{i} + 2\mathbf{j} + 3\mathbf{k}$. What are the directions of maximum and minimum increase at $(1, -1, 2)$? ☐

Solution   We have $\partial\phi/\partial x = y^2 - 8xy$, $\partial\phi/\partial y = 2xy - 4x^2$ and $\partial\phi/\partial z = 2z$ so that

$$\nabla\phi(x, y, z) = (y^2 - 8xy)\mathbf{i} + (2xy - 4x^2)\mathbf{j} + 2z\mathbf{k}$$
$$\nabla\phi(1, -1, 1) = 9\mathbf{i} - 6\mathbf{j} + 4\mathbf{k}.$$

Since $|6\mathbf{i} + 2\mathbf{j} + 3\mathbf{k}| = 7$, then $\mathbf{U} = \frac{6}{7}\mathbf{i} + \frac{2}{7}\mathbf{j} + \frac{3}{7}\mathbf{k}$ is a unit vector in the indicated direction. Then we have

$$D_{\mathbf{U}}\phi(1, -1, 2) = \nabla\phi \cdot \mathbf{U} = (9\mathbf{i} - 6\mathbf{j} + 4\mathbf{k}) \cdot \left(\frac{6}{7}\mathbf{i} + \frac{2}{7}\mathbf{j} + \frac{3}{7}\mathbf{k}\right) = \frac{54}{7}.$$

The maximum value of the directional derivative at ϕ at $(1, -1, 2)$ is $|\nabla\phi(1, -1, 2)| = \sqrt{133}$. The minimum value of $D_{\mathbf{U}}\phi(1, -1, 2)$ is then $-\sqrt{133}$. ∎

The previous two statements concerning the maximum and minimum value of the directional derivative can be another way. The gradient vector $\nabla\phi$ points in the direction in which ϕ increases most rapidly, whereas $-\nabla\phi$ points in the direction of most rapid decrease of ϕ.

Example 4.5   Each year in Los Angeles there is a bicycle race up to the top of a hill by a road known to be the steepest in the city. To understand why a bicyclist will zigzag up the road, let us suppose the graph of $f(x, y) = 4 - \frac{2}{3}\sqrt{x^2 + y^2}$, $0 \le z \le 4$ shown in Fig. 4.3a is a mathematical model of the hill. □

Solution   The gradient of f is

$$\nabla f(x, y) = \frac{2}{3}\left(\frac{-x}{\sqrt{x^2 + y^2}}\mathbf{i} + \frac{-y}{\sqrt{x^2 + y^2}}\mathbf{j}\right) = \frac{\frac{2}{3}}{\sqrt{x^2 + y^2}}\mathbf{R},$$

where $\mathbf{R} = -x\mathbf{i} - y\mathbf{j}$ is a vector pointing to the centre of the circular base. Thus, the steepest ascent up the hill is a straight road whose projection in the xy-plane is a radius of the circular base. Obviously, this direction will cause the bicyclist to expend an enormous amount of energy to arrive at the top of the hill. Because of this, a bicyclist will zigzag, or seek a direction $\mathbf{U}$ other than ∇f, in order to reduce this component. ∎

Example 4.6   The temperature in a rectangular box is approximated by $T(x, y, z) = xyz(1 - x)(2 - y)(3 - z)$, $0 \le x \le 1$, $0 \le y \le 2$, $0 \le z \le 3$. If a mosquito is located at $(\frac{1}{2}, 1, 1)$, in which direction should it fly to cool off as rapidly as possible? □

Solution   The gradient of T is

$$\nabla T(x, y, z) = yz(2 - y)(3 - z)(1 - 2x)\mathbf{i} + xz(1 - x)(3 - z)(2 - 2y)\mathbf{j}$$
$$+ xy(1 - x)(2 - y)(3 - 2z)\mathbf{k}.$$

Therefore,

$$\nabla T\left(\frac{1}{2}, 1, 1\right) = \frac{1}{4}\mathbf{k}.$$

To cool off most rapidly, the mosquito should fly in the direction of $-\frac{1}{4}\mathbf{k}$, i.e. it should dive for the floor of the box, where the temperature is $T(x, y, z) = 0$.   ∎

4.4 Tangent planes and normal lines

4.4.1 Geometric interpretation of the gradient – functions of two variables

Suppose $f(x, y) = c$ is the level curve of the differentiable function $z = f(x, y)$ that passes through a specified point $P(x_0, y_0)$, i.e. $f(x_0, y_0) = c$. If this level curve is parameterized by the differentiable functions $x = g(t), y = h(t)$ such that

$$x_0 = gt(t_0), \quad y_0 = h(t_0),$$

then the derivative of $f(g(t), h(t)) = c$ with respect to t is

$$\frac{\partial f}{\partial x}\frac{dx}{dt} + \frac{\partial f}{\partial y}\frac{dy}{dt} = 0. \tag{4.9}$$

When we introduce the vectors

$$\nabla f(x, y) = \frac{\partial f}{\partial x}\mathbf{i} + \frac{\partial f}{\partial y}\mathbf{j} \quad \text{and} \quad \mathbf{R}'(t) = \frac{dx}{dt}\mathbf{i} + \frac{dy}{dt}\mathbf{j}$$

equation (4.9) becomes $\nabla f \cdot \mathbf{R}'$. Specifically, at $t = t_0$, we have

$$\nabla f(x_0, y_0) \cdot \mathbf{R}'(t_0) = 0. \tag{4.10}$$

Thus, if $\mathbf{R}'(t_0) \neq 0$, the vector $\nabla f(x_0, y_0)$ is orthogonal to the tangent vector $\mathbf{R}'(t_0)$ at $P(x_0, y_0)$. This is interpreted to mean ∇f is perpendicular to the level curve at P (see Fig. 4.4).

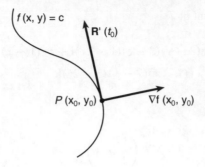

Figure 4.4 $\nabla f(x_0, y_0)$ shown perpendicular to a level curve.

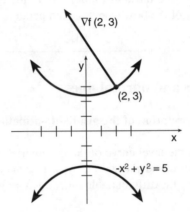

Figure 4.5   Level curve and gradient at (2, 3).

Example 4.7   Find the level curve of $f(x, y) = -x^2 + y^2$ passing through $(2, 3)$. Graph the gradient at the point.   □

Solution   Since $f(2, 3) = -4 + 9 = 5$, the level curve is the hyperbola $-x^2 + y^2 = 5$. Now,

$$\nabla f(x, y) = -2x\mathbf{i} + 2y\mathbf{j} \quad \text{and} \quad \nabla f(2, 3) = -5\mathbf{i} + 6\mathbf{j}.$$

Figure 4.5 shows the level curve and $\nabla f(2, 3)$.   ∎

4.4.2 Geometric interpretation of the gradient – functions of three variables

Here again, if we let $F(x, y, z) = c$ be the level surface of a differentiable function $w = F(x, y, z)$ that passes through $P(x_0, y_0, z_0)$. If the differentiable functions

$$x = f(t), \quad y = g(t) \quad \text{and} \quad z = h(t)$$

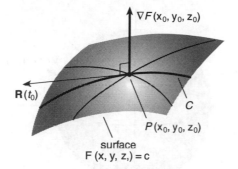

Figure 4.6 ∇F shown perpendicular to level surface at P.

are the parametric equations of a curve C on the surface for which $x_0 = f(t_0)$, $y_0 = g(t_0)$, $z_0 = h(t_0)$, then the derivative of $F(f(t), g(t), h(t)) = 0$ implies that

$$\frac{\partial F}{\partial x}\frac{dx}{dt} + \frac{\partial F}{\partial y}\frac{dy}{dt} + \frac{\partial F}{\partial z}\frac{dz}{dt} = 0$$

or

$$\left(\frac{\partial F}{\partial x}\mathbf{i} + \frac{\partial F}{\partial y}\mathbf{j} + \frac{\partial F}{\partial z}\mathbf{k}\right) \cdot \left(\frac{dx}{dt}\mathbf{i} + \frac{dy}{dt}\mathbf{j} + \frac{dz}{dt}\mathbf{k}\right) = 0. \tag{4.11}$$

In particular, at $t = t_0$ (4.11) becomes

$$\nabla F(x_0, y_0, z_0) \cdot \mathbf{R}'(t_0) = 0. \tag{4.12}$$

Thus, when $\mathbf{R}'(t_0) \neq 0$, the vector $\nabla F(x_0, y_0, z_0)$ is orthogonal to the tangent vector $\mathbf{R}'(t_0)$. Since this argument holds for any differentiable curve through $P(x_0, y_0, z_0)$ on the surface, we conclude that ∇F is perpendicular (normal) to the level surface at P (see Fig. 4.6).

Example 4.8   Find the level surface of $F(x, y, z) = x^2 + y^2 + z^2$ passing through $(1, 1, 1)$. Graph the gradient at the point.   □

Solution   Since $F(1, 1, 1) = 3$, the level surface passing through $(1, 1, 1)$ is the sphere $x^2 + y^2 + z^2 = 3$. The gradient of the function is

$$\nabla F(x, y, z) = 2x\mathbf{i} + 2y\mathbf{j} + 2z\mathbf{k}$$

and so, at the given point

$$\nabla F(1, 1, 1) = 2\mathbf{i} + 2\mathbf{j} + 2\mathbf{k}.$$

The level surface and $\nabla F(1, 1, 1)$ are illustrated in Fig. 4.7.   ∎

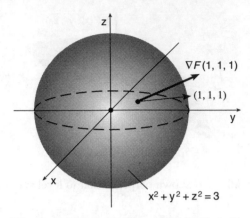

Figure 4.7   Level surface and $\nabla F(1, 1, 1)$.

4.5 Tangent plane

In the study of differential calculus a basic problem is finding an equation of a tangent line to the graph of a function. In 3-space, the analogous problem is finding an equation of a *tangent plane* to a surface. We assume again that $w = F(x, y, z)$ is a differentiable function and that the surface is given by $F(x, y, z) = c$.

Definition 4.2   *Let $P(x_0, y_0, z_0)$ be a point on the graph of $F(x, y, z) = c$ where ∇F is not equal to the zero vector. The tangent plane at P is that plane through P that is perpendicular to ∇F evaluated at P.*

Thus, if $P(x, y, z)$ and $P_1(x_0, y_0, z_0)$ are points on the tangent plane and $\mathbf{R}$ and $\mathbf{R}_0$ are their respective position vectors, then a vector equation of the tangent plane is

$$\nabla F(x_0, y_0, z_0) \cdot (\mathbf{R} - \mathbf{R}_0) = 0. \tag{4.13}$$

Example 4.9   Find an equation of the tangent plane to the graph of $x^2 - 4y^2 + z^2 = 16$ at $(2, 1, 4)$.   □

Solution   By defining $\phi(x, y, z) = x^2 - 4y^2 + z^2$, the given surface is the level surface $\phi(x, y, z) = \phi(2, 1, 4) = 16$ passing through $(2, 1, 4)$. Now, $\phi_x(x, y, z) = 2x$, $\phi_y(x, y, z) = -8y$ and $\phi_z(x, y, z) = 2z$, so that

$$\nabla \phi(x, y, z) = 2x\mathbf{i} - 8y\mathbf{j} + 2z\mathbf{k} \quad \text{and} \quad \nabla \phi(2, 1, 4) = 4\mathbf{i} - 8\mathbf{j} + 8\mathbf{k}.$$

It follows that an equation of the tangent plane is $\nabla \phi \cdot (\mathbf{R} - \mathbf{R}_0) = 0$, i.e.

$$4(x - 2) - 8(y - 1) + 8(z - 4) = 0 \quad \text{or} \quad x - 2y + 2z = 8.$$

■

4.6 Normal lines

Let $P(x_0, y_0, z_0)$ be a point on the graph of $\phi(x, y, z) = c$, where $\nabla\phi$ is not $\mathbf{0}$. The line containing $P(x_0, y_0, z_0)$ that is parallel to $\nabla\phi(x_0, y_0, z_0)$ is called the *normal line* to the surface at P. The normal line is perpendicular to the tangent plane to the surface at P. This equation is given by (in parametric form)

$$\mathbf{R} = \mathbf{R}_0 + t(\nabla\phi)_{\mathbf{R}_0}.$$

Example 4.10   Find the parametric equations for the normal line to the surface $\phi = \frac{1}{2}x^2 + \frac{1}{2}y^2 + 4$ at $(1, -1, 5)$.   □

Solution   A normal line equation is given by $\mathbf{R} = \mathbf{R}_0 + t(\nabla\phi)$. A direction vector for the normal line at $(1, -1, 5)$ is $\nabla\phi(1, -1, 5) = \mathbf{i} - \mathbf{j} - \mathbf{k}$. It follows that the parametric equations for the normal line are $x = 1 + t$, $y = -1 - t$ and $z = 5 - t$.   ∎

Example 4.11   Find the equations of the tangent plane and the normal line to the surface of the paraboloid of revolution $x^2 + y^2 - z = 1$ at the point $(1, 2, 4)$.   □

Solution   This example will illustrate how to obtain the normal line and tangent plane. Firstly, we can consider $\phi(x, y, z) \equiv x^2 + y^2 - z = 1$. Then we have

$$\nabla\phi = 2x\mathbf{i} + 2y\mathbf{j} - \mathbf{k}$$
$$= 2\mathbf{i} + 4\mathbf{j} - \mathbf{k} \quad \text{at the given point } (1, 2, 4).$$

Therefore, the direction numbers of the normal are simply $(2, 4, -1)$. From this, it follows that the tangent plane has the equation $2x + 4y - z = d$, where d is to be determined. Since we know that the point $(1, 2, 4)$ is in the plane, we can then find the value of d, i.e. $2(1) + 4(2) - 1(4) = d$. Thus the tangent plane is given by $2x + 4y - z = 6$. We can write the equation of the normal line in symmetric form as

$$\vec{PP_0} \times \nabla\phi\Big|_{(1,2,4)} = \mathbf{0}$$

$$= \begin{vmatrix} \mathbf{i} & \mathbf{j} & \mathbf{k} \\ x - 1 & y - 2 & z - 4 \\ 2 & 4 & -1 \end{vmatrix}$$

$$= \mathbf{i}[-(y - 2) - 4(z - 4)] - \mathbf{j}[-(x - 1) - 2(z - 4)]$$
$$+ \mathbf{k}[4(x - 1) - 2(y - 2)] = \mathbf{0}.$$

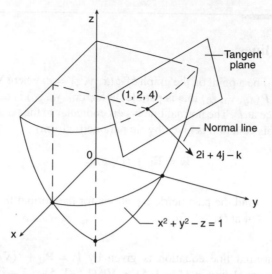

Figure 4.8   Paraboloid of revolution $x^2 + y^2 - z = 1$.

The normal line will then be given by the following equations

$$\frac{(y-2)}{4} = \frac{(z-4)}{-1},$$

$$\frac{(x-1)}{2} = \frac{(z-4)}{-1},$$

$$\frac{(x-1)}{2} = \frac{(y-2)}{4}.$$

These can be written in symmetric form in the following manner

$$\frac{x-1}{2} = \frac{y-2}{4} = \frac{z-4}{-1}.$$

Hence, the normal line is given by the equation above. It is worth noting that the direction numbers $(2, 4, -1)$ apply to the outward-pointing normal at the given point as shown in Fig. 4.8. If we redefine $\phi(x, y, z)$ as $\phi(x, y, z) \equiv -x^2 - y^2 + z$, we will then obtain the direction numbers for the inward-pointing normal.

4.7 Divergence and curl of a vector field

The vector character of the operator ∇ suggests that we also consider dot and cross products in which it appears as a factor. Let $\mathbf{F}(x, y, z)$ be a differentiable function, where x, y, z are Cartesian coordinates, and let F_1, F_2, F_3 be the components of $\mathbf{F}$. Then the function

$$\nabla \cdot \mathbf{F} = \left(\mathbf{i} \frac{\partial}{\partial x} + \mathbf{j} \frac{\partial}{\partial y} + \mathbf{k} \frac{\partial}{\partial z} \right) \cdot (F_1 \mathbf{i} + F_2 \mathbf{j} + F_3 \mathbf{k})$$

$$= \frac{\partial F_1}{\partial x} + \frac{\partial F_2}{\partial y} + \frac{\partial F_3}{\partial z} \tag{4.14}$$

is called the *divergence* of $\mathbf{F}$. This is commonly denoted as div $\mathbf{F}$. Note that $\nabla \cdot \mathbf{F}$ means the scalar div $\mathbf{F}$, whereas ∇f means the vector grad f defined by equation (4.7).

In a similar manner, we can define the cross product of ∇ with $\mathbf{F}$ as

$$\nabla \times \mathbf{F} = \begin{vmatrix} \mathbf{i} & \mathbf{j} & \mathbf{k} \\ \frac{\partial}{\partial x} & \frac{\partial}{\partial y} & \frac{\partial}{\partial z} \\ F_1 & F_2 & F_3 \end{vmatrix}$$

$$= \left(\frac{\partial F_3}{\partial y} - \frac{\partial F_2}{\partial z} \right) \mathbf{i} + \left(\frac{\partial F_1}{\partial z} - \frac{\partial F_3}{\partial x} \right) \mathbf{j} + \left(\frac{\partial F_2}{\partial x} - \frac{\partial F_1}{\partial y} \right) \mathbf{k}. \tag{4.15}$$

This expression is known as the *curl* of $\mathbf{F}$ or the curl of the vector field defined by $\mathbf{F}$.

4.8 Physical interpretation of divergence

Consider a region of space filled with a moving fluid and let

$$\mathbf{V} = v_1 \mathbf{i} + v_2 \mathbf{j} + v_3 \mathbf{k}$$

be a vector function representing the instantaneous velocity of a particle of the fluid at each point. If we fix our attention on the infinitesimal volume (Fig. 4.9) in the region occupied by the fluid, there will be flow through each of its faces, and as a result the amount of fluid within the element may vary. We can compute the loss of fluid from the element in the time Δt to measure this variation.

The volume of the fluid which passes through one face of the element ΔV in time Δt is approximately equal to the product of the component of the fluid velocity normal to the face, the area of the face and Δt, and the corresponding mass flow is the product of this volume and the density of the fluid ρ.

Therefore, computing the loss of fluid through each face in turn (remembering that since the fluid is not assumed to be incompressible, the density as well as the

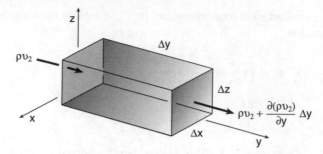

Figure 4.9   Typical volume element in a region filled with moving fluid.

velocity may vary from point to point), we have

Right face: $\left(\rho v_2 + \dfrac{\partial(\rho v_2)}{\partial y}\right)\Delta x \Delta z \Delta t,$

Left face: $-\rho v_2 \Delta x \Delta z \Delta t,$

Front face: $\left(\rho v_1 + \dfrac{\partial(\rho v_1)}{\partial x}\Delta x\right)\Delta y \Delta z \Delta t,$

Rear face: $-\rho v_1 \Delta y \Delta z \Delta t,$

Top face: $\left(\rho v_3 + \dfrac{\partial(\rho v_3)}{\partial z}\Delta z\right)\Delta x \Delta y \Delta t,$

Bottom face: $-\rho v_3 \Delta x \Delta y \Delta t.$

If these are added and the resulting estimate of absolute loss of fluid from ΔV in the interval Δt is converted into loss per unit volume per unit time, dividing by $\Delta V \Delta t \equiv \Delta x \Delta y \Delta z \Delta t$, we obtain the limit:

$$\text{Rate of loss per unit volume} = \frac{\partial(\rho v_1)}{\partial x} + \frac{\partial(\rho v_2)}{\partial y} + \frac{\partial(\rho v_3)}{\partial z}, \qquad (4.16)$$

which is precisely the divergence of the vector $(\rho \mathbf{V})$. From fluid mechanics we know that a possible interpretation of the divergence is the rate of loss or gain of fluid per unit volume. The divergence of a vector field can also be interpreted as a measure of the rate of change of the density of the fluid at a point. In other words, div $\mathbf{F}$ is a measure of the fluid's compressibility. If $\nabla \cdot \mathbf{F} = 0$, the fluid is said to be *incompressible*. In electromagnetic theory, if $\nabla \cdot \mathbf{F} = 0$, the vector field $\mathbf{F}$ is said to be *solenoidal*.

If the fluid is incompressible, then there is no loss or gain of fluid. In a general element where $\rho = $ constant,

$$\nabla \cdot (\rho \mathbf{V}) = 0,$$
$$\rho(\nabla \cdot \mathbf{V}) = 0, \qquad (4.17)$$
$$\nabla \cdot \mathbf{V} = 0,$$

which is known as the *equation of continuity*.

4.9 Physical interpretation of the curl

Consider a body rotating with uniform angular speed Ω about an axis l (Fig. 4.10). We define the *vector angular velocity* $\boldsymbol{\Omega}$ to be a vector of length ω extending along l in the direction in which a right-handed screw would advance if subject to the same rotation as the body. Let $\mathbf{R}$ be the vector drawn from any point O on the axis l to an arbitrary point P in the body. From Fig. 4.10, it is evident that the radius at which P rotates is $|\mathbf{R}| \cdot |\sin\theta|$. Therefore, the linear speed of P is

$$|\mathbf{V}| = \Omega|\mathbf{R}| \cdot |\sin\theta| = |\boldsymbol{\Omega}||\mathbf{R}||\sin\theta| = |\boldsymbol{\Omega} \times \mathbf{R}|.$$

Furthermore, the vector velocity $\mathbf{V}$ is directed perpendicular to the plane of $\boldsymbol{\Omega}$ and $\mathbf{R}$ so that $\boldsymbol{\Omega}$, $\mathbf{R}$ and $\mathbf{V}$ form a right-handed system. Hence, the cross product $\boldsymbol{\Omega} \times \mathbf{R}$ gives not only the magnitude of $\mathbf{V}$ but the direction as well.

If we take the point O as the origin of the coordinates, we can write

$$\mathbf{R} = x\mathbf{i} + y\mathbf{j} + z\mathbf{k} \quad \text{and} \quad \boldsymbol{\Omega} = \Omega_1\mathbf{i} + \Omega_2\mathbf{j} + \Omega_3\mathbf{k}.$$

Therefore, the equation $\mathbf{V} = \boldsymbol{\Omega} \times \mathbf{R}$ can be written in the form

$$\mathbf{V} = (\Omega_2 z - \Omega_3 y)\mathbf{i} - (\Omega_1 z - \Omega_3 z)\mathbf{j} + (\Omega_1 y - \Omega_2 x)\mathbf{k}. \tag{4.18}$$

If we take the curl of $\mathbf{V}$, we have

$$\nabla \times \mathbf{V} = \begin{vmatrix} \mathbf{i} & \mathbf{j} & \mathbf{k} \\ \dfrac{\partial}{\partial x} & \dfrac{\partial}{\partial y} & \dfrac{\partial}{\partial z} \\ (\Omega_2 z - \Omega_3 y) & -(\Omega_1 z - \Omega_3 x) & (\Omega_1 y - \Omega_2 x) \end{vmatrix}. \tag{4.19}$$

We can expand this, remembering that $\boldsymbol{\Omega}$ is a constant vector, to get

$$\nabla \times \mathbf{V} = 2\Omega_1\mathbf{i} + 2\Omega_2\mathbf{j} + 2\Omega_3\mathbf{k} = 2\boldsymbol{\Omega},$$

$$\boldsymbol{\Omega} = \frac{1}{2}\nabla \times \mathbf{V}. \tag{4.20}$$

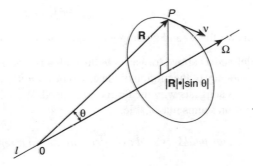

Figure 4.10   A physical interpretation of the curl.

The angular velocity of a uniformly rotating body is equal to one-half of the curl of the linear velocity of any point of the body. The aptness of the name *curl* in this connection is apparent.

Example 4.12   Calculate the curl of the vector field if

(a) $\mathbf{F} = 2xy\mathbf{i} + z\mathbf{j} + x^2 \cos(yz)\mathbf{k}$.
(b) $\mathbf{F} = (qQ/4\pi\epsilon_0|\mathbf{R}|^3)\mathbf{R}$, where $\mathbf{R} = x\mathbf{i} + y\mathbf{j} + z\mathbf{k}$. $\square$

Solution

(a)
$$\nabla \times \mathbf{F} = \begin{vmatrix} \mathbf{i} & \mathbf{j} & \mathbf{k} \\ \dfrac{\partial}{\partial x} & \dfrac{\partial}{\partial y} & \dfrac{\partial}{\partial z} \\ 2xy & z & x^2 \cos(yz) \end{vmatrix}$$

$$= (-x^2 z \sin(yz) - 1)\mathbf{i} + (-2x \cos(yz))\mathbf{j} + (-2x)\mathbf{k}.$$

(b)
$$\nabla \times \mathbf{F} = \frac{qQ}{4\pi\epsilon_0} \begin{vmatrix} \mathbf{i} & \mathbf{j} & \mathbf{k} \\ \dfrac{\partial}{\partial x} & \dfrac{\partial}{\partial y} & \dfrac{\partial}{\partial z} \\ \dfrac{x}{|\mathbf{R}|^3} & \dfrac{y}{|\mathbf{R}|^3} & \dfrac{z}{|\mathbf{R}|^3} \end{vmatrix}.$$

The x-component of $\nabla \times \mathbf{F}$ is $qQ/(4\pi\epsilon_0)$ multiplied by

$$\frac{\partial}{\partial y}\left(\frac{z}{|\mathbf{R}|^3}\right) - \frac{\partial}{\partial z}\left(\frac{y}{|\mathbf{R}|^3}\right) = \frac{\partial}{\partial y}\left(\frac{z}{(x^2+y^2+z^2)^{3/2}}\right)$$

$$- \frac{\partial}{\partial z}\left(\frac{y}{(x^2+y^2+z^2)^{3/2}}\right)$$

$$= \frac{-3yz}{(x^2+y^2+z^2)^{5/2}}$$

$$+ \frac{3yz}{(x^2+y^2+z^2)^{5/2}} = 0.$$

Similar results for the y- and z-components give $\nabla \times \mathbf{F} = 0$.   ■

If a vector $\rho\mathbf{V}$ is defined as above for gas flow through a region D, then the curl of $\rho\mathbf{V}$ describes the tendency of the motion of the gas to be circular rather than flowing in a straight line. This suggests why the term *curl* is used. With this interpretation, the following definition seems reasonable.

Definition 4.3   *A vector field* $\mathbf{F}$ *is said to be irrotational in a region* D *if in* D

$$\nabla \times \mathbf{F} = 0. \tag{4.21}$$

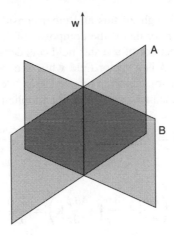

Figure 4.11   Paddle device.

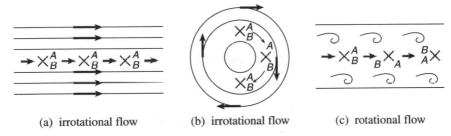

(a) irrotational flow   (b) irrotational flow   (c) rotational flow

Figure 4.12   Fluid flow.

The word *curl* was introduced by Maxwell in his studies of electromagnetic fields. However, the curl is easily understood in connection with the flow of fluids. If a paddle device (Fig. 4.11) is inserted in a flowing fluid, then the curl of the velocity field **F** is a measure of the tendency of the fluid to turn the device about its vertical axis w. If $\nabla \times \mathbf{F} = \mathbf{0}$, then the flow of the fluid is *irrotational* (from Definition 4.3), which means that it is free of vortices or whirlpools that would cause the paddle to rotate.

For this reason, in some science texts the word *rotation* is sometimes used instead of *curl*. The symbol curl **F** is then replaced by rot **F**. In Fig. 4.12, the axis w points straight out of the page. Note from Fig. 4.12b that *irrotational* does *not* mean that the fluid does not rotate.

4.10 The Laplacian operator

In the applied field of electrostatics, the gradient of the potential is a scalar multiple of the electric field intensity. The divergence of the electric intensity is related to the charge density. In the light of this and other considerations, it is convenient to introduce a single operator that is the composite of the gradient (grad) and the divergence (div). This operator of a scalar field ϕ is defined to be div(grad ϕ). It must be noted that grad ϕ is a vector field while the divergence of grad ϕ is a scalar field. The Laplacian of ϕ is often referred to as 'Del squared ϕ'. This can be written in the following manner, using $\nabla^2\phi$, for simplicity

$$\nabla^2\phi = \nabla \cdot (\nabla\phi) = \frac{\partial^2\phi}{\partial x^2} + \frac{\partial^2\phi}{\partial y^2} + \frac{\partial^2\phi}{\partial z^2},$$

since we have

$$\nabla \cdot (\nabla\phi) = \nabla \cdot \left(\frac{\partial\phi}{\partial x}\mathbf{i} + \frac{\partial\phi}{\partial y}\mathbf{j} + \frac{\partial\phi}{\partial z}\mathbf{k}\right) = \frac{\partial^2\phi}{\partial x^2} + \frac{\partial^2\phi}{\partial y^2} + \frac{\partial^2\phi}{\partial z^2}.$$

From the above, we see that the symbol ∇^2 can be considered an abbreviation for

$$\frac{\partial^2}{\partial x^2} + \frac{\partial^2}{\partial y^2} + \frac{\partial^2}{\partial z^2}.$$

This equation occurs so often that we abbreviate it as $\nabla^2\phi$. The operator

$$\nabla^2 = \nabla \cdot \nabla$$

is called the *Laplace operator* because of its relation to *Laplace's equation*

$$\nabla^2\phi = \frac{\partial^2\phi}{\partial x^2} + \frac{\partial^2\phi}{\partial y^2} + \frac{\partial^2\phi}{\partial z^2} = 0.$$

Thus, the Laplacian of ϕ is just the sum of the second partial derivatives of ϕ. The Laplacian can also act on a vector quantity. In this case, the result is a vector whose components are the Laplacians of the components of the original vector

$$\nabla^2\mathbf{U} = (\nabla^2 u_1, \nabla^2 u_2, \nabla^2 u_3).$$

It can further be stated that any function satisfying this equation in a given region is said to be *harmonic* in that region. For example, in any region where the charge density is zero, the electric potential of a static distribution of charges is harmonic. A function describing the steady-state temperature distribution of a homogeneous material is harmonic in the interior of the region occupied by the material. Laplace's equation is seen to occur mainly in gravitation, electrostatics, steady-state heat flow and fluid flow. The Laplacian operator is by far the most important differential operator in mathematical physics.

If ϕ is a scalar, then $\nabla^2\phi(x, y, z)$ denotes the value of $\nabla^2\phi$ at the point (x, y, z). This is a number that tells us something about the behaviour of the scalar field in the neighbourhood of (x, y, z). Roughly speaking, it provides a measure of the *difference between the average value of the field in the immediate neighbourhood of the point and the precise value of the field at the point.* It must be noted that the word 'average' here refers to an average over a region of space, not to a time average. Variations with time t do not enter into the calculation of $\nabla^2 \phi$.

Thus, if $\nabla^2 \phi$ is positive at a point, and ϕ denotes the temperature, this means that the temperature in the vicinity of the point is, on an average, greater than the temperature at the point itself. In particular, if the temperature takes its minimum value at a certain point in space, it is reasonable to expect that the value of $\nabla^2 \phi$ will *not* be negative at that point. From this viewpoint, the Laplacian can be viewed as a sort of three-dimensional generalization of the ordinary differential operator d^2/dx^2, which is used in elementary calculus to test extreme points to see if they represent maxima or minima.

4.11 Vector identities

Although we continue to use the Del notation, the practice of formally manipulating

$$\nabla = \mathbf{i}\frac{\partial}{\partial x} + \mathbf{j}\frac{\partial}{\partial y} + \mathbf{k}\frac{\partial}{\partial z}$$

as though it were a vector has certain hazards. Keep in mind that the derivative operators appearing in the Del operator act only on functions appearing to the right of the Del operator. For example, suppose that

$$\mathbf{F} = x^3 y\mathbf{i} + y^2\mathbf{j} + x^2 z\mathbf{k} \quad \text{and} \quad \mathbf{R} = x\mathbf{i} + y\mathbf{j} + z\mathbf{k}.$$

Let us now compare the expressions $(\nabla \cdot \mathbf{R})\mathbf{F}$ and $(\mathbf{R} \cdot \nabla)\mathbf{F}$. For the first expression we have

$$(\nabla \cdot \mathbf{R})\mathbf{F} = 3\mathbf{F} = 3x^3 y\mathbf{i} + 3y^2\mathbf{j} + 3x^2 z\mathbf{k}.$$

On the other hand, in the second expression, $\mathbf{R}$ is to the left of ∇ and, therefore, the derivatives in the Del operator do not act on $\mathbf{R}$. In this case, we have

$$
\begin{aligned}
(\mathbf{R} \cdot \nabla)\mathbf{F} &= \left(x\frac{\partial}{\partial x} + y\frac{\partial}{\partial y} + z\frac{\partial}{\partial z} \right)(x^3 y\mathbf{i} + y^2\mathbf{j} + x^2 z\mathbf{k}) \\
&= x(3x^2 y\mathbf{i} + 2xz\mathbf{k}) + y(x^3\mathbf{i} + 2y\mathbf{j}) + z(x^2\mathbf{k}) \\
&= 4x^3 y\mathbf{i} + 2y^2\mathbf{j} + 3x^2 z\mathbf{k}.
\end{aligned}
$$

To further confuse matters, parentheses are sometimes omitted so that the reader should supply such parentheses as necessary to make the expression meaningful. For example, $\nabla \cdot \mathbf{RF}$ and $\mathbf{R} \cdot \nabla\mathbf{F}$ must mean $(\nabla \cdot \mathbf{R})\mathbf{F}$ and $(\mathbf{R} \cdot \nabla)\mathbf{F}$, respectively, since $\nabla \cdot (\mathbf{RF})$ and $\mathbf{R} \cdot (\nabla\mathbf{F})$ do not make sense within the framework of vector analysis. Similarly, $\nabla \cdot f\mathbf{F}$ means $\nabla \cdot (f\mathbf{F})$, simply the divergence of $f\mathbf{F}$, since $\nabla \cdot f$, and, hence, $(\nabla \cdot f)\mathbf{F}$, is meaningless. Because of these problems of interpretation, it is vitally important to preserve the working order with ∇. We now list a number of identities. Here $\mathbf{U}$ and $\mathbf{V}$ denote vector fields, while ϕ denotes a scalar field.

$$\nabla \cdot \phi \mathbf{V} = \phi \nabla \cdot \mathbf{V} + \mathbf{V} \cdot \nabla \phi \tag{4.22}$$

$$\nabla \times \phi \mathbf{V} = \phi \nabla \times \mathbf{V} + (\nabla \phi) \times \mathbf{V} \tag{4.23}$$

$$\nabla \cdot (\mathbf{U} \times \mathbf{V}) = \mathbf{V} \cdot \nabla \times \mathbf{U} - \mathbf{U} \cdot \nabla \times \mathbf{V} \tag{4.24}$$

$$\nabla \times (\mathbf{U} \times \mathbf{V}) = (\mathbf{V} \cdot \nabla)\mathbf{U} - (\mathbf{U} \cdot \nabla)\mathbf{V} + \mathbf{U}(\nabla \cdot \mathbf{V}) - \mathbf{V}(\nabla \cdot \mathbf{U}) \tag{4.25}$$

$$\nabla(\mathbf{U} \cdot \mathbf{V}) = (\mathbf{U} \cdot \nabla)\mathbf{V} + (\mathbf{V} \cdot \nabla)\mathbf{U}$$

$$+ \mathbf{U} \times (\nabla \times \mathbf{V}) + \mathbf{V} \times (\nabla \times \mathbf{U}) \tag{4.26}$$

$$\nabla \times \nabla \phi = \mathbf{0} \tag{4.27}$$

$$\nabla \cdot \nabla \times \mathbf{V} = 0 \tag{4.28}$$

$$\nabla \times (\nabla \times \mathbf{V}) = \nabla(\nabla \cdot \mathbf{V}) - \nabla \cdot \nabla \mathbf{V}$$

$$= \nabla(\nabla \cdot \mathbf{V}) - \nabla^2 \mathbf{V} \tag{4.29}$$

These identities can all be verified by direct expansion.

Proof of identity (4.22)   For example, to prove identity (4.22), we have

$$\nabla \cdot (\phi \mathbf{V}) = \nabla \cdot (\phi(v_1\mathbf{i} + v_2\mathbf{j} + v_3\mathbf{k}))$$

$$= \frac{\partial(\phi v_1)}{\partial x} + \frac{\partial(\phi v_2)}{\partial y} + \frac{\partial(\phi v_3)}{\partial z}$$

$$= \left(\phi\frac{\partial v_1}{\partial x} + v_1\frac{\partial \phi}{\partial x}\right) + \left(\phi\frac{\partial v_2}{\partial y} + v_2\frac{\partial \phi}{\partial y}\right) + \left(\phi\frac{\partial v_3}{\partial z} + v_3\frac{\partial \phi}{\partial z}\right)$$

$$= \phi\left(\frac{\partial v_1}{\partial x} + \frac{\partial v_2}{\partial y} + \frac{\partial v_3}{\partial z}\right) + v_1\frac{\partial \phi}{\partial x} + v_2\frac{\partial \phi}{\partial y} + v_3\frac{\partial \phi}{\partial z}$$

$$= \phi\nabla \cdot \mathbf{V} + \mathbf{V} \cdot \nabla \phi.$$

To establish these formulae, it is generally easier if ∇ is treated as a vector. This leads to the manipulation of the expressions according to the appropriate formulae from vector algebra and finally giving ∇ its operational meaning. Since ∇ is a linear combination of scalar differential operators which obey the usual product rule of differentiation, it is clear that ∇ itself has this property.

We can apply ∇ to products of various sorts by assuming that each of the factors, in turn, is the only one which is variable and adding the partial results so obtained. As a notation to help us determine these partial results, it is useful to attach to ∇, whenever it is followed by more than one factor, a subscript indicating the one factor upon which it is currently allowed to operate.   ∎

Proof of identity (4.23) To prove identity (4.23), we first suppose that the scalar function ϕ is constant. Then we can write

$$\nabla_v \times (\phi \mathbf{V}) = \phi \nabla \times \mathbf{V}.$$

Similarly, if we regard $\mathbf{V}$ as constant and ϕ as variable, we have

$$\nabla_\phi \times (\phi \mathbf{V}) = (\nabla \phi) \times \mathbf{V},$$

where the parenthesis now restrict the effect of ∇ to the factor ϕ alone. Adding these two partial results gives

$$\nabla_v \times (\phi \mathbf{V}) + \nabla_\phi \times (\phi \mathbf{V}) \equiv \nabla \times (\phi \mathbf{V}) = \phi \nabla \times \mathbf{V} + (\nabla \phi) \times \mathbf{V}.$$

∎

Proof of identity (4.24) To prove identity (4.24), from the cyclic properties of scalar triple products, we have the following,

$$\nabla_u \cdot (\mathbf{U} \times \mathbf{V}) = \nabla \cdot \mathbf{V} \times \mathbf{U} \quad \text{and}$$
$$\nabla_v \cdot (\mathbf{U} \times \mathbf{V}) = -\mathbf{U} \cdot \nabla \times \mathbf{V}.$$

Therefore, adding these two partial results gives

$$\nabla_u \cdot (\mathbf{U} \times \mathbf{V}) + \nabla_v (\mathbf{U} \times \mathbf{V}) \equiv \nabla \cdot (\mathbf{U} \times \mathbf{V})$$
$$= \mathbf{V} \cdot \nabla \times \mathbf{U} - \mathbf{U} \cdot \nabla \times \mathbf{V}.$$

∎

Proof of identity (4.25) To prove identity (4.25), we have

$$\nabla_u \times (\mathbf{U} \times \mathbf{V}) = (\nabla_u \cdot \mathbf{V})\mathbf{U} - (\nabla_u \cdot \mathbf{U})\mathbf{V} = \mathbf{V} \cdot \nabla \mathbf{U} - \mathbf{V} \nabla \cdot \mathbf{U}$$

and

$$\nabla_v \times (\mathbf{U} \times \mathbf{V}) = (\nabla_v \cdot \mathbf{V})\mathbf{U} - (\nabla_v \cdot \mathbf{U})\mathbf{V} = \mathbf{U} \nabla \cdot \mathbf{V} - \mathbf{U} \cdot \nabla \mathbf{V}.$$

Adding these gives

$$\nabla_u \times (\mathbf{U} \times \mathbf{V}) + \nabla_v \times (\mathbf{U} \times \mathbf{V}) \equiv \nabla \times (\mathbf{U} \times \mathbf{V})$$
$$= \mathbf{V} \cdot \nabla \mathbf{U} - \mathbf{U} \cdot \nabla \mathbf{V} + \mathbf{U} \nabla \cdot \mathbf{V} - \mathbf{V} \nabla \cdot \mathbf{U}.$$

∎

Proof of identity (4.26) To prove identity (4.26), we note that

$$\mathbf{U} \times (\nabla \times \mathbf{V}) \equiv \mathbf{U} \times (\nabla_v \times \mathbf{V}) = (\mathbf{U} \cdot \mathbf{V})\nabla_v - (\mathbf{U} \cdot \nabla)\mathbf{V}$$
$$= \nabla_v (\mathbf{U} \cdot \mathbf{V}) - \mathbf{U} \cdot \nabla \mathbf{V}$$

and

$$\mathbf{V} \times (\nabla \times \mathbf{U}) \equiv \mathbf{V} \times (\nabla_u \times \mathbf{U}) = (\mathbf{V} \cdot \mathbf{U})\nabla_u - (\mathbf{V} \cdot \nabla)\mathbf{U}$$
$$= \nabla_u(\mathbf{U} \cdot \mathbf{V}) - \mathbf{V} \cdot \nabla\mathbf{U}.$$

Hence, transposing and adding these, we find

$$\nabla_u(\mathbf{U} \cdot \mathbf{V}) + \nabla_v(\mathbf{U} \cdot \mathbf{V}) \equiv \nabla(\mathbf{U} \cdot \mathbf{V})$$
$$= \mathbf{U} \times (\nabla \times \mathbf{V}) + \mathbf{V} \times (\nabla \times \mathbf{U})$$
$$+ \mathbf{U} \cdot \nabla\mathbf{V} + \mathbf{V} \cdot \nabla\mathbf{U}.$$

∎

Proof of identity (4.27) The fact that the operational coefficient $\nabla \times \nabla$ appears in identity (4.27) as the cross product of identical factors suggests that $\nabla \times \nabla\phi$ is zero and can be easily verified. ∎

Proof of identity (4.28) Similarly, the fact that the left-hand side of formula (4.28) appears as a scalar triple product with two identical factors suggests the truth of the formula, and again it is not difficult to show that $\nabla \cdot \nabla \times \mathbf{V}$ is always equal to zero. ∎

Proof of identity (4.29) Formula (4.29) can be established by direct expansion or verified by applying the usual rule for expanding a vector triple product.

$$\nabla \times (\nabla \times \mathbf{V}) = (\nabla \cdot \mathbf{V})\nabla - (\nabla \cdot \nabla)\mathbf{V} = \nabla(\nabla \cdot \mathbf{V}) - \nabla^2\mathbf{V},$$

where the conventional symbol ∇^2 has been substituted for the second-order operator

$$\nabla \cdot \nabla = \left(\mathbf{i}\frac{\partial}{\partial x} + \mathbf{j}\frac{\partial}{\partial y} + \mathbf{k}\frac{\partial}{\partial z}\right) \cdot \left(\mathbf{i}\frac{\partial}{\partial x} + \mathbf{j}\frac{\partial}{\partial y} + \mathbf{k}\frac{\partial}{\partial z}\right)$$
$$= \frac{\partial^2}{\partial x^2} + \frac{\partial^2}{\partial y^2} + \frac{\partial^2}{\partial z^2}.$$

∎

Example 4.13 Calculate $\nabla \cdot \mathbf{R}$, $\nabla \cdot \mathbf{R}f(r)$ and then show that

$$\nabla \cdot \mathbf{R}r^{n-1} = (n + 2)r^{n-1}.$$

Note that here the magnitude of the position vector is denoted as r, i.e. $|\mathbf{R}| = r$. □

Solution We know that

$$\nabla \cdot \mathbf{R} = \left(\mathbf{i} \frac{\partial}{\partial x} + \mathbf{j} \frac{\partial}{\partial y} + \mathbf{k} \frac{\partial}{\partial z} \right) \cdot (\mathbf{i}x + \mathbf{j}y + \mathbf{k}z)$$

$$= \frac{\partial x}{\partial x} + \frac{\partial y}{\partial y} + \frac{\partial z}{\partial z},$$

or

$$\nabla \cdot \mathbf{R} = 3.$$

For the second part, we have

$$\nabla \cdot \mathbf{R} \, f(r) = \frac{\partial}{\partial x}[xf(r)] + \frac{\partial}{\partial y}[yf(r)] + \frac{\partial}{\partial x}[zf(r)]$$

$$= 3f(r) + \frac{x^2}{r}\frac{df}{dr} + \frac{y^2}{r}\frac{df}{dr} + \frac{z^2}{r}\frac{df}{dr}$$

$$= 3f(r) + r\frac{df}{dr}.$$

Now to show the identity, we make use of the above taking

$$f(r) = r^{n-1},$$

$$\nabla \cdot \mathbf{R} \, r^{n-1} = \nabla \cdot \mathbf{R}_0 \, r^n$$

$$= 3r^{n-1} + (n-1)r^{n-1}$$

$$= (n+2)r^{n-1}.$$

It is worth noting that this divergence vanishes for $n = -2$.   ∎

Example 4.14 Given a vector field function

$$\mathbf{F} = (3y - c_1 z)\mathbf{i} + (c_2 x - 2z)\mathbf{j} - (c_3 y + z)\mathbf{k}.$$

(a) Determine the constants c_1, c_2 and c_3 if $\mathbf{F}$ is irrotational.
(b) Determine the scalar potential ϕ whose negative gradient equals $\mathbf{F}$.   ☐

Solution (a) For $\mathbf{F}$ to be irrotational, $\nabla \times \mathbf{F} = 0$, i.e.

$$\nabla \times \mathbf{F} = \begin{vmatrix} \mathbf{i} & \mathbf{j} & \mathbf{k} \\ \dfrac{\partial}{\partial x} & \dfrac{\partial}{\partial y} & \dfrac{\partial}{\partial z} \\ 3y - c_1 z & c_2 x - 2z & -(c_3 y + z) \end{vmatrix}$$

$$= (-c_3 + 2)\mathbf{i} - c_1 \mathbf{j} + (c_2 - 3)\mathbf{k} = 0.$$

Each component of $\nabla \times \mathbf{F}$ must vanish. Hence, $c_1 = 0$, $c_2 = 3$ and $c_3 = 2$.

(b) Since **F** is irrotational, it can be expressed as the negative gradient of a scalar function ϕ; that is,

$$\mathbf{F} = -\nabla\phi = -\mathbf{i}\frac{\partial\phi}{\partial x} - \mathbf{j}\frac{\partial\phi}{\partial y} - \mathbf{k}\frac{\partial\phi}{\partial z}$$
$$= 3y\mathbf{i} + (3x - 2z)\mathbf{j} - (2y + z)\mathbf{k}.$$

Three partial differential equations are obtained:

$$\frac{\partial\phi}{\partial x} = -3y,$$

$$\frac{\partial\phi}{\partial y} = -3x + 2z,$$

$$\frac{\partial\phi}{\partial z} = 2y + z.$$

Integrating the first one with respect to x partially, we obtain

$$\phi = -3xy + f(y, z),$$

where $f(y, z)$ is an arbitrary function of y and z. Now differentiating this ϕ solution with respect to y partially, and equating with the second equation, we have

$$\frac{\partial\phi}{\partial y} = -3x + 2z = -3x + \frac{\partial f}{\partial y},$$

and after reduction we have $\partial f/\partial y = 2z$. Now integrating partially with respect to y, $f(y, z) = 2yz + g(z)$, where g is an arbitrary function of z alone. Thus, we have

$$\phi = -3xy + 2yz + g(z).$$

Next, we differentiate this ϕ with respect to z and equate with the third equation, which yields

$$\frac{\partial\phi}{\partial z} = 2y + z = 2y + g'(z),$$

and this reduces to $g'(z) = z$. Integrating with respect to z, we obtain $g(z) = (z^2/2) + c$, where c is a pure constant. Hence the solution for ϕ is obtained as

$$\phi(x, y, z) = -3xy + 2yz + \frac{z^2}{2} + c.$$

The constant c is to be determined by the boundary condition or the condition at infinity. ∎

4.12 Summary

- The *gradient* of a scalar field ϕ is the vector field

$$\nabla\phi = \frac{\partial\phi}{\partial x}\mathbf{i} + \frac{\partial\phi}{\partial y}\mathbf{j} + \frac{\partial\phi}{\partial z}\mathbf{k},$$

  where

$$\nabla \equiv \mathbf{i}\frac{\partial}{\partial x} + \mathbf{j}\frac{\partial}{\partial y} + \mathbf{k}\frac{\partial}{\partial z}.$$

- If ϕ is a function of a single variable u, which in turn is a function of (x, y) and z, then

$$\nabla\phi(u) = \frac{d\phi}{du} \cdot \nabla u.$$

- If $\phi = \phi(x, y, z)$ is a differentiable function of x, y and z and $\mathbf{U}$ is any unit vector, then

$$D_{\mathbf{U}}\phi = \nabla\phi \cdot \mathbf{U}$$

  is defined as the directional derivative of $\phi = \phi(x, y, z)$ in the direction of the unit vector $\mathbf{U}$.

- *The divergence of a vector* $\mathbf{F}$

  The vector character of the operator ∇ suggests that we also consider dot and cross products in which it appears as a factor. Let $\mathbf{F}(x, y, z)$ be a differentiable function, where x, y, z are Cartesian coordinates, and let F_1, F_2, F_3 be the components of $\mathbf{F}$. Then the function

$$\nabla \cdot \mathbf{F} = \left(\mathbf{i}\frac{\partial}{\partial x} + \mathbf{j}\frac{\partial}{\partial y} + \mathbf{k}\frac{\partial}{\partial z}\right) \cdot (F_1\mathbf{i} + F_2\mathbf{j} + F_3\mathbf{k})$$

$$= \frac{\partial F_1}{\partial x} + \frac{\partial F_2}{\partial y} + \frac{\partial F_3}{\partial z}$$

  is called the *divergence* of $\mathbf{F}$. This is commonly denoted as div $\mathbf{F}$. Note that $\nabla \cdot \mathbf{F}$ means the scalar div $\mathbf{F}$, whereas ∇f means the vector grad f defined by equation (4.7).

- *The curl of a vector* $\mathbf{F}$

In a similar manner, we can define the cross product of ∇ with $\mathbf{F}$ as

$$\nabla \times \mathbf{F} = \begin{vmatrix} \mathbf{i} & \mathbf{j} & \mathbf{k} \\ \dfrac{\partial}{\partial x} & \dfrac{\partial}{\partial y} & \dfrac{\partial}{\partial z} \\ F_1 & F_2 & F_3 \end{vmatrix}$$

$$= \left(\frac{\partial F_3}{\partial y} - \frac{\partial F_2}{\partial z} \right) \mathbf{i} + \left(\frac{\partial F_1}{\partial z} - \frac{\partial F_3}{\partial x} \right) \mathbf{j} + \left(\frac{\partial F_2}{\partial x} - \frac{\partial F_1}{\partial y} \right) \mathbf{k}.$$

This expression is known as the *curl* of $\mathbf{F}$ or the curl of the vector field defined by $\mathbf{F}$.

- *The vector identities*

We list eight vector identities. Here, $\mathbf{U}$ and $\mathbf{V}$ denote vector fields, while ϕ denotes a scalar field. They are all functions of space variables.

$$\nabla \cdot \phi \mathbf{V} = \phi \nabla \cdot \mathbf{V} + \mathbf{V} \cdot \nabla \phi \tag{4.22}$$

$$\nabla \times \phi \mathbf{V} = \phi \nabla \times \mathbf{V} + (\nabla \phi) \times \mathbf{V} \tag{4.23}$$

$$\nabla \cdot (\mathbf{U} \times \mathbf{V}) = \mathbf{V} \cdot \nabla \times \mathbf{U} - \mathbf{U} \cdot \nabla \times \mathbf{V} \tag{4.24}$$

$$\nabla \times (\mathbf{U} \times \mathbf{V}) = (\mathbf{V} \cdot \nabla)\mathbf{U} - (\mathbf{U} \cdot \nabla)\mathbf{V} + \mathbf{U}(\nabla \cdot \mathbf{V}) - \mathbf{V}(\nabla \cdot \mathbf{U}) \tag{4.25}$$

$$\nabla(\mathbf{U} \cdot \mathbf{V}) = (\mathbf{U} \cdot \nabla)\mathbf{V} + (\mathbf{V} \cdot \nabla)\mathbf{U}$$

$$+ \mathbf{U} \times (\nabla \times \mathbf{V}) + \mathbf{V} \times (\nabla \times \mathbf{U}) \tag{4.26}$$

$$\nabla \times \nabla \phi = 0 \tag{4.27}$$

$$\nabla \cdot \nabla \times \mathbf{V} = 0 \tag{4.28}$$

$$\nabla \times (\nabla \times \mathbf{V}) = \nabla(\nabla \cdot \mathbf{V}) - \nabla \cdot \nabla \mathbf{V}$$

$$= \nabla(\nabla \cdot \mathbf{V}) - \nabla^2 \mathbf{V} \tag{4.29}$$

- The Laplacian ∇^2 is given by

$$\nabla^2 = \nabla \cdot \nabla = \left(\mathbf{i}\frac{\partial}{\partial x} + \mathbf{j}\frac{\partial}{\partial y} + \mathbf{k}\frac{\partial}{\partial z} \right) \cdot \left(\mathbf{i}\frac{\partial}{\partial x} + \mathbf{j}\frac{\partial}{\partial y} + \mathbf{k}\frac{\partial}{\partial z} \right)$$

$$= \frac{\partial^2}{\partial x^2} + \frac{\partial^2}{\partial y^2} + \frac{\partial^2}{\partial z^2}.$$

Exercises

1. Find the directional derivative of f at the given point in the direction indicated by the angle θ.
   (a) $f(x, y) = x^2 y^3 + 2x^4 y$, $(1, -2)$, $\theta = \pi/3$.
   (b) $f(x, y) = (x^2 - y)^3$ $(3, 1)$, $\theta = 3\pi/4$.

(c) $f(x, y) = y^x$, $(1, 2)$, $\theta = \pi/2$.

(d) $f(x, y) = \sin(x + 2y)$, $(4, -2)$, $\theta = -2\pi/3$.

For the following problems,

(a) find the gradient of f.

(b) evaluate the gradient at the point P.

(c) find the rate of change of f at P in the direction of the vector $\mathbf{U}$.

2. $f(x, y) = x^3 - 4x^2y + y^2$, $P(0, -1)$, $\mathbf{U} = (3/5, 4/5)$.

3. $f(x, y) = e^x \sin y$, $P(1, \pi/4)$, $\mathbf{U} = (-1/\sqrt{5}, 2/\sqrt{5})$.

4. $f(x, y, z) = xy^2z^3$, $P(1, -2, 1)$, $\mathbf{U} = (1/\sqrt{3}, -1/\sqrt{3}, 1/\sqrt{3})$.

5. $f(x, y, z) = xy + yz^2 + xz^3$, $P(2, 0, 3)$, $\mathbf{U} = (-2/3, -1/3, 2/3)$.

6. Find the directional derivative of the function at the given point in the direction of the vector $\mathbf{V}$.

(a) $f(x, y) = \sqrt{x - y}$, $(5, 1)$, $\mathbf{V} = (12, 5)$.

(b) $f(x, y) = xe^{xy}$ $(-3, 0)$, $\mathbf{V} = 2\mathbf{i} + 3\mathbf{j}$.

(c) $f(x, y, z) = \sqrt{xyz}$ $(2, 4, 2)$, $\mathbf{V} = (4, 2, -4)$.

(d) $f(x, y) = x \tan^{-1}(y/z)$ $(1, 2, -2)$, $\mathbf{V} = \mathbf{i} + \mathbf{j} - \mathbf{k}$.

(e) $f(x, y) = xe^{yz} + xye^z$ $(-2, 1, 1)$, $\mathbf{V} = \mathbf{i} - 2\mathbf{j} + 3\mathbf{k}$.

7. Find the maximum rate of change of f at the given point and the direction in which it occurs.

(a) $f(x, y) = xe^{-y} + 3y$, $(1, 0)$.

(b) $f(x, y) = \ln(x^2 + y^2)$, $(1, 2)$.

(c) $f(x, y) = \sqrt{x^2 + 2y}$, $(4, 10)$.

(d) $f(x, y) = x + y/z$, $(4, 3, -1)$.

(e) $f(x, y) = \cos(3x + 2y)$, $(\pi/6, -\pi/8)$.

(f) $f(x, y) = x/y + y/z$, $(4, 2, 1)$.

8. The temperature at a point (x, y, z) is given by

$$T(x, y, z) = 200e^{-x^2 - 3y^2 - 9z^2},$$

where T is measured in degrees Celsius and x, y, z in metres.

(a) Find the rate of change of temperature at the point $P(2, -1, 2)$ in the direction toward the point $(3, -3, 3)$.

(b) In which direction does the temperature increase fastest at P?

(c) Find the maximum rate of increase of T.

9. Suppose that you are climbing a hill whose shape is given by the equation $z = 1000 - 0.01x^2 - 0.02y^2$ and you are standing at a point with coordinates $(60, 100, 764)$.

(a) In which direction should you proceed initially to reach the top of the hill as fast as possible?

(b) If you climb in that direction, at what angle above the horizontal will you be climbing initially?

10. Given $\mathbf{F}(x, y, z) = x^2y\mathbf{i} + z\mathbf{j} - (x + y - z)\mathbf{k}$, find

(a) $\nabla \cdot \mathbf{F}$.

(b) $\nabla \times \mathbf{F}$.

(c) $\nabla(\nabla \cdot \mathbf{F})$.

11. Which of the following functions satisfy Laplace's equation?
    (a) $f(x, y, z) = e^z \sin y$.
    (b) $f(x, y, z) = \sin x \sinh y \cos \cosh z$.
    (c) $f(x, y, z) = \sin px \sinh qy$ (p and q are constants).

12. Given that $f(x, y, z) = 2x^2 + y$ and $\mathbf{R} = x\mathbf{i} + y\mathbf{j} + z\mathbf{k}$, find
    (a) ∇f.
    (b) $\nabla \cdot \mathbf{R}$.
    (c) $\nabla^2 f$.
    (d) $\nabla \times (f\mathbf{R})$.

13. Evaluate $\nabla^2[(\mathbf{i} + \mathbf{j} + \mathbf{k}) \times \nabla(\mathbf{R} \cdot \mathbf{R})^2]$ where $\mathbf{R} = x\mathbf{i} + y\mathbf{j} + z\mathbf{k}$.

14. Evaluate $\mathbf{A} \cdot \nabla\mathbf{R} + \nabla(\mathbf{A} \cdot \mathbf{R}) + \mathbf{A} \cdot \nabla \times \mathbf{R}$ where $\mathbf{A}$ is a constant vector field and $\mathbf{R} = x\mathbf{i} + y\mathbf{j} + z\mathbf{k}$.

15. If $r^2 = x^2 + y^2 + z^2$, $\mathbf{R} = x\mathbf{i} + y\mathbf{j} + z\mathbf{k}$ and $\mathbf{A}$ is a constant vector field, find
    (a) $\nabla \cdot (r^2\mathbf{A})$.
    (b) $\nabla \times (r^2\mathbf{A})$.
    (c) $\mathbf{R} \cdot \nabla(r^2\mathbf{A})$.
    (d) $\nabla(\mathbf{A} \cdot \mathbf{R})^4$.
    (e) $\nabla \cdot (r\mathbf{A})$.
    (f) $\mathbf{R} \cdot \nabla(\mathbf{A} \cdot \mathbf{R}\mathbf{A})$.
    (g) $\nabla \cdot (\mathbf{A} \times \mathbf{R})$.
    (h) $\nabla^2(\mathbf{R} \cdot \mathbf{R})$.

16. If r is the distance from the origin to the point (x, y, z) and $\mathbf{A}$ is a constant vector, evaluate

$$\nabla\left(\mathbf{A} \cdot \nabla\frac{1}{r}\right) + \nabla \times \left(\mathbf{A} \times \nabla\frac{1}{r}\right).$$

17. Given $\phi = \tan^{-1} x + \tan^{-1} y$ and $\psi = (x + y)/(1 - xy)$, show that

$$\nabla\phi \times \nabla\psi = \mathbf{0}.$$

18. Find the divergence of the field

$$\mathbf{F} = \frac{x\mathbf{i} + y\mathbf{j} + z\mathbf{k}}{(x^2 + y^2 + z^2)^{3/2}}.$$

    Is the divergence of this field defined at every point in space?

19. Given that $\mathbf{R} = x\mathbf{i} + y\mathbf{j} + z\mathbf{k}$ and $r = |\mathbf{R}| = \sqrt{x^2 + y^2 + z^2}$, prove that

$$\nabla r^n = nr^{n-2}\mathbf{R}.$$

20. Using the same definition of r as in the previous question, show that

$$\nabla \times \left[\frac{1}{r}(\mathbf{A} \times \mathbf{R})\right] = \frac{1}{r}\mathbf{A} + \frac{\mathbf{A} \cdot \mathbf{R}}{r^3}\mathbf{R}$$

    for any constant vector $\mathbf{A}$.

21. (a) Show that a necessary and sufficient condition that $u(x, y, z)$ and $v(x, y, z)$ are related by some function $f(u, v) = 0$ is that $(\nabla u) \times (\nabla v) = \mathbf{0}$.

    (b) If $u = u(x, y)$ and $v = v(x, y)$, show that the condition $(\nabla u) \times (\nabla v) = \mathbf{0}$ leads to the two-dimensional Jacobian

$$J \left(\frac{u, v}{x, y} \right) = \begin{vmatrix} \dfrac{\partial u}{\partial x} & \dfrac{\partial u}{\partial y} \\ \dfrac{\partial v}{\partial x} & \dfrac{\partial v}{\partial y} \end{vmatrix} = 0.$$

    The functions u and v are assumed differentiable.

22. If a force $\mathbf{F}$ is given by

$$\mathbf{F} = (x^2 + y^2 + z^2)^n (\mathbf{i}x + \mathbf{j}y + \mathbf{k}z),$$

    find

    (a) $\nabla \cdot \mathbf{F}$.
    (b) $\nabla \times \mathbf{F}$.
    (c) A scalar potential $\phi(x, y, z)$ so that $\mathbf{F} = -\nabla \phi$.
    (d) For what value of the exponent n does the scalar potential diverge at both the origin and infinity?

23. The origin of the Cartesian coordinates is at the earth's centre. The moon is on the z-axis, a fixed distance R away (centre-to-centre distance). The tidal force exerted by the moon on a particle at the earth's surface (point x, y, z) is given by

$$F_x = -GMm\frac{x}{R^3}, \quad F_y = -GMm\frac{y}{R^3}, \quad F_z = -GMm\frac{z}{R^3}.$$

    Find the potential that yields this tidal force.

24. Vector $\mathbf{B}$ is formed by the product of two gradients

$$\mathbf{B} = (\nabla u) \times (\nabla v),$$

    where u and v are scalar functions.

    (a) Show that $\mathbf{B}$ is solenoidal.
    (b) Show that

$$\mathbf{A} = \frac{1}{2}(u\nabla v - v\nabla u)$$

    is a vector potential for $\mathbf{B}$, where

$$\mathbf{B} = \nabla \times \mathbf{A}.$$

    [Hint: When a vector $\mathbf{B}$ is solenoidal, i.e. $\nabla \cdot \mathbf{B} = 0$, a vector potential exists such that $\mathbf{B} = \nabla \times \mathbf{A}$.]

25. Find the potential function for the gravitational field of a uniform sphere of radius a and mass M. Show that the attraction of the sphere at a point P a distance r from the centre of the sphere is

$$\mathbf{F} = \begin{cases} -\dfrac{M\mathbf{R}}{a^3} & r \leq a, \\[2ex] -\dfrac{M\mathbf{R}}{r^3} & r \geq a. \end{cases}$$

Chapter 5

Line, surface and volume integrals

5.1 Introduction

The notion of the definite integral $\int_a^b f(x)\mathrm{d}x$, i.e. integration of a function defined over an interval, can be generalized to integration of a function defined along a curve. With this in mind, we now introduce some terminology about curves. Suppose C is a curve parameterized by $x = f(t)$, $y = g(t)$, $a \le t \le b$ and A and B are the points $(f(a), g(a))$ and $(f(b), g(b))$, respectively. We say that:

(i) C is a *smooth curve* if f' and g' are continuous on the closed interval $[a, b]$ and not simultaneously zero on the open interval (a, b).

(ii) C is *piecewise smooth* if it consists of a finite number of smooth curves $C_1, C_2, \ldots, C_n$ joined end to end – i.e. $C = C_1 \cup C_2 \cup \cdots \cup C_n$.

(iii) C is a *closed curve* if $A = B$.

(iv) C is a *simple closed curve* if $A = B$ and the curve does not cross itself.

(v)   If C is not a closed curve, then the *positive direction* on C is the direction corresponding to increasing values of t.

The graphs in Fig. 5.1a, b, c and d illustrate each type of curve defined in (i)–(iv).

This same terminology carries over in a natural manner to curves in space. For example, a curve C defined by $x = f(t)$, $y = g(t)$, $z = h(t)$, $a \le t \le b$, is smooth if f', g' and h' are continuous on $[a, b]$ and not simultaneously zero.

If we partition the interval $[a, b]$ into n subintervals $[x_{k-1}, x_k]$ of length $\Delta x_k = x_k - x_{k-1}$, we can denote this partition to be P. We can then take $\|P\|$ to be the length of the longest subinterval and call it the *norm* of the partition P. Then from

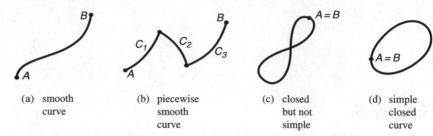

Figure 5.1 Categories of curves.

calculus we can define the definite integral of a single variable to be given by the limit of the sum:

$$\int_a^b f(x)\mathrm{d}x = \lim_{\|P\|\to 0} \sum_{k=1}^n f(x_k)\Delta x_k. \tag{5.1}$$

Alternatively, we could let Δs_i be the length of each sub-arc of the curve. Then the limit shown in equation (5.1) would become

$$\int_a^b f(x)\mathrm{d}x = \lim_{\Delta s_i\to 0} \sum_{k=1}^n f(x_k)\Delta x_k. \tag{5.2}$$

Let $F(x, y)$ be a function of x and y, and let C be a continuous, piecewise smooth curve joining the points A and B. Furthermore, let the arc of C between A and B be divided into n segments Δs_i whose projections on the x and y axes are, respectively, Δx_i and Δy_i and let (ξ_i, η_i) be the coordinates of an arbitrary point in segment Δs_i (see Fig. 5.2).

If we evaluate the given function $F(x, y)$ at each of the points (ξ_i, η_i) and form the products

$$F(\xi_i, \eta_i)\Delta x_i, \quad F(\xi_i, \eta_i)\Delta y_i, \quad F(\xi_i, \eta_i)\Delta s_i,$$

and then sum over all the subdivisions of the arc AB, we have the three sums

$$\sum_{i=1}^n F(\xi_i, \eta_i)\Delta x_i, \quad \sum_{i=1}^n F(\xi_i, \eta_i)\Delta y_i, \quad \sum_{i=1}^n F(\xi_i, \eta_i)\Delta s_i.$$

The limits of these sums, as n becomes infinite in such a way that the length of each Δs_i approaches zero, are known as *line integrals* and are written, respectively, as

$$\int_C F(x, y)\mathrm{d}x, \quad \int_C F(x, y)\mathrm{d}y, \quad \int_C F(x, y)\mathrm{d}s.$$

A more appropriate name might be curvilinear integral instead of line integral. The word 'line' is regarded as meaning 'curved line' as opposed to a straight line. The following definition summarizes the preceding points.

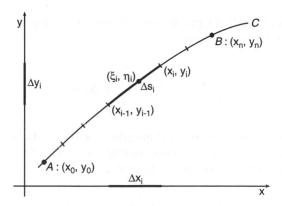

Figure 5.2   Subdivision of an arc preparatory to defining a line integral.

Definition 5.1   *Let G be a function of two variables x and y defined on a region of the plane containing a smooth curve C and let (x_k^*, y_k^*) be the coordinates of any arbitrary point in the segment Δs_k.*

(i)   The line integral of G along C from A to B with respect to x is

$$\int_C G(x, y)\mathrm{d}x = \lim_{\|P\|\to 0} \sum_{k=1}^{n} G(x_k^*, y_k^*)\Delta x_k$$

(ii)   The line integral of G along C from A to B with respect to y is

$$\int_C G(x, y)\mathrm{d}y = \lim_{\|P\|\to 0} \sum_{k=1}^{n} G(x_k^*, y_k^*)\Delta y_k.$$

(iii)   The line integral of G along C from A to B with respect to arc length is

$$\int_C G(x, y)\mathrm{d}s = \lim_{\|P\|\to 0} \sum_{k=1}^{n} G(x_k^*, y_k^*)\Delta s_k.$$

It can be shown that if $G(x, y)$ is continuous on C, then the integrals defined in (i), (ii) and (iii) exist. We shall assume continuity as a matter of course.

In these definitions, Δx_l and Δy_l are signed quantities whereas Δs_i is intrinsically positive. Thus, the following properties of ordinary definite integrals hold:

(a) $\int_A^B c\phi(t)\mathrm{d}t = c\int_A^B \phi(t)\mathrm{d}t$, where c is a constant;

(b) $\int_A^B [\phi_1(t) \pm \phi_2(t)]\mathrm{d}t = \int_A^B \phi_1(t)\mathrm{d}t \pm \int_A^B \phi_2(t)\mathrm{d}t$;

(c) $\int_A^B \phi(t)\mathrm{d}t = -\int_B^A \phi(t)\mathrm{d}t$;

(d) $\int_A^P \phi(t)\mathrm{d}t + \int_P^B \phi(t)\mathrm{d}t = \int_A^B \phi(t)\mathrm{d}t$.

These properties are equally valid for line integrals of types (i) and (ii), provided that throughout each formula the curve joining A and B remains the same. On the other hand, line integrals of type (iii), although they have properties (a) and (b), do not have property (c) since, in fact,

$$\int_A^B F(x, y)ds = \int_B^A F(x, y)ds.$$

Moreover, property (d) holds for these integrals if and only if P is between A and B on the path of integration. In general, we shall be more interested in integrals of the first two types than in those of the third type.

Remark 5.1   In many problems the path of integration C will consist of one or more simple closed curves enclosing the boundary of a region R. Clearly, integration can be performed in either of two directions around a closed curve, and it is important that we be able to distinguish between them. This is done by defining the *positive direction* around a closed curve as the direction in which an observer would move if he traversed the curve in such a way that the area of R was always on his left. According to this definition, if R is the interior of a simple closed curve C, then the positive direction around C is the counterclockwise direction. If R is the region exterior to a simple closed curve C, then the positive direction around C is clockwise. If R is the region interior to a closed curve C_1 and exterior to another simple closed curve C_2, then the positive direction of traversing the entire boundary of R, namely $C = C_1 \cup C_2$, is counterclockwise around C_1 and clockwise around C_2.   ∎

5.1.1 Method of evaluation – curve defined parametrically

The line integrals in Definition 5.1 can be evaluated in two ways, depending on whether the curve C is defined parametrically or by an explicit function. In either case, the basic idea is to convert the line integral to a definite integral in a single variable. If C is a smooth curve parameterized by $x = f(t)$, $y = g(t)$, $a \le t \le b$, then we simply replace x and y in the integral by the functions $f(t)$ and $g(t)$, and the appropriate differential dx, dy or ds by $f'dt$, $g'dt$, or $\sqrt{(f'(t))^2 + (g'(t))^2}dt$. The expression $ds = \sqrt{(f'(t))^2 + (g'(t))^2}dt$ is called the *differential of arc length*. The integration is carried out with respect to the variable t in the usual manner:

$$\int_C G(x, y)dx = \int_a^b G(f(t), g(t))f'(t)dt, \tag{5.3}$$

$$\int_C G(x, y)dy = \int_a^b G(f(t), g(t))g'(t)dt, \tag{5.4}$$

$$\int_C G(x, y)ds = \int_a^b G(f(t), g(t))\sqrt{(f'(t))^2 + (g'(t))^2}dt. \tag{5.5}$$

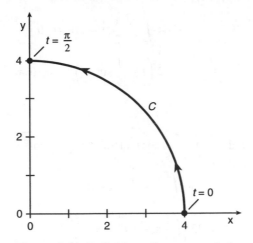

Figure 5.3   Definition of a quarter-circle.

Example 5.1   Evaluate

(a) $\int_C xy^2 dx$,

(b) $\int_C xy^2 dy$ and

(c) $\int_C xy^2 ds$

on the quarter circle C defined by $x = 4\cos t, y = 4\sin t, 0 \le t \le \frac{\pi}{2}$ (see Fig. 5.3).
$\square$

Solution

(a)   From (5.3), we have

$$\int_C xy^2 dx = \int_0^{\pi/2} (4\cos t)(16\sin^2 t)(-4\sin t dt)$$

$$= -256 \int_0^{\pi/2} \sin^3 t \cos t dt$$

$$= -256 \left(\frac{1}{4}\sin^4 t\right)_0^{\pi/2} = -64.$$

(b)   From (5.4),

$$\int_C xy^2 dy = \int_0^{\pi/2} (4\cos t)(16\sin^2 t)(4\cos t dt)$$

$$= 256 \int_0^{\pi/2} \sin^2 t \cos^2 t dt$$

$$= 256 \int_0^{\pi/2} \frac{1}{4}\sin^2 2t dt$$

$$= 64 \int_0^{\pi/2} \frac{1}{2}(1 - \cos 4t)dt$$

$$= 32 \left(t - \frac{1}{4} \sin 4t \right)_0^{\pi/2} = 16\pi.$$

(c)   From (5.5),

$$\int_C xy^2 ds = \int_0^{\pi/2} (4\cos t)(16\sin^2 t)\sqrt{16(\cos^2 t + \sin^2 t)}dt$$

$$= 256 \int_0^{\pi/2} \sin^2 t \cos t\, dt$$

$$= 256 \left(\frac{1}{3}\sin^3 t \right)_0^{\pi/2} = \frac{256}{3}.$$

■

5.1.2 Method of evaluation – curve defined by an explicit function

If the curve C is defined by an explicit function $y = f(x), a \le x \le b$, we can use x as a parameter with $dy = f'(x)dx$ and $ds = \sqrt{1 + (f'(x))^2}dx$, the foregoing line integrals become, in turn,

$$\int_C G(x, y)dx = \int_a^b G(x, f(x))dx, \tag{5.6}$$

$$\int_C G(x, y)dy = \int_a^b G(x, f(x))f'(x)dx, \tag{5.7}$$

$$\int_C G(x, y)ds = \int_a^b G(x, f(x))\sqrt{1 + (f'(x))^2}dx. \tag{5.8}$$

A line integral along a piecewise smooth curve is defined as the sum of the integrals over the various smooth curves whose union comprises C. For example, if C is composed of smooth curves C_1 and C_2 then

$$\int_C G(x, y)ds = \int_{C_1} G(x, y)ds + \int_{C_2} G(x, y)ds.$$

It is common practice to write this sum as one integral without parenthesis as

$$\int_C P(x, y)dx + Q(x, y)dy \quad \text{or simply} \quad \int_C Pdx + Qdy. \tag{5.9}$$

A line integral along a *closed* curve C is very often denoted by

$$\oint_C Pdx + Qdy.$$

Example 5.2   Evaluate $\int_C xy\,dx + x^2\,dy$, where C is given by $y = x^3$, $-1 \le x \le 2$.   □

Solution   The curve C is illustrated in Fig. 5.4 and is defined by the explicit function $y = x^3$. Hence, we can use x as the parameter. Using $dy = 3x^2\,dx$, we get

$$\int_C xy\,dx + x^2\,dy = \int_{-1}^{2} x(x^3)\,dx + x^2(3x^2\,dx)$$

$$= \int_{-1}^{2} 4x^4\,dx$$

$$= \left(\frac{4}{5}x^5\right)_{-1}^{2} = \frac{132}{5}.$$

■

Example 5.3   Evaluate $\oint_C x\,dx$, where C is the circle $x = \cos t$, $y = \sin t$, $0 \le t \le 2\pi$.   □

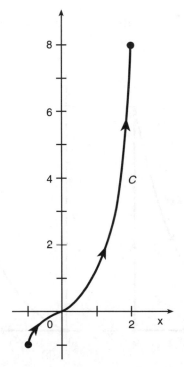

Figure 5.4   The curve $y = x^3$.

Solution   From (5.3), we have

$$\oint_C x\,dx = \int_0^{2\pi} \cos t(-\sin t\,dt) = \left(\frac{1}{2}\cos^2 t\right)_0^{2\pi} = \frac{1}{2}(1-1) = 0.$$

∎

Example 5.4   Evaluate $\oint_C y^2 dx - x^2 dy$ on the closed curve C that is shown in Fig. 5.5a. □

Solution   Since C is piecewise smooth, we can express the integral as a sum of integrals. Symbolically, we write

$$\oint_C = \int_{C_1} + \int_{C_2} + \int_{C_3},$$

where C_1, C_2 and C_3 are the curves shown in Fig. 5.5b. On C_1, we use x as a parameter. Since $y = 0$, $dy = 0$; therefore,

$$\int_{C_1} y^2 dx - x^2 dy = \int_0^2 (0)dx - x^2(0) = 0.$$

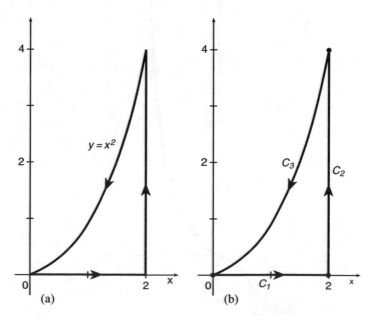

Figure 5.5   Integration of closed curve using integral sums.

On C_2, we use y as a parameter. From $x = 2$, $dx = 0$, we have

$$\int_{C_2} y^2 dy - x^2 dy = \int_0^4 y^2(0) - 4dy$$

$$= -\int_0^4 4dy = (-4y)_0^4 = -16.$$

Finally, on C_3, we again use x as a parameter. From $y = x^2$, we get $dy = 2x\, dx$ and so

$$\int_{C_3} y^2\, dx - x^2\, dy = \int_2^0 x^4 dx - x^2(2x\, dx)$$

$$= \int_2^0 (x^4 - 2x^3)\, dx$$

$$= \left(\frac{1}{5}x^5 - \frac{1}{2}x^4\right)_2^0 = \frac{8}{5}.$$

Therefore,

$$\oint_C y^2 dx - x^2 dy = 0 - 16 + \frac{8}{5} = -\frac{72}{5}.$$

$\blacksquare$

It is important to note that a line integral is independent of the parameterization of the curve C, provided C is given the same orientation by all sets of parametric equations defining the curve. Also, recall for definite integrals that

$$\int_b^a f(x)dx = -\int_a^b f(x)dx.$$

Line integrals possess a similar property. Suppose, as shown in Fig. 5.6, that $-C$ denotes the curve having the opposite orientation of C. Then it can be shown that

$$\int_{-C} P dx + Q dy = -\int_C P dx + Q dy$$

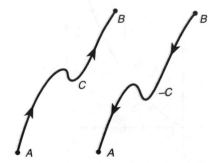

Figure 5.6   Line integrals along C and $-C$.

or equivalently,

$$\int_{-C} P\,dx + Q\,dy + \int_{C} P\,dx + Q\,dy = 0. \tag{5.10}$$

Example 5.5 What is the value of $\int_B^A dx/(x+y)$ along each of the paths shown in Fig. 5.7? □

Solution Before this integral can be evaluated, y must be expressed in terms of x. To do this, recall from the definition of a line integral that the integrand is always to be evaluated *along the path of integration.*

(i) Along the parabolic arc joining A and B, we have $y = x^2$ and making this substitution in the given line integral yields the ordinary definite integral

$$\int_1^2 \frac{dx}{x+x^2} = \int_1^2 \left(\frac{1}{x} - \frac{1}{1+x}\right) dx = \ln(4/3).$$

(ii) Similarly, along the straight-line path from A to B, we have $y = 3x - 2$, and making this substitution in the integrand of the given integral, we obtain the ordinary definite integral

$$\int_1^2 \frac{dx}{x+(3x-2)} = \frac{1}{4}\ln(4x-2)\big|_1^2 = \frac{1}{4}(\ln 6 - \ln 2) = \frac{1}{4}\ln(3).$$

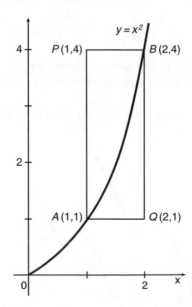

Figure 5.7 Possible paths for line integration from $A(1, 1)$ to $B(2, 4)$.

(iii) To compute the line integral along the path APB, we must perform two integrations, one along AP and the other along PB, since the relation expressing y in terms of x is different on these two segments. Along AP the integral is obviously zero, since x remains constant and, therefore, in the sum leading to the integral each Δx_i is zero. Along PB, on which $y = 4$, we have the integral

$$\int_1^2 \frac{dx}{x+4} = \ln(x+4)|_1^2 = \ln(6/5),$$

which is thus the value of the integral along the entire path APB.

(iv) Along the path AQB we again have two integrations to perform. Along AQ, on which $y - 1$, we have the integral

$$\int_1^2 \frac{dx}{x+1} - \ln(x+1)|_1^2 = \ln(3/2).$$

Along the vertical segment QB the integral is again zero. Hence, for the entire path AQB the value of the given integral is $\ln(3/2)$. ∎

This example illustrates the computational details of line integration and also shows that in general a line integral depends not only on the end points of the integration but also on the particular path which joins them.

5.2 Line integrals and vector functions

Line integrals of a function G of three variables, $\int_C G(x, y, z)dx$, $\int_C G(x, y, z)dy$, and $\int_C G(x, y, z)ds$, are defined in a manner analogous to Definition 5.1. However, to that list we add a fourth *line integral along a space curve C with respect to z:*

$$\int_C G(x, y, z)dz = \lim_{\|P\| \to 0} \sum_{k=1}^n G(x_k^*, y_k^*, z_k^*)\Delta z_k. \qquad (5.11)$$

If C is a smooth curve in 3-space defined by the parametric equations

$$x = f(t), \quad y = g(t), \quad z = h(t), \quad a \le t \le b,$$

then the integral in (5.11) can be evaluated by using

$$\int_C G(x, y, z)dz = \int_a^b G(f(t), g(t), h(t))h'(t)dt.$$

The integrals $\int_C G(x, y, z)dz$ and $\int_C G(x, y, z)dy$ are evaluated in a similar manner. The line integral with respect to arc length is

$$\int_C G(x, y, z)ds = \int_a^b G(f(t), g(t), h(t))\sqrt{(f'(t))^2 + (g'(t))^2 + (h'(t))^2}dt.$$

As in (5.9), in 3-space we are often concerned with line integrals in the form of a sum

$$\int_C P(x, y, z)dx + Q(x, y, z)dy + R(x, y, z)dz.$$

Example 5.6   Evaluate $\int_C ydx + xdy + zdz$, where C is the helix $x = \cos t$, $y = 2\sin t, z = t, 0 \le t \le 2\pi$.   □

Solution   Substituting the expressions for x, y and z along with $dx = -2\sin t\,dt$, $dy = 2\cos t\,dt$, $dz = dt$, we obtain

$$\int_C y\,dx + x\,dy + z\,dz = \int_0^{2\pi} (-4\sin^2 t + 4\cos^2 t)\,dt + t\,dt$$

$$= \int_0^{2\pi} (4\cos t + t)\,dt$$

$$= \left(2\sin 2t + \frac{t^2}{2}\right)\Bigg|_0^{2\pi} = 2\pi^2.$$ ■

Example 5.7   Find the value of the line integral $\int_C (xy + z^2)ds$, where C is the arc of the helix $x = \cos t, y = \sin t$ and $z = t$, which joins the point $A(1, 0, 0)$ and $B(-1, 0, \pi)$. Here ds is the elementary arc length of the given helix.   □

Solution   We know that

$$(ds)^2 = (dx)^2 + (dy)^2 + (dz)^2,$$

and since

$$dx = -\sin t\,dt, \quad dy = \cos t\,dt, \quad dz = dt,$$

we have at once that

$$ds = \sqrt{\sin^2 t + \cos^2 t + 1}|dt| = \sqrt{2}|dt|.$$

Furthermore, it is clear that the point $A(1, 0, 0)$ corresponds to the parametric value $t = 0$ and that the point $B(-1, 0, \pi)$ corresponds to the parametric value $t = \pi$.

Hence, when the integrand is expressed in terms of the parameter t, the required integral becomes

$$\int_C (xy + z^2)ds = \int_0^\pi (\cos t \sin t + t^2)\sqrt{2}dt$$

$$= \sqrt{2}\left\{\frac{\cos^2 t}{2} + \frac{t^3}{3}\right\}_0^\pi$$

$$= \frac{\sqrt{2}\pi^3}{3}.$$

This is the required solution. ∎

The concept of a vector function of several variables can be used to write a general line integral in a compact fashion. For example, suppose the vector-valued function $\mathbf{F}(x, y) = P(x, y)\mathbf{i} + Q(x, y)\mathbf{j}$ is defined along a curve $C: x = f(t)$, $y = g(t)$, $a \le t \le b$, and suppose $\mathbf{r}(t) = f(t)\mathbf{i} + g(t)\mathbf{j}$ is the position vector of points on C. Then the derivative of $\mathbf{r}(t)$,

$$\frac{d\mathbf{r}}{dt} = f'(t)\mathbf{i} + g'(t)\mathbf{j} = \frac{dx}{dt}\mathbf{i} + \frac{dy}{dt}\mathbf{j}$$

prompts us to define the following:

$$d\mathbf{r} = \frac{d\mathbf{r}}{dt}dt = dx\mathbf{i} + dy\mathbf{j}.$$

Since,

$$\mathbf{F}(x, y) \cdot d\mathbf{r} = P(x, y)dx + Q(x, y)dy,$$

we can write

$$\int_C P(x, y)dx + Q(x, y)dy = \int_C \mathbf{F} \cdot d\mathbf{r}. \tag{5.12}$$

Similarly, for a line integral on a space curve,

$$\int_C P(x, y, z)dx + Q(x, y, z)dy + R(x, y, z)dz = \int_C \mathbf{F} \cdot d\mathbf{r}, \tag{5.13}$$

where $\mathbf{F}(x, y, z) = P(x, y, z)\mathbf{i} + Q(x, y, z)\mathbf{j} + R(x, y, z)\mathbf{k}$ and $d\mathbf{r} = dx\mathbf{i} + dy\mathbf{j} + dz\mathbf{k}$.

It may well be asked what physical result is gained by computing a line integral. The answer is: you have a number. Depending on the type of problem, this number may represent work done, change in potential energy, total heat flow, change in entropy, circulation of a fluid and so on. In other words, the physical application determines how a line integral result is interpreted.

5.3 Work: interpretation of line integrals as ubiquitous

In the beginning courses in calculus or physics it is shown that the work done in moving an object from $x = a$ to $x = b$ by a force $F(x)$, which varies in magnitude but not in direction, is given by the definite integral $W = \int_a^b F(x)\,dx$. From the physical interpretation of the dot product, the work done in moving an object a distance d by a force of magnitude F in the same direction as the force is

$$W = Fd.$$

When a constant force $\mathbf{F}$ is applied to a body acting at an angle θ to the direction of motion, then the work done by $\mathbf{F}$ is defined to be

$$W = (|\mathbf{F}|\cos\theta)|\mathbf{d}| = |\mathbf{F}||\mathbf{d}|\cos\theta. \tag{5.14}$$

This implies that if $\mathbf{F}$ causes a displacement $\mathbf{d}$ of a body, then the work done is

$$W = \mathbf{F} \cdot \mathbf{d} \tag{5.15}$$

In general, a force $\mathbf{F}(x, y) = P(x, y)\mathbf{i} + Q(x, y)\mathbf{j}$ acting at each point on a smooth curve $C: x = f(t)$, $y = g(t)$, $a \le t \le b$, varies in both magnitude and direction (see Fig. 5.8a).

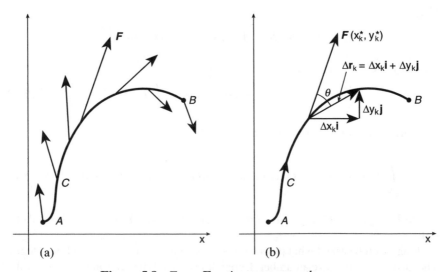

Figure 5.8 Force $\mathbf{F}$ acting on a smooth curve.

If A and B are the points $(f(a), g(a))$ and $(f(b), g(b))$, respectively, we need to know what the work done by $\mathbf{F}$ is as its point of application moves along C from A to B. To answer this, suppose C is divided into n sub-arcs of length Δs_k. Let (x_k^*, y_k^*) be the coordinates of any arbitrary point in segment Δs_k. On each sub-arc $\mathbf{F}(x_k^*, y_k^*)$ is a constant force.

If, as shown in Fig. 5.8b, the length of the vector $\Delta \mathbf{R}_k = (x_k - x_{k-1})\mathbf{i} + (y_k - y_{k-1})\mathbf{j} = \Delta x_k \mathbf{i} + \Delta y_k \mathbf{j}$ is an approximation to the length of the k-th sub-arc, then the approximate work done by $\mathbf{F}$ over the sub-arc is

$$(|\mathbf{F}(x_k^*, y_k^*)| \cos\theta)|\Delta \mathbf{R}_k| = \mathbf{F}(x_k^*, y_k^*) \cdot \Delta \mathbf{R}_k$$
$$= P(x_k^*, y_k^*)\Delta x_k + Q(x_k^*, y_k^*)\Delta y_k.$$

By summing these elements of work and passing to the limit, we naturally define *work done by* $\mathbf{F}$ *along* C as the line integral

$$W = \int_c P(x, y)dx + Q(x, y)dy \quad \text{or} \quad W = \int_C \mathbf{F} \cdot d\mathbf{R}. \tag{5.16}$$

Equation (5.16) extends to force fields acting at points on a space curve. In this case, work $\int_C \mathbf{F} \cdot d\mathbf{R}$ is defined as in (5.13). Now, since

$$\frac{d\mathbf{R}}{dt} = \frac{d\mathbf{R}}{dt}\frac{ds}{dt},$$

we let $d\mathbf{R} = \mathbf{T}ds$, where $\mathbf{T} = d\mathbf{R}/ds$ is a unit tangent to C. Hence, we have

$$W = \int_C \mathbf{F} \cdot d\mathbf{R} = \int_C \mathbf{F} \cdot \mathbf{T}ds. \tag{5.17}$$

In other words, the work done by a force $\mathbf{F}$ along a curve C is due entirely to the tangential component of $\mathbf{F}$.

Example 5.8   Find the work done by (a) $\mathbf{F} = x\mathbf{i} + y\mathbf{j}$ and (b) $\mathbf{F} = \frac{3}{4}\mathbf{i} + \frac{1}{2}\mathbf{j}$ along the curve C traced by $\mathbf{R}(t) = \cos t\mathbf{i} + \sin t\mathbf{j}$ from $t = 0$ to $t = \pi$.   □

Solution

(a)   The vector function $\mathbf{R}(t)$ gives the parametric equations $x = \cos t$, $y = \sin t$, $0 \le t \le \pi$ which we recognize as a half circle. As seen in Fig. 5.9, the force field **F** is perpendicular to C at every point. Since the tangential components of **F** are zero, the work done along C is zero. To see this we use (5.16):

$$W = \int_C \mathbf{F} \cdot d\mathbf{R} = \int_C (x\mathbf{i} + y\mathbf{j}) \cdot d\mathbf{R}$$

$$= \int_0^\pi (\cos t\mathbf{i} + \sin t\mathbf{j}) \cdot (-\sin t\mathbf{i} + \cos t\mathbf{j}) dt$$

$$= \int_0^\pi (-\cos t \sin t + \sin t \cos t) dt = 0.$$

(b) In Fig. 5.10 the vectors tangent to the semi-circle are the projections of **F** on the unit tangent vectors. The work done by **F** is

$$W = \int_c \mathbf{F} \cdot d\mathbf{R} = \int_C \left(\frac{3}{4}\mathbf{i} + \frac{1}{2}\mathbf{j}\right) \cdot d\mathbf{r}$$

$$= \int_0^\pi \left(\frac{3}{4}\mathbf{i} + \frac{1}{2}\mathbf{j}\right) \cdot (-\sin t\mathbf{i} + \cos t\mathbf{j}) dt$$

$$= \int_0^\pi \left(-\frac{3}{4}\sin t + \frac{1}{2}\cos t\right) dt$$

$$= \left(\frac{3}{4}\cos t + \frac{1}{2}\sin t\right)\Big|_0^\pi = -\frac{3}{2}.$$

The units of work depend on the units of $|\mathbf{F}|$ and on the units of distance. ∎

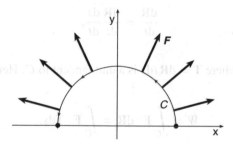

Figure 5.9 Half-circle C, with force **F** perpendicular to C.

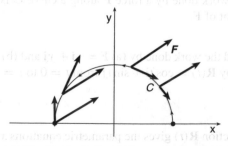

Figure 5.10 Projections of **F** on unit tangent vectors.

5.4 Line integrals independent of path

The following concepts are important in the discussion that follows:

Functions of two variables: The *differential* of a function of two variables $\phi(x, y)$ is

$$d\phi = \frac{\partial \phi}{\partial x} dx + \frac{\partial \phi}{\partial y} dy.$$

A differential expression $P(x, y)dx + Q(x, y)dy$ is said to be an *exact differential* if there exists a function $\phi(x, y)$ such that

$$d\phi = P(x, y)dx + Q(x, y)dy.$$

For example, the expression $x^2 y^3 dx + x^3 y^2 dy$ is an exact differential since it is the differential of $\phi(x, y) = \frac{1}{3} x^3 y^3$. On the other hand, $(2y^2 - 2y)dx + (2xy - x)dy$ is not an exact differential.

Functions of three variables: The differential of a function of three variables $\phi(x, y, z)$ is

$$d\phi = \frac{\partial \phi}{\partial x} dx + \frac{\partial \phi}{\partial y} dy + \frac{\partial \phi}{\partial z} dz$$

A differential expression $P(x, y, z) dx + Q(x, y, z) dy + R(x, y, z) dz$ is an exact differential if there exists a function $\phi(x, y, z)$ such that

$$d\phi = P(x, y, z)dx + Q(x, y, z)dy + R(x, y, z)dz$$

The value of a line integral generally depends on the curve or *path* between two points A and B. There are, however, exceptions. A line integral whose value is the same for *every* curve connecting A and B is said to be *independent of path*.

The following theorem establishes an important relationship between the seemingly disparate notions of path independence and exact differentials. In addition, it provides a means of evaluating path-independent line integrals in a manner analogous to the fundamental theorem of calculus:

$$\int_a^b f'(x)dx = f(b) - f(a).$$

Theorem 5.1 *Suppose there exists a function $\phi(x, y)$ such that $d\phi = Pdx + Qdy$; i.e. $Pdx + Qdy$ is an exact differential. Then $\int_C Pdx + Qdy$ depends on only the endpoints A and B of the path C and*

$$\int_C Pdx + Qdy = \phi(B) - \phi(A).$$

Proof Let C be a smooth path defined parametrically by $x = f(t)$, $y = g(t)$, $a \leq t \leq b$, and let the coordinate of A and B be $(f(a), g(a))$ and $(f(b), g(b))$, respectively. Then, by the chain rule,

$$
\begin{aligned}
\int_C P dx + Q dy &= \int_a^b \left(\frac{\partial \phi}{\partial x} \frac{dx}{dt} + \frac{\partial \phi}{\partial y} \frac{dy}{dt} \right) dt \\
&= \int_a^b \frac{d\phi}{dt} dt = (\phi(f(t), g(t)))|_a^b \\
&= \phi(f(b), g(b)) - \phi(f(a), g(a)) \\
&= \phi(B) - \phi(A).
\end{aligned}
$$

∎

The theorem is also valid for piecewise smooth curves. In addition, the converse of the theorem is also true. Hence, $\int_C P dx + Q dy$ is independent of path if and only if $P dx + Q dy$ is an exact differential. We now state without proof the following theorem.

Theorem 5.2 *Let P and Q have continuous first partial derivatives in an open simply connected region. Then $\int_C P dx + Q dy$ is independent of the path C if and only if*

$$
\frac{\partial P}{\partial y} = \frac{\partial Q}{\partial x}
$$

for all (x, y) in the region.

Example 5.9 Show that the integral $\int_C (x^2 - 2y^3) dx + (x + 5y) dy$ is not independent of the path C. □

Solution From $P = x^2 - 2y^3$ and $Q = x + 5y$, we find

$$
\frac{\partial P}{\partial y} = -6y^2 \quad \text{and} \quad \frac{\partial Q}{\partial x} = 1.
$$

Since $\partial P/\partial y \neq \partial Q/\partial x$ then by Theorem 5.2 the integral is not independent of the path. In other words, the differential expression $(x^2 - 2y^3) dx + (x + 5y) dy$ is not an exact differential. ∎

Example 5.10 Show that $\int_C (y^2 - 6xy + 6) dx + (2xy - 3x^2) dy$ is independent of any path C between $(-1, 0)$ and $(3, 4)$. Evaluate the integral. □

Solution Identifying $P = y^2 - 6xy + 6$ and $Q = 2xy - 3x^2$ yields

$$
\frac{\partial P}{\partial y} = 2y - 6x \quad \text{and} \quad \frac{\partial Q}{\partial x} = 2y - 6x.
$$

Since $\partial P/\partial y = \partial Q/\partial x$, the integral is independent of path and so there exists a function ϕ such that

$$\frac{\partial \phi}{\partial x} = y^2 - 6xy + 6 \quad \text{and} \quad \frac{\partial \phi}{\partial y} = 2xy - 3x^2.$$

To find the function ϕ we can integrate either $\partial \phi/\partial x$ or $\partial \phi/\partial y$. Integrating $\partial \phi/\partial x$ with respect to x partially gives

$$\phi = y^2 x - 3x^2 y + 6x + g(y),$$

where $g(y)$ is the 'constant' of integration. Taking the partial derivative of the last expression with respect to y and setting the result equal to Q (i.e. $\partial \phi/\partial y$) then gives

$$\frac{\partial \phi}{\partial y} = 2yx - 3x^2 + g'(y) = 2yx - 3x^2,$$

which implies $g'(y) = 0$ and so $g(y) = c$, a constant. But since the differentials

$$d(y^2 x - 3x^2 y + 6x + c) \quad \text{and} \quad d(y^2 x - 3x^2 y + 6x),$$

both yield $(y^2 - 6xy + 6)dx + (2xy - 3x^2)dx$, we can drop the constant C and take

$$\phi = xy^2 - 3x^2 y + 6x.$$

It follows from the fundamental theorem for line integrals (Theorem 5.1) that

$$\int_{(-1,0)}^{(3,4)} (y^2 - 6xy + 6)dx + (2xy - 3x^2)dy = \int_{(-1,0)}^{(3,4)} d(xy^2 - 3x^2 y + 6x)$$

$$= (xy^2 - 3x^2 y + 6x)\Big|_{(-1,0)}^{(3,4)}$$

$$= (48 - 108 + 18) - (-6)$$

$$= -36.$$

■

Alternative solution   Because the integral is independent of the path, we can integrate along any convenient curve connecting the given points. In particular, $y = x + 1$ is such a curve. Using x as a parameter then gives

$$\int_C (y^2 - 6xy + 6)dx + (2xy - 3x^2)dy = \int_{-1}^{3} (-6x^2 - 2x + 7)dx$$

$$= -36.$$

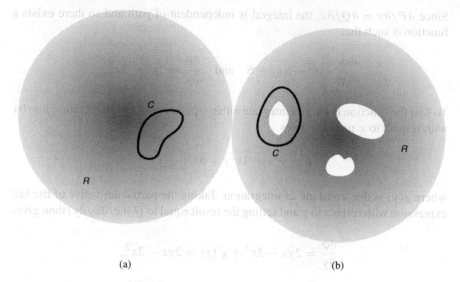

(a)   (b)

Figure 5.11   Simple connected regions.

In order to be able to state the next theorem, we need to consider a particular kind of region in the plane. We say that a region is *simply connected* if:

(a) R is connected, i.e. every pair of points in the region can be joined by a piecewise smooth curve that lies entirely in R, and

(b)   every single closed curve C lying entirely within R can be shrunk, or contracted to a point without leaving R.

The last statement means that if C is any simple closed curve lying entirely in R, then the region in the interior C also lies entirely in R. Roughly put, a simply connected region has no holes in it. In Fig. 5.11a we have illustrated a simply connected region.

The representative simple closed curve C could be shrunk to a point without leaving the region. In Fig. 5.11b the region has three distinct holes in it; since the representative curve C surrounds one of the holes, it could not be shrunk to a point without leaving the region.

This last region is known as a *multiply connected* one. A simply connected region is said to be *open* if it contains no boundary points. We now state without proof the following theorem giving a test for path independence.   ■

Theorem 5.3   *Let D be a domain in which $P(x, y, z)$, $Q(x, y, z)$ and $R(x, y, z)$ have continuous first derivatives. If the line integral $\int \mathbf{F} \cdot d\mathbf{r} = \int P\,dx + Q\,dy + R\,dz$ is independent of path in D, then $\nabla \times \mathbf{F} = 0$ in D. Conversely, if D is simply connected, the $\nabla \times \mathbf{F} = 0$ in D, then the line integral is independent of path in D.*

The preceding theorem gives a simple test to determine whether a given line integral is independent of path: we see whether the curl of $\mathbf{F}$ is zero. Evaluation of a line integral that is independent of path still requires the function $\phi(x, y, z)$, but it is at least useful to know that ϕ exists before searching for it.

Example 5.11 Evaluate $\int_C 2xye^z dx + (x^2e^z + y)dy + (x^2ye^z - z)dz$ along the straight line C from $(0, 1, 2)$ to $(2, 1, -8)$. ◻

Solution *Method 1:* It is evident that

$$\nabla\left(x^2ye^z + \frac{y^2}{2} - \frac{z^2}{2}\right) = 2xye^z\mathbf{i} + (x^2e^z + y)\mathbf{j} + (x^2ye^z - z)\mathbf{k}$$

and, hence, the line integral is independent of path. Its value is, therefore,

$$I = \left(x^2ye^z + \frac{y^2}{2} - \frac{z^2}{2}\right)\Big|_{(0,1,2)}^{(2,1,-8)} = \left(4e^{-8} + \frac{1}{2} - 32\right) - \left(\frac{1}{2} - 2\right).$$

$$= 4e^{-8} - 30.$$

Method 2: Since

$$\nabla \times (2xye^z\mathbf{i} + (x^2e^z + y)\mathbf{j} + (x^2ye^z - z)\mathbf{k})$$

$$= \begin{vmatrix} \mathbf{i} & \mathbf{j} & \mathbf{k} \\ \dfrac{\partial}{\partial x} & \dfrac{\partial}{\partial y} & \dfrac{\partial}{\partial z} \\ 2xye^z & x^2e^z + y & x^2ye^z - z \end{vmatrix}$$

$$= (x^2e^z - x^2e^z)\mathbf{i} + (2xye^z - 2xye^z)\mathbf{j}$$

$$+ (2xe^z - 2xe^z)\mathbf{k} = \mathbf{0},$$

the line integral is independent of path. Thus, there exists a function $\phi(x, y, z)$ such that

$$\nabla\phi = 2xye^z\mathbf{i} + (x^2e^z + y)\mathbf{j} + (x^2ye^z - z)\mathbf{k}$$

or

$$\frac{\partial\phi}{\partial x} = 2xye^z, \quad \frac{\partial\phi}{\partial y} = x^2e^z + y, \quad \frac{\partial\phi}{\partial z} = x^2ye^z - z.$$

Integration of the first of these equations yields

$$\phi(x, y, z) = x^2ye^z + f(y, z).$$

Substitution of this function into the left-hand side of the second equation gives

$$x^2e^z + \frac{\partial f}{\partial y} = x^2e^z + y,$$

that implies

$$\frac{\partial f}{\partial y} = y,$$

consequently,

$$f(y, z) = \frac{y^2}{2} + g(z),$$

We know both the x- and y-dependence of ϕ

$$\phi(x, y, z) = x^2 + ye^z + \frac{y^2}{2} + g(z).$$

To obtain the z-dependence in $g(z)$, we substitute into the left-hand side of the third equation,

$$x^2 ye^z + \frac{dg}{dz} = x^2 ye^z - z,$$

from which we have

$$\frac{dg}{dz} = -z.$$

Hence, $g(z) = -z^2/2 + c$ (c a constant), and

$$\phi(x, y, z) = x^2 ye^z + \frac{y^2}{2} - \frac{z^2}{2} + c.$$

Finally, we have

$$I = \left(x^x ye^z + \frac{y^2}{2} - \frac{z^2}{2} \right)\Big|_{(0,1,2)}^{(2,1,-8)} = 4e^{-8} - 30.$$

■

5.5   Conservative vector fields

If $\int_C P dx + Q dy$ is independent of the path C, we know there exists a function ϕ such that

$$d\phi = \frac{\partial \phi}{\partial x} dx + \frac{\partial \phi}{\partial y} dy = P dx + Q dy$$

$$= (P\mathbf{i} + Q\mathbf{j}) \cdot (dx\mathbf{i} + dy\mathbf{j}) = \mathbf{F} \cdot d\mathbf{r},$$

where $\mathbf{F} = P\mathbf{i} + Q\mathbf{j}$ is a vector field and $P = \partial\phi/\partial x$, $Q = \partial\phi/\partial y$. In other words, the vector field $\mathbf{F}$ is a gradient of the function ϕ. Since $\mathbf{F} = \nabla\phi$, $\mathbf{F}$ is said to be a *gradient field* and the function ϕ is said to be a *potential function* for $\mathbf{F}$.

In a gradient force field $\mathbf{F}$, the work done by the force upon a particle moving from position A to position B is the same for all paths between these points. Moreover, the work done by the force along a closed path is zero. For this reason, such a force field is also said to be *conservative*. In a conservative field $\mathbf{F}$ the law of conservation of mechanical energy holds: for a particle moving along a path in a conservative field,

$$\text{Kinetic energy} + \text{Potential energy} = \text{constant.}$$

In a simply connected region the hypotheses of Theorem 5.3 imply that a force field $\mathbf{F}(x, y) = P(x, y)\mathbf{i} + Q(x, y)\mathbf{j}$ is a gradient (i.e. conservative) if and only if

$$\frac{\partial P}{\partial y} = \frac{\partial Q}{\partial x}.$$

Example 5.12   Show that the vector field $\mathbf{F} = (y^2 + 5)\mathbf{i} + (2xy - 8)\mathbf{j}$ is a gradient field. Find a potential function for $\mathbf{F}$. □

Solution   By identifying $P = y^2 + 5$ and $Q = 2xy - 8$, we see that

$$\frac{\partial P}{\partial y} = \frac{\partial Q}{\partial x} = 2y.$$

Hence, $\mathbf{F}$ is a gradient field and so there exists a potential function ϕ satisfying

$$\frac{\partial \phi}{\partial x} = y^2 + 5 \quad \text{and} \quad \frac{\partial \phi}{\partial y} = 2xy - 8.$$

Proceeding in the manner of the previous example, we find that $\phi = xy^2 - 8y + 5x$. ■

Note: $\nabla\phi = (\partial\phi/\partial x)\mathbf{i} + (\partial\phi/\partial y)\mathbf{j} = (y^2 + 5)\mathbf{i} + (2xy - 8)\mathbf{j}$. In general, we can state that in a simply connected region of space, a force field $\mathbf{F}(x, y, z) = P(x, y, z)\mathbf{i} + Q(x, y, z)\mathbf{j} + R(x, y, z)\mathbf{k}$ is conservative if and only if the curl of $\mathbf{F} = \mathbf{0}$ (i.e. $\nabla \times \mathbf{F} = \mathbf{0}$).

5.6 Surface integrals

Surface area in the plane the length of an arc of the graph of $y = f(x)$ from $x = a$ to $x = b$ is given by

$$s = \int_a^b \left(\sqrt{1 + \left(\frac{dy}{dx}\right)^2} \right) dx. \tag{5.18}$$

The problem in three dimensions is to find the area $A(s)$ of that portion of the surface S given by a function $z = f(x, y)$ having continuous first partial derivatives on a closed region R in the xy-plane.

Suppose, as shown in Fig. 5.12a, that an inner partition P of R is formed using lines parallel to the x- and y-axes. P then consists of n rectangular elements R_k of area $\Delta A_k = \Delta x_k \Delta y_k$ that lie entirely within R. Let $(x_k, y_k, 0)$ denote any point in R_k. As we see in Fig. 5.12a, by projecting the sides of R_k upward, we determine two quantities: a portion S_k of the surface and a portion T_k, a tangent plane at $(x_k, y_k, f(x_k, y_k))$. It seems reasonable to assume that when R_k is small, the area ΔT_k of T_k is approximately the same as the area ΔS_k of S_k.

To find the area of T_k let us choose $(x_k, y_k, 0)$ at a corner of R_k as shown in Fig. 5.12b. The indicated vectors $\mathbf{U}$ and $\mathbf{V}$, which form two sides of T_k, are given by

$$\mathbf{U} = \Delta x_k \mathbf{i} + F_x(x_k, y_k) \Delta x_k \mathbf{k},$$
$$\mathbf{V} = \Delta y_k \mathbf{j} + f_y(x_k, y_k) \Delta y_k \mathbf{k},$$

where $f_x(x_k, y_k)$ and $f_y(x_k, y_k)$ are slopes of the lines containing $\mathbf{U}$ and $\mathbf{V}$, respectively. We know that

$$\Delta T_k = |\mathbf{U} \times \mathbf{V}|,$$

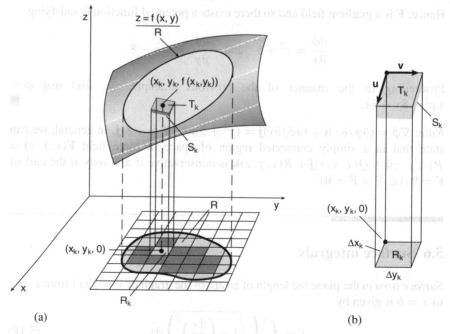

Figure 5.12 Inner partition P of region R.

where

$$U \times V = \begin{vmatrix} \mathbf{i} & \mathbf{j} & \mathbf{k} \\ \Delta x_k & 0 & f_x(x_k, y_k)\Delta x_k \\ 0 & \Delta y_k & f_y(x_k, y_k)\Delta y_k \end{vmatrix}$$

$$= (-f_x(x_k, y_k)\mathbf{i} - f_y(x_k, y_k)\mathbf{j} + \mathbf{k})\Delta x_k \Delta y_k.$$

In other words,

$$\Delta T_k = \left(\sqrt{(f_x(x_k, y_k))^2 + (f_y(x_k, y_k))^2 + 1} \right) \Delta x_k \Delta y_k.$$

Consequently, the area A is approximately

$$\sum_{k=1}^{n} \left(\sqrt{1 + (f_x(x_k, y_k))^2 + (f_y(x_k, y_k))^2} \right) \Delta x_k \Delta y_k.$$

Taking the limit of the above sum as $|P| \to 0$ leads us to the following definition.

Definition 5.2 *Let f be a function for which the first partial derivatives f_x and f_y are continuous on a closed region R. Then the area of the surface over R is given by*

$$A(s) = \int_R \int \left(\sqrt{1 + (f_x(x, y))^2 + (f_y(x, y))^2} \right) dA. \tag{5.19}$$

Example 5.13 Find the surface area of that portion of the sphere $x^2 + y^2 + z^2 = a^2$ that is above the xy-plane and within the cylinder $x^2 + y^2 = b^2, 0 < b < a$. □

Solution If we define $z = f(x, y)$ by $f(x, y) = \sqrt{a^2 - x^2 - y^2}$, then

$$f_x(x, y) = \frac{-x}{\sqrt{a^2 - x^2 - y^2}} \quad \text{and} \quad f_y(x, y) = \frac{-y}{\sqrt{a^2 - x^2 - y^2}},$$

and so $1 + (f_x(x, y))^2 + (f_y(x, y))^2 = a^2/(a^2 - x^2 - y^2)$. Hence, (5.19) is

$$A(s) = \int\int_R \frac{a}{\sqrt{a^2 - x^2 - y^2}} dA$$

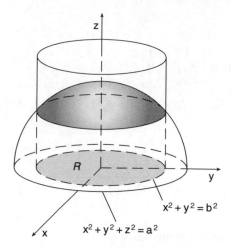

Figure 5.13   Integration region R for Example 5.10.

where R is indicated in Fig. 5.13. To evaluate this double integral, we change to polar coordinates

$$A(s) = a \int_0^{2\pi} \int_0^b (a^2 - r^2)^{-1/2} r \, dr \, d\theta,$$

$$= a \int_0^{2\pi} \left(-(a^2 - r^2)^{1/2} \right) \Big|_0^b \, d\theta = a \left(a - \sqrt{a^2 - b^2} \right) \int_0^{2\pi} d\theta,$$

$$= 2\pi a \left(a - \sqrt{a^2 - b^2} \right) \quad \text{square units.}$$

■

As we have seen, double and triple integrals

$$\int_R \int f(x, y) \, dA \qquad \iint_D \int f(x, y, z) \, dV$$

are generalizations of the definite integral $\int_a^b f(x) \, dx$. The surface area integral (5.19) is a generalization of the arc length integral (5.18). We now consider a generalization of the line integral $\int_C G(x, y) \, ds$. This generalization is called the *surface integral*.

Definition 5.3   *Let G be a function of three variables defined over a region of space containing the surface S. Then the surface integral of G over S is given by*

$$\iint_S G(x, y, z) \, dS = \lim_{\|P\| \to 0} \sum_{k=1}^n G(x_k^*, y_k^*, z_k^*) \Delta S_k \tag{5.20}$$

If G, f, f_x and f_y are continuous throughout a region containing S, then equation (5.20) can be evaluated by means of a double integral. The left-hand side of (5.20) becomes

$$\iint_S G(x, y, z)\,dS = \iint_R G(x, y, f(x, y))$$
$$\times \sqrt{1 + (f_x(x, y))^2 + (f_y(x, y))^2}\,dA. \qquad (5.21)$$

If $y = g(x, z)$ is the equation of a surface S that projects onto a region R of the xz-plane, then

$$\iint_S G(x, y, z)\,dS = \iint_R G(x, g(x, z), z)$$
$$\times \sqrt{1 + (g_x(x, z))^2 + (g_z(x, z))^2}\,dA. \qquad (5.22)$$

Similarly, if $x = h(y, z)$ is the equation of a surface that projects onto the yz-plane, then the analogue of (5.21) is

$$\iint_S G(x, y, z)\,dS = \iint_R G(h(y, z), y, z)$$
$$\times \sqrt{1 + (h_y(y, z))^2 + (h_z(y, z))^2}\,dA. \qquad (5.23)$$

Example 5.14 Evaluate $\iint_S xz^2\,dS$ where S is that portion of the cylinder $y = 2x^2 + 1$ in the first octant bounded by $x = 0$, $x = 2$, $z = 4$, and $z = 8$. $\square$

Solution We shall use (5.22) with $g(x, z) = 2x^2 + 1$ and R the rectangular region in the xz-plane shown in Fig. 5.14.

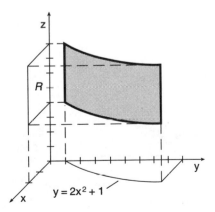

Figure 5.14 Rectangular region R in xz-plane.

Since $g_x(x, z) = 4x$ and $g_z(x, z) = 0$, it follows that

$$\iint_S xz^2 \, dS = \int_0^2 \int_4^8 xz^2 \sqrt{1 + 16x^2} \, dz dx$$

$$= \int_0^2 \left(\frac{z^3}{3} x \sqrt{1 + 16x^2} \right) \Big|_4^8 \, dx$$

$$= \frac{448}{3} \int_0^2 x(1 + 16x^2)^{1/2} dx$$

$$= \frac{28}{9} \left((1 + 16^2)^{3/2} \right) \Big|_0^2 = \frac{28}{9} (65^{3/2} - 1)$$

$$= 1627.3.$$

$\blacksquare$

5.7 Orientation of a surface

Roughly, an orientable surface S, such as that given in Fig. 5.15a, has two sides that could be painted different colours. The Möbius strip (named after August Ferdinand Möbius (1790–1868), German mathematician, student of Gauss, professor of astronomy at Leipzig, who is known for his important work in the theory of surfaces, projective geometry and mechanics, and who also contributed to number theory) shown in Fig. 5.15b is not an orientable surface since it is one-sided. A person who starts to paint the surface of a Möbius strip at a point will paint the entire surface and return to the starting point.

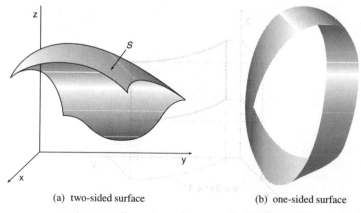

(a) two-sided surface   (b) one-sided surface

Figure 5.15 Orientable surfaces.

Specifically, we say a smooth surface S is *orientable* or is an *orientable surface* if there exists a continuous unit normal vector function $\mathbf{n}$ defined at each point (x, y, z) on the surface.

The vector field $\mathbf{n}(x, y, z)$ is called the *orientation* of S. But since a unit normal to the surface S at (x, y, z) can be either $\mathbf{n}(x, y, z)$ or $-\mathbf{n}(x, y, z)$, an orientable surface has two orientations (see Fig. 5.16a–c).

The Möbius strip shown again in Fig. 5.15d is not an oriented surface since if a unit normal $\mathbf{n}$ starts at P on the surface and moves once around the strip on the curve C, it ends up on the 'opposite side' of the strip at P and so points in the opposite direction.

A surface S defined by $z = f(x, y)$ has an *upward orientation* (Fig. 5.16b) when the unit normals are directed upward – i.e. have positive $\mathbf{k}$ components, and it has a *downward orientation* (Fig. 5.16c) when the unit normals are directed downward – i.e. have negative $\mathbf{k}$ components.

If a smooth surface is defined by $g(x, y, z) = 0$, then a unit normal is

$$\mathbf{n} = \frac{1}{|\nabla g|} \nabla g, \tag{5.24}$$

where $\nabla g = (\partial g/\partial x)\mathbf{i} + (\partial g/\partial y)\mathbf{j} + (\partial g/\partial z)\mathbf{k}$ is the gradient of g. If S is defined by $z = f(x, y)$, then we can use either $g(x, y, z) = z - f(x, y) = 0$ or $g(x, y, z) = f(x, y) - z = 0$ depending on the orientation of S.

As an example, consider the sphere of radius $a > 0$: $x^2 + y^2 + z^2 = a^2$. If we define $g(x, y, z) = x^2 + y^2 + z^2 - a^2$, then

$$\nabla g = 2x\mathbf{i} + 2y\mathbf{j} + 2z\mathbf{k} \quad \text{and} \quad |\nabla g| = \sqrt{4x^2 + 4y^2 + 4z^2} = 2a.$$

Then the two orientations of the surface are

$$\mathbf{n} = \frac{x}{a}\mathbf{i} + \frac{y}{a}\mathbf{j} + \frac{z}{a}\mathbf{k} \quad \text{and} \quad \mathbf{n}_1 = -\mathbf{n} = -\frac{x}{a}\mathbf{i} - \frac{y}{a}\mathbf{j} - \frac{z}{a}\mathbf{k}.$$

The vector field $\mathbf{n}$ defines an outward orientation, whereas $\mathbf{n}_1 = -\mathbf{n}$ defines an inward orientation (see Fig. 5.17).

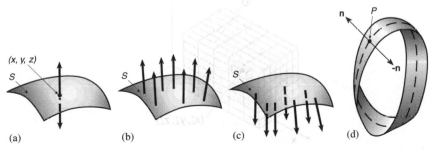

(a)   (b)   (c)   (d)

Figure 5.16 Orientable surfaces with unit normals.

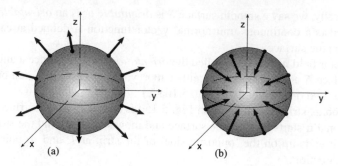

Figure 5.17   Sphere with inward and outward **n**.

5.8 Volume integration

The steps leading to the definition of the three-dimensional definite integral or *triple integral* $\iiint_D F(x, y, z)\mathrm{d}V$ are similar to those of the double integral. A sum of the form $\sum_{k=1}^{n} F(x_k^*, y_k^*, z_k^*)\Delta V_k$, where (x_k^*, y_k^*, z_k^*) is an arbitrary point within each D_k and ΔV_k denotes the volume of each D_k, is called a *Riemman sum*. The type of partition used, where all D_k lie completely within D, is called an *inner partition* of D (Fig. 5.18).

Definition 5.4   *Let F be a function of three variables defined over a closed region D of space. Then the* triple integral *of F over D is given by*

$$\iiint_D F(x, y, z)\mathrm{d}V = \lim_{|P|\to 0} \sum_{k=1}^{n} F(x_k^*, y_k^*, z_k^*)\Delta V_k. \qquad (5.25)$$

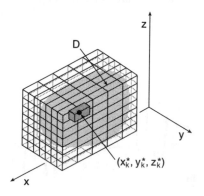

Figure 5.18   Inner partition of D.

If the region D is bounded above by the graph $z = f_2(x, y)$ and bounded below by the graph $z = f_1(x, y)$, then it can be shown that the triple integral in (5.25) can be expressed as a double integral of the partial integral as follows:

$$\iiint_D F(x, y, z) dV = \iint_R \left(\int_{f_1(x,y)}^{f_2(x,y)} F(x, y, z) dz \right) dA$$

where R is the orthogonal projection of D onto the xy-plane. If R is the region shown in Fig. 5.19, the triple integral of F over D can be rewritten as the iterated integral

$$\iiint_D F(x, y, z) \, dV = \int_{x=a}^{b} \int_{y=g_1(x)}^{g_2(x)} \int_{z=f_1(x,y)}^{f_2(x,y)} F(x, y, z) dz \, dy \, dx. \quad (5.26)$$

To evaluate the integral in (5.26) we begin by evaluating the partial integral

$$\int_{f_1(x,y)}^{f_2(x,y)} F(x, y, z) \, dz$$

in which both x and y are held fixed. In a double integral there are only two possible orders of integration, $dy \, dx$ and $dx \, dy$. The triple integral in (5.26) illustrates one of six possible orders of integration:

$$\begin{array}{ccc} dz \, dy \, dx & dz \, dx \, dy & dy \, dx \, dz \\ dx \, dy \, dz & dx \, dz \, dy & dy \, dz \, dx. \end{array}$$

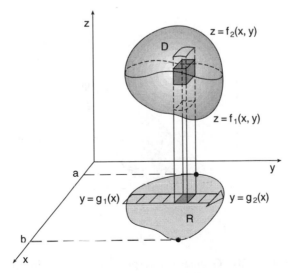

Figure 5.19   Type I region R.

The last two differentials tell the coordinate plane in which the region R is situated. As an example, the iterated integral corresponding to the order of integration $dx\, dz\, dy$ must have the form

$$\int_{y=c}^{d} \int_{z=k_1(y)}^{k_2(y)} \int_{x=h_1(y,z)}^{h_2(y,z)} F(x, y, z)dx\, dz\, dy.$$

The geometric interpretation of this integral and the region R of integration in the yz-plane are shown in Fig. 5.20.

A list of some of the standard applications of the triple integral follows:

- **Mass**: If $\rho(x, y, z)$ is density, then the *mass* of the solid D is given by

$$m = \iint_{D} \int \rho(x, y, z)\, dV.$$

- *First moments*: The *first moments* of the solid about the coordinate planes indicated by the subscripts are given by

$$M_{xy} = \iint_{D} \int z\rho(x, y, z)\, dV,$$

$$M_{xz} = \iint_{D} \int y\rho(x, y, z)\, dV,$$

$$M_{yz} = \iint_{D} \int x\rho(x, y, z)\, dV.$$

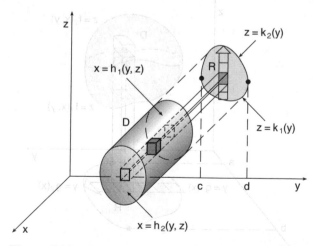

Figure 5.20   Geometric interpretation of triple integral.

- **Second moments**: The *second moments*, or *moments of inertia* of D about the coordinate axes indicated by the subscripts, are given by

$$I_x = \iint_D\!\int (y^2 + z^2)\rho(x, y, z)\,dV,$$

$$I_y = \iint_D\!\int (x^2 + z^2)\rho(x, y, z)\,dV,$$

$$I_z = \iint_D\!\int (x^2 + y^2)\rho(x, y, z)\,dV.$$

Example 5.15 Find the volume of the solid in the first octant bounded by the graphs of $z = 1 - y^2$, $y = 2x$, and $x = 3$. $\square$

Solution As indicated in Fig. 5.21a, the first integration with respect to z is from 0 to $1 - y^2$. Figure 5.21b shows the projection of the solid D in the xy-plane. Hence, we next integrate with respect to x, from $y/2$ to 3. The last integration is with respect to y from 0 to 1. Thus, we have

$$V = \iiint_D dV = \int_0^1\!\int_{y/2}^3\!\int_0^{1-y^2} dz\,dx\,dy$$

$$= \int_0^1\!\int_{y/2}^3 (1 - y^2)\,dx\,dy$$

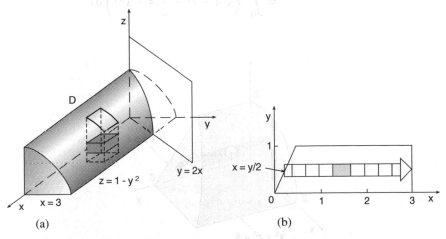

Figure 5.21 Integration region in first octant.

$$= \int_0^1 \left(x - xy^2\right)\Big|_{y/2}^3 dy$$

$$= \int_0^1 \left(3 - 3y^3 - \frac{1}{2}y + \frac{1}{2}y^3\right) dy$$

$$= \left(3y - y^3 - \frac{1}{4}y^2 + \frac{1}{8}y^4\right)\Big|_0^1$$

$$= \frac{15}{8} \text{ cubic units.}$$

■

Example 5.16   Find the volume in the first octant cut from the cylinder $x^2 + z^2 = 4$ by the plane $y + z = 6$.   □

Solution   From Fig. 5.22, we have

$$V = \iiint_D dV = \int_0^2 \int_0^{\sqrt{4-x^2}} \int_0^{6-z} dy \, dz \, dx$$

$$= \int_0^2 \int_0^{\sqrt{4-x^2}} (6 - z) dz \, dx$$

$$= \int_0^2 \left(6z - \frac{z^2}{2}\right)\Big|_0^{\sqrt{4-x^2}} dx$$

$$= \int_0^2 \left(\sqrt{4 - x^2} - \frac{1}{2}(4 - x^2)\right) dx.$$

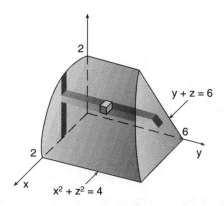

Figure 5.22   Region of integration in first octant.

Polar coordinates can now be used in the first term of the preceding integral by setting $x = 2 \sin \theta$ so that $dx = 2 \cos \theta d\theta$ which leads to

$$V = 6 \int_0^{\pi/2} (2 \cos \theta) 2 \cos \theta d\theta + \left(-2x + \frac{x^3}{6} \right)_0^2$$

$$= 12 \int_0^{\pi/2} (1 + \cos \theta) d\theta - \frac{8}{3}$$

$$= 12 \left(\theta + \frac{1}{2} \sin \theta \right) \Big|_0^{\pi/2} - \frac{8}{3} = 6\pi - \frac{8}{3} \text{ cubic units.} \quad \blacksquare$$

5.9 Triple integrals in cylindrical coordinates

When evaluating double integrals, polar coordinates are often used for convenience and simplicity. In a similar manner, other coordinate systems can simplify the evaluation of triple integrals. Two of the most common are cylindrical and spherical coordinates. Cylindrical coordinates are useful in problems involving an axis of symmetry. They are based on a Cartesian coordinate along the axis of symmetry and polar coordinates in a plane perpendicular to the axis of symmetry. If the z-axis is the axis of symmetry and polar coordinates are defined in the xy-plane with the origin as pole and the positive x-axis as polar axis, then cylindrical coordinates and Cartesian coordinates are related by the equations (see Fig. 5.23)

$$x = r \cos \theta, \quad y = r \sin \theta, \quad z = z. \tag{5.27}$$

Recall that r can be expressed in terms of x and y by

$$r = \sqrt{x^2 + y^2}$$

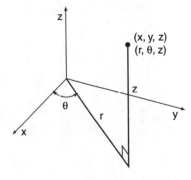

Figure 5.23   Cylindrical coordinate system.

and θ is defined implicitly by the equations

$$\cos \theta = \frac{x}{\sqrt{x^2 + y^2}}, \quad \sin \theta = \frac{y}{\sqrt{x^2 + y^2}}.$$

To use cylindrical coordinates in the evaluation of triple integrals, we must express equations of surfaces in terms of these coordinates. This is very simple. If $F(x, y, z) = 0$ is the equation of a surface in Cartesian coordinates, then to express this equation in cylindrical coordinates we substitute from equation (5.27):

$$F(r \cos \theta, r \sin \theta, z) = 0.$$

For example, the right circular cylinder $x^2 + y^2 = 9$, which has the z-axis as its axis of symmetry, has the very simple equation $r = 3$ in cylindrical coordinates.

Suppose we are to evaluate the triple integral of a continuous function $F(x, y, z)$ over some region V of space. The choice of triple iterated integral in cylindrical coordinates implies a subdivision of V into small volumes dV by means of coordinate surfaces $r = $ constant, $\theta = $ constant, and $z = $ constant (Fig. 5.24).

Surfaces $r = $ constant are right circular cylinders coaxial with the z-axis; surfaces $\theta = $ constant are planes containing the z-axis and, therefore, perpendicular to the xy-plane; surfaces $z = $ constant are planes parallel to the xy-plane. If we denote small variations in r, θ and z for the element dV by $dr, d\theta$, and dz (see Fig. 5.25), then the volume of the element is approximately $(r \, dr \, d\theta) \, dz$, where $r \, dr \, d\theta$ is the polar cross-sectional area parallel to the xy-plane. Hence, in cylindrical coordinates, we get

$$dV = r \, dz \, dr \, d\theta. \qquad (5.28)$$

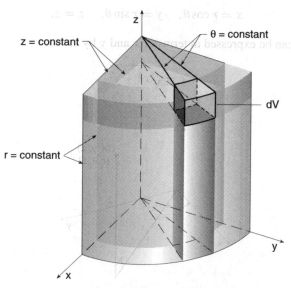

Figure 5.24   First quadrant portion of cylinder.

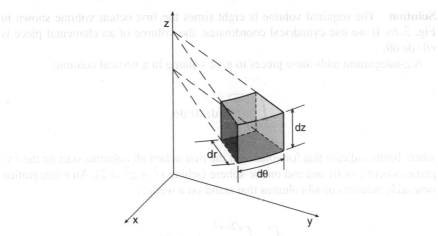

Figure 5.25 Volume element.

The integrand $F(x, y, z)$ is expressed in cylindrical coordinates as

$$F(r\cos\theta, r\sin\theta, z).$$

It remains only to affix appropriate limits to the triple integral. The most commonly used triple iterated integral is with respect to z, r, and θ and in this case the z-integration adds the quantities

$$F(r\cos\theta, r\sin\theta, z)r\,dz\,dr\,d\theta$$

in a vertical column, where r and θ are constant. The limits identify surfaces in which each and every column starts and stops and generally depend on r and θ:

$$\int_{h_1(r,\theta)}^{h_2(r,\theta)} F(r\cos\theta, r\sin\theta, z)r\,dz\,dr\,d\theta.$$

The remaining integrations with respect to r and θ perform additions over the area in the xy-plane onto which all vertical columns project. Since r and θ are simply polar coordinates, the r-integration adds over small areas in a wedge and the θ-integration adds over all wedges. The triple integral with respect to z, r and θ has the form

$$\int_{\theta=a}^{b} \int_{r=g_1(\theta)}^{g_2(\theta)} \int_{z=h_1(\theta)}^{h_2(\theta)} F(r\cos\theta, r\sin\theta, z)r\,dz\,dr\,d\theta.$$

Example 5.17 Find the volume inside both the sphere $x^2 + y^2 + z^2 = 2$ and the cylinder $x^2 + y^2 = 1$. ☐

Solution   The required volume is eight times the first octant volume shown in Fig. 5.26. If we use cylindrical coordinates, the volume of an elemental piece is $r\,dz\,dr\,d\theta$.

A z-integration adds these pieces to give volume in a vertical column:

$$\int_{z=0}^{\sqrt{2-r^2}} r\,dz\,dr\,d\theta,$$

where limits indicate that for volume in the first octant all columns start on the xy-plane (where $z = 0$) and end on the sphere (where $r^2 + z^2 = 2$). An r-integration now adds volumes of all columns that stand on a wedge:

$$\int_{r=0}^{1}\int_{z=0}^{\sqrt{2-r^2}} r\,dz\,dr\,d\theta,$$

where limits indicate that all wedges start at the origin (where $r = 0$) and end on the curve $x^2 + y^2 = 1$ (or $r = 1$) in the xy-plane. This integration yields the volume of a slice (Fig. 5.26). Finally, θ-integration adds volumes of all such slices

$$\int_{\theta=0}^{\pi/2}\int_{r=0}^{1}\int_{z=0}^{\sqrt{2-r^2}} r\,dz\,dr\,d\theta,$$

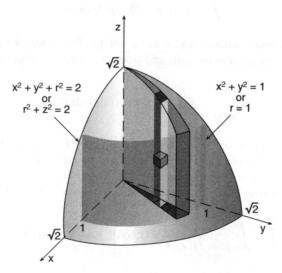

Figure 5.26   First octant volume element.

where limits 0 and $\pi/2$ identify positions of first and last wedges, respectively, in the first quadrant. The required volume is then

$$8 \int_{\theta=0}^{\pi/2} \int_{r=0}^{1} \int_{z=0}^{\sqrt{2-r^2}} r \, dz \, dr \, d\theta = 8 \int_{0}^{\pi/2} \int_{0}^{1} r\sqrt{2-r^2} dr \, d\theta$$

$$= \int_{0}^{\pi/2} \left(-\frac{1}{3}(2-r^2)^{3/2} \right)_{0}^{1} d\theta$$

$$= \frac{8}{3}(2\sqrt{2}-1) \int_{0}^{\pi/2} d\theta = \frac{4\pi}{3}(2\sqrt{2}-1).$$

■

5.10 Triple integrals in spherical coordinates

Spherical coordinates are useful in solving problems where figures are symmetric about a point. If the origin is that point, then spherical coordinates (ρ, θ, ϕ) in Fig. 5.27 are related to Cartesian coordinates (x, y, z) by the equations

$$x = \rho \sin \phi \cos \theta,$$
$$y = \rho \sin \phi \sin \theta, \tag{5.29}$$
$$z = \rho \cos \phi.$$

As is the case for polar and cylindrical coordinates, without restrictions on ρ, θ and ϕ, each point in space has many sets of spherical coordinates. The positive value of its spherical coordinate ρ is given by

$$\rho = \sqrt{x^2 + y^2 + z^2}.$$

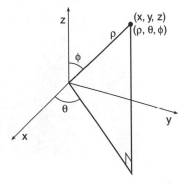

Figure 5.27   Spherical coordinate system.

The θ-coordinate in cylindrical and spherical coordinates is identical so no simple formula for θ in terms of x, y, and z exists. The value of θ lies in the range $0 \le \theta \le 2\pi$.

The ϕ-coordinate is the angle between the positive z-axis and the line joining the origin to the point (x, y, z), and that value of ϕ in the range $0 \le \phi \le \pi$ is determined by the formula

$$\phi = \cos^{-1}\left(\frac{z}{\sqrt{x^2 + y^2 + z^2}}\right).$$

To transform equations of surfaces, $F(x, y, z) = 0$, from Cartesian to spherical coordinates, we substitute from equations (5.29):

$$F(\rho \sin\phi \cos\theta, \rho \sin\phi \sin\theta, \rho \cos\phi) = 0.$$

For example, the sphere $x^2 + y^2 + z^2 = 4$ is symmetric about its centre, and its equation in spherical coordinates is simply $\rho = 2$.

The choice of a triple iterated integral in spherical coordinates implies a subdivision of V into small volumes by means of coordinate surfaces $\rho = $ constant, $\theta = $ constant, and $\phi = $ constant (see Fig. 5.28). Surfaces $\rho = $ constant are spheres

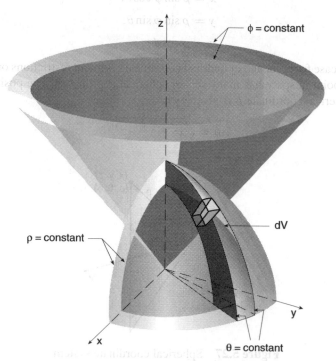

Figure 5.28   Triple integration using spherical coordinates.

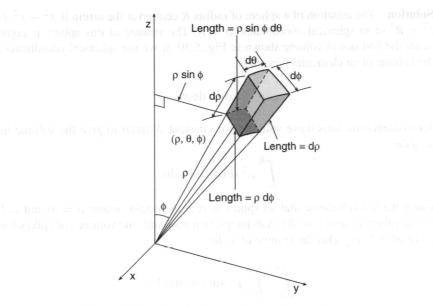

Figure 5.29   Spherical coordinate volume element.

centred at the origin; surfaces θ = constant are planes containing the z-axis; and surfaces ϕ = constant are right circular cones symmetric about the z-axis with the origin as apex.

If we denote small variations in ρ, θ and ϕ for the element dV by $d\rho, d\theta$ and $d\phi$ (Fig. 5.29) and approximate dV by a rectangular parallelepiped with dimensions $d\rho, \rho d\phi$ and $\rho \sin \phi d\theta$, then

$$dV = (\rho \sin \phi d\theta)(\rho d\phi)d\rho = \rho^2 \sin \phi d\rho \, d\theta \, d\phi. \tag{5.30}$$

The integrand $F(x, y, z)$ is expressed in spherical coordinates as

$$F(\rho \sin \phi \cos \theta, \rho \sin \phi \sin \theta, \rho \cos \phi).$$

Affixing the appropriate limits is the only remaining duty. The limits depend on which of the six possible triple iterated integrals in spherical coordinates we choose. If we use a triple iterated integral with respect to ρ, ϕ, and θ, then its form is

$$\iiint_V F(x, y, z) dV$$

$$= \int_{\theta=a}^{b} \int_{\phi=g_1(\theta)}^{g_2(\theta)} \int_{\rho=h_1(\theta,\phi)}^{h_2(\theta,\phi)} F(\rho \sin \phi \cos \theta, \rho \sin \phi \sin \theta, \rho \cos \phi)\rho^2 \sin \phi d\rho \, d\phi \, d\theta.$$

$$\tag{5.31}$$

Example 5.18   Find the volume of a sphere.   □

Solution The equation of a sphere of radius R centred at the origin is $x^2 + y^2 + z^2 = R^2$ or in spherical coordinates, $\rho = R$. The volume of this sphere is eight times the first octant volume shown in Fig. 5.30. If we use spherical coordinates, the volume of an elemental piece is

$$\rho^2 \sin\phi d\rho \, d\phi \, d\theta.$$

An ρ-integration adds these volumes for constant ϕ and θ to give the volume in a 'spike'

$$\int_0^R \rho^2 \sin\phi d\rho \, d\phi \, d\theta,$$

where the limits indicate that all spikes start at the origin (where $\rho = 0$) and end on the sphere (where $\rho = R$). A ϕ-integration now adds the volumes of spikes for constant θ. This yields the volume of a slice

$$\int_0^{\pi/2} \int_0^R \rho^2 \sin\phi d\rho \, d\phi \, d\theta,$$

where the limits indicate that all slices in the first octant start on the z-axis (where $\phi = 0$) and end on the xy-plane (where $\phi = \frac{\pi}{2}$). Finally, the θ-integration adds the volumes of all such slices

$$\int_0^{\pi/2} \int_0^{\pi/2} \int_0^R \rho^2 \sin\phi d\rho \, d\phi \, d\theta,$$

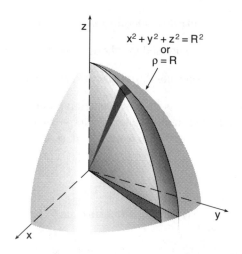

Figure 5.30 First octant volume element for Example 5.17.

where the limits 0 and $\pi/2$ identify positions of first and last slices, respectively, in the first octant. The volume of the sphere is given by

$$8 \int_0^{\pi/2} \int_0^{\pi/2} \int_0^R \rho \sin \phi \, d\rho \, d\phi \, d\theta = 8 \int_0^{\pi/2} \int_0^{\pi/2} \left(\frac{\rho^3}{3} \sin \phi \right)_0^R d\phi \, d\theta$$

$$= \frac{8R^3}{3} \int_0^{\pi/2} (-\cos \phi)_0^{\pi/2} \, d\theta$$

$$= \frac{8R^3}{3} (\theta)_0^{\pi/2} = \frac{4\pi R^3}{3}.$$

■

Example 5.19   Find the centre of mass of a solid object of constant density if it is in the shape of a right circular cone.   ☐

Solution   Let the cone have altitude h and base radius R. Then its mass is

$$M = \frac{1}{3} \pi R^2 h \rho,$$

where ρ is the density of the object. If we place axes as shown in Fig. 5.31, then $\bar{x} = \bar{y} = 0$; i.e., the centre of mass is on the axis of symmetry of the cone. To find $\bar{z}$ we offer three solution methods.

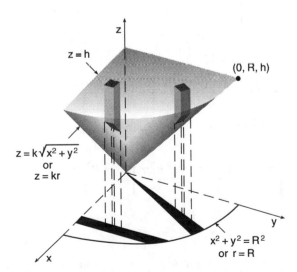

Figure 5.31   First quadrant element for Example 5.18.

Method 1: If we use Cartesian coordinates, the equation of the surface of the cone is of the form $z = k\sqrt{x^2 + y^2}$. Since $(0, R, h)$ is a point on the cone, $h = kR$, and, therefore, $k = H/R$. Now

$$M\bar{z} = 4 \int_0^R \int_0^{\sqrt{R^2-x^2}} \int_{k\sqrt{x^2+y^2}}^h z\rho \, dz \, dy \, dx$$

$$= 4\rho \int_0^R \int_0^{\sqrt{R^2-x^2}} \left\{ \frac{z^2}{2} \right\}_{k\sqrt{x^2+y^2}}^h dy \, dx$$

$$= 2\rho \int_0^R \int_0^{\sqrt{R^2-x^2}} \left\{ h^2 - k^2(x^2 + y^2) \right\} dy \, dx$$

$$= 2\rho \int_0^R \left\{ h^2 y - k^2 x^2 y - \frac{k^2 y^3}{3} \right\}_0^{\sqrt{R^2-x^2}} dx$$

$$= 2\rho \int_0^R \left\{ h^2 \sqrt{R^2 - x^2} - k^2 x^2 \sqrt{R^2 - x^2} - \frac{k^2}{3}(R^2 - x^2)^{3/2} \right\} dx.$$

If we now set $x = R \sin\theta$, then $dx = R \cos\theta \, d\theta$, and

$$M\bar{z} = 2\rho \int_0^{\pi/2} \left\{ h^2 R \cos\theta - k^2(R^2 \sin^2\theta) R \cos\theta - \frac{k^2}{3} \cos^3\theta \right\} R \cos\theta \, d\theta$$

$$= 2\rho R^2 \int_0^{\pi/2} \left\{ \frac{h^2}{2}(1 + \cos 2\theta) - \frac{k^2 R^2}{8}(1 - \cos 4\theta) \right.$$

$$\left. - \frac{k^2 R^2}{12}\left(1 + 2\cos 2\theta + \frac{1 + \cos 4\theta}{2}\right) \right\} d\theta$$

$$= \rho R^2 \left\{ h^2 \left(\theta + \frac{\sin 2\theta}{2}\right) - \frac{k^2 R^2}{4}\left(\theta - \frac{\sin 4\theta}{4}\right) \right.$$

$$\left. - \frac{k^2 R^2}{6}\left(\frac{3\theta}{2} + \sin 2\theta + \frac{\sin 4\theta}{8}\right) \right\}_0^{\pi/2}$$

$$= \rho R^2 \left\{ \frac{\pi h^2}{2} - \frac{k^2 R^2 \pi}{8} - \frac{k^2 R^2 \pi}{8} \right\} = \frac{\pi R^2 h^2 \rho}{4}.$$

Therefore,

$$\bar{z} = \frac{\pi R^2 h^2 \rho}{4} \frac{3}{\pi R^2 h \rho} = \frac{3h}{4}.$$

Method 2: If cylindrical coordinates are used, the equation of the surface of the cone is $z = kr$. From Fig. 5.31, we see that

$$M\bar{z} = 4 \int_0^{\pi/2} \int_0^R \int_{kr}^h z\rho\, r\, dz\, dr\, d\theta = 4\rho \int_0^{\pi/2} \int_0^R \left\{ \frac{rz^2}{2} \right\}_{kr}^h dr\, d\theta$$

$$= 2\rho \int_0^{\pi/2} \int_0^R r(h^2 - k^2 r^2)\, dr\, d\theta = 2\rho \int_0^{\pi/2} \left\{ \frac{r^2 h^2}{2} - \frac{k^2 r^4}{4} \right\}_0^R d\theta$$

$$= \rho \left(R^2 h^2 - \frac{k^2 R^4}{2} \right) \{\theta\}_0^{\pi/2} = \rho \left(R^2 h^2 - \frac{k^2 R^4}{2} \right) \frac{\pi}{2} = \frac{\pi R^2 h^2 \rho}{4}.$$

Here again, we have

$$\bar{z} = \frac{3h}{4}.$$

Method 3: In using spherical coordinates, we take $\rho = \gamma$ to avoid confusion with the density. The equation of the surface of the cone then becomes

$$\phi = \cos^{-1} \left(\frac{h}{\sqrt{h^2 + R^2}} \right) = \phi_1.$$

From Fig. 5.32, we see that

$$M\bar{z} = 4 \int_0^{\pi/2} \int_0^{\phi_1} \int_0^{h\sec\phi} (\gamma\cos\phi)\rho\, \gamma^2 \sin\phi\, d\gamma\, d\phi\, d\theta$$

$$= 4\rho \int_0^{\pi/2} \int_0^{\phi_1} \left\{ \frac{\gamma^4}{4} \sin\phi\cos\phi \right\}_0^{h\sec\phi} d\phi\, d\theta$$

$$= \rho h^4 \int_0^{\pi/2} \int_0^{\phi_1} \left(\frac{\sin\phi}{\cos^3\phi} \right) d\phi\, d\theta = \rho h^4 \int_0^{\pi/2} \left\{ \frac{1}{2\cos^2\phi} \right\}_0^{\phi_1} d\theta$$

$$= \frac{\rho h^4}{2} \left\{ \frac{1}{\cos^2\phi} - 1 \right\} \{\theta\}_0^{\pi/2}$$

$$= \frac{\pi\rho h^4}{4} \left\{ \frac{h^2 + R^2}{h^2} - 1 \right\} = \frac{\pi h^2 R^2 \rho}{4}.$$

Once again, $\bar{z} = 3h/4$ as before. ∎

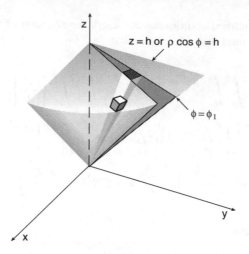

Figure 5.32   Spherical coordinate volume element for Example 5.18.

5.11 Summary

- Line integrals, surface integrals (flux integrals) and volume integrals yield only scalar quantities. They are all discrete (limiting) cases of summations — along curves, over surfaces and throughout volumes, respectively.
- A line integral is usually defined mathematically in the form $\int_C \mathbf{F} \cdot d\mathbf{R}$, where $\mathbf{F}$ is a given vector field and C is a given path between points P and Q. If P and Q coincide, that is, if C is a *closed path*, we write $\oint_C \mathbf{F} \cdot d\mathbf{R}$. A line integral can be evaluated by using the equation of the path to convert it to a definite integral.
- If $\oint_C \mathbf{F} \cdot d\mathbf{R} = 0$ for all paths in a region D, we say that the vector field $\mathbf{F}$ is conservative in D. A necessary and sufficient condition for $\mathbf{F}$ to be conservative is the vanishing of its *curl* everywhere in that region, that is, $\nabla \times \mathbf{F} = 0$, which implies that $\mathbf{F} = \nabla \Phi$ in which Φ is a scalar function (potential function).
- A conservative vector field $\mathbf{F}$ can be expressed as the gradient of a scalar function Φ, called the potential. In this situation,

$$\int_P^Q \mathbf{F} \cdot d\mathbf{R} = \int_P^Q \nabla \Phi \cdot d\mathbf{R} = \int_P^Q d\Phi = \Phi|_P^Q = \Phi(Q) - \Phi(P).$$

- The form of a surface integral or *flux integral* of a vector field across a surface S is $\iint_S \mathbf{F} \cdot \mathbf{n}\,dS$, or equivalently, $\iint_S \mathbf{F} \cdot d\mathbf{S}$, where $\mathbf{n}$ is an outward unit normal vector at a point S. For a closed surface we write $\oiint_S \mathbf{F} \cdot \mathbf{n}\,dS$.
- The *unit normal vector* $\mathbf{n}$ to a surface, if not obvious by inspection, can be calculated using the formula $\mathbf{n} = \nabla\Phi/|\nabla\Phi|$ where the equation of the surface is written in the form $\Phi = $ constant.

- Flux integrals over cylinders may be evaluated using *cylindrical coordinates* (r, θ, z). The surface elements are
  $dS = ad\theta dz$ on the cylinder surface on $r = a$, and
  $dS = rd\theta dr$ on an end surface $z = $ constant.
- Flux integrals over a spherical surface may be evaluated using *spherical coordinates* (r, θ, ϕ). The surface element on a sphere $r = a$ is

$$dS = (ad\theta)(a \sin \theta d\phi) = a^2 \sin \theta d\phi d\theta.$$

- Projecting a surface onto a coordinate plane can be a useful method for evaluating the surface integral.
- The form of the *volume integral* of a scalar field is $\iiint_V f dV$, where V is the given volume.
- The basic volume elements are
  (a) In Cartesian coordinates: $dV = dx dy dz$.
  (b) In cylindrical coordinates: $dV = (dr)(rd\theta)(dz)$.
  (c) In spherical coordinates: $dV = (dr)(rd\theta)(r \sin \theta d\phi)$.

Remark 5.2   Volume integrals are evaluated as triple integrals using limits, and a coordinate system, that are appropriate to the volume of integration.   ∎

Exercises

1. Evaluate the line integral, where C is the given curve.
    (a) $\int_C x ds, \; C: x = t^3, \; y = t, \; 0 \le t \le 1$.
    (b) $\int_C xy^4 ds, \; C$ is the right half of the circle $x^2 + y^2 = 16$.
    (c) $\int_C (x - 2y^2) dy, \; C$ is the arc of the parabola $y = x^2$ from $(-2, 4)$ to $(2, 3)$.
    (d) $\int_C xy dx + (x - y) dy, \; C$ consists of line segments from $(0, 0)$ to $(2, 0)$ and from $(2, 0)$ to $(3, 2)$.
    (e) $\int_C x \sqrt{y} dx + 2y \sqrt{x} dy, \; C$ consists of the arc of the circle $x^2 + y^2 = 1$ from $(1, 0)$ to $(0, 1)$ and the line segment from $(0, 1)$ to $(4, 3)$.
2. Evaluate the line integral in the following questions.
    (a) $\int_C xz ds$ along the first octant part of $y = x^2, \; z + y = 1$ from $(0, 0, 1)$ to $(1, 1, 0)$.
    (b) $\int_C x \sqrt{y + z} ds$ where C is that part of the curve $3x + 2y + 3z = 6$, $x - 2y + 4z = 5$ from $(1, 0, 1)$ to $(0, 9/14, 11/7)$.
    (c) $\int_C (x + y) z ds$ where C is the curve $y = x, \; z = 1 + y^4$ from $(-1, -1, 2)$ to $(1, 1, 2)$.
    (d) $\int_C (2y + yz) ds$ where C is the curve $z = xy, \; x = y^2$ from $(0, 0, 0)$ to $(4, -2, -8)$.
3. Find the work done by a force $\mathbf{F} = x\mathbf{i} + y\mathbf{j}$ on a particle as it moves once counter clockwise around the ellipse $b^2 x^2 + a^2 y^2 = a^2 b^2, \; z = 0$.
4. Evaluate the line integral

$$\int_C \frac{x^3}{(1 + x^4)^3} dx + y^2 e^y dy + \frac{z}{\sqrt{1 + z^2}} dz$$

where C consists of line segments joining successively the points $(0, -1, 1)$, $(1, -1, 1)$, $(1, 0, 1)$ and $(1, 0, 2)$.

5. The cycloid $x = R(\theta - \sin\theta)$, $y = R(1 - \cos\theta)$ is the curve traced out by a fixed point on the circumference of a circle of radius R rolling along the x-axis. Suppose the point is acted on by a force of unit magnitude directed toward the centre of the rolling circle.
    (a) Find the work done by the force as the point from $\theta = 0$ to $\theta = \pi$.
    (b) How much of the work in (a) is done by the vertical component of the force?

6. Evaluate the line integral $\int_C \mathbf{F} \cdot d\mathbf{r}$, where C is given by the vector function $\mathbf{r}(t)$.
    (a) $\mathbf{F}(x, y) = x^2 y\mathbf{i} - xy\mathbf{j}$, $\mathbf{r}(t) = t^3\mathbf{i} + t^4\mathbf{j}$, $0 \le t \le 1$.
    (b) $\mathbf{F}(x, y, z) = (y + z)\mathbf{i} - x^2\mathbf{j} - 4y^2\mathbf{k}$ $\mathbf{r}(t) = t\mathbf{i} + t^2\mathbf{j} + t^4\mathbf{k}$, $0 \le t \le 1$.
    (c) $\mathbf{F}(x, y, z) = \sin x\mathbf{i} + \cos y\mathbf{j} + xz\mathbf{k}$, $\mathbf{r}(t) = t^3\mathbf{i} - t^2\mathbf{j} + t\mathbf{k}$, $0 \le t \le 1$.
    (d) $\mathbf{F}(x, y, z) = x^2\mathbf{i} + xy\mathbf{j} + z^2\mathbf{k}$, $\mathbf{r}(t) = \sin t\mathbf{i} + \cos t\mathbf{j} + t^2\mathbf{k}$, $0 \le t \le \pi/2$.

7. Evaluate the line integral $\int_C \mathbf{F} \cdot d\mathbf{r}$ where $\mathbf{F}(x, y) = e^{x-1}\mathbf{i} + xy\mathbf{j}$ and C is given by $\mathbf{r}(t) = t^2\mathbf{i} + t^3\mathbf{j}$, $0 \le t \le 1$.

8. Find the centre of mass of a wire in the shape of the helix $x = 2\sin t$, $y = 2\cos t$ $z = 3t$, $0 \le t \le 2\pi$ if the density is a constant k.

9. Find the mass and centre of mass of a wire in the shape of the helix $x = t$, $y = \cos t$, $z = \sin t$, $0 \le t \le 2\pi$, if the density at any point is equal to the square of the distance from the origin.

10. Determine whether or not $\mathbf{F}$ is a conservative field. If it is, find a function f such that $\mathbf{F} = \nabla f$.
    (a) $\mathbf{F}(x, y) = (3x^2 - 4y)\mathbf{i} + (4y^2 - 2x)\mathbf{j}$.
    (b) $\mathbf{F}(x, y) = (x^2 + y)\mathbf{i} + (y^2 + x)\mathbf{j}$.
    (c) $\mathbf{F}(x, y) = (1 + 4x^3 y^3)\mathbf{i} + 3x^4 y^2\mathbf{j}$.
    (d) $\mathbf{F}(x, y) = (y\cos x - \cos y)\mathbf{i} + (\sin x + x\sin y)\mathbf{j}$.
    (e) $\mathbf{F}(x, y) = (ye^x + \sin y)\mathbf{i} + (e^x + x\cos y)\mathbf{j}$.

In the next two questions show that the line integral is independent of path and evaluate the integral.

11. $\int_C 2x\sin y dx + (x^2\cos y - 3y^2)dy$ C is any path from $(-1, 0)$ to $(5, 1)$.

12. $\int_C (2y^2 - 12x^3 y^3)dx + (4xy - 9x^4 y^2)dy$, C is any path from $(1, 1)$ to $(3, 2)$.

13. Let $\mathbf{F}(x, y) = (-y\mathbf{i} + x\mathbf{j})/(x^2 + y^2)$
    (a) Show that $\partial P/\partial y = \partial Q/\partial x$.
    (b) Show that $\int_C \mathbf{F} \cdot d\mathbf{r}$ is not independent of path. [*Hint:* Show that $\int_{C_1} \mathbf{F} \cdot d\mathbf{r}$ and $\int_{C_2} \mathbf{F} \cdot d\mathbf{r}$, where C_1 and C_2 are the upper and lower halves of the circle $x^2 + y^2 = 1$ from $(1, 0)$ to $(-1, 0)$]. Does this contradict Theorem 5.3?

14. (a) Suppose that $\mathbf{F}$ is an inverse square force field, i.e.
$$\mathbf{F}(\mathbf{r}) = \frac{c\mathbf{r}}{|\mathbf{r}|^3}$$

for some constant c, where $\mathbf{r} = x\mathbf{i} + y\mathbf{j} + z\mathbf{k}$. Find the work done by $\mathbf{F}$ in moving an object from a point P_1 along a path to a point P_2 in terms of the distances d_1 and d_2 from these points to the origin.

(b) An example of an inverse square field is the gravitational field $\mathbf{F} = -(mMG)/|\mathbf{r}|^3$. Use part (a) to find the work done by the gravitational field when the earth moves from aphelion (at a maximum distance of 1.52×10^8 km from the sun) to perihelion (at a minimum distance of 1.47×10^8 km). Use the values $m = 5.97 \times 10^{24}$ kg, $M = 1.99 \times 10^{30}$ kg and $G = 6.67 \times 10^{-11}$ Nm^2/kg^2.

(c) Another example of an inverse square field is the electric field $\mathbf{E} = \varepsilon q\, Q\mathbf{r}/|\mathbf{r}|^3$. Suppose an electron with a charge of -1.6×10^{-19} C is at the origin. A positive charge is at a distance 10^{-12} m from the electron. Use part (a) to find the work done by the electric field. (Use the value $\varepsilon = 8.985 \times 10^{10}$.)

15. Evaluate the given surface integral
   (a) $\iint_S y\,ds$, S is the part of the plane $3x + 2y + z = 6$ that lies on the first octant.
   (b) $\iint_S xz\,ds$, S is the triangle with vertices $(1, 0, 0)$, $(0, 1, 0)$ and $(0, 0, 1)$.
   (c) $\iint_S x\,ds$, S is the surface $y = x^2 + 4z$, $0 \le x \le 2, 0 \le z \le 2$.
   (d) $\iint_S (y^2 + z^2)\,ds$, S the part of the paraboloid $x = 4 - y^2 - z^2$ that lies in front of the plane $x = 0$.

16. A fluid with density 1200 flows with velocity $\mathbf{V} = y\mathbf{i} + x\mathbf{j} + 2z\mathbf{k}$. Find the rate of flow upward through the paraboloid $z = 9 - (x^2 + y^2)/4$, $x^2 + y^2 \le 36$.

17. Evaluate the given surface integral
   (a) $\iint_S (yz^2\mathbf{i} + ye^x\mathbf{j} + x\mathbf{k}) \cdot \mathbf{n}ds$, where S is defined by $y = x^2, 0 \le y \le 4$, $0 \le z \le 1$ and $\mathbf{n}$ is the unit normal to S with positive y-component.
   (b) $\iint_S (yz\mathbf{i}_x z\mathbf{j} + xy\mathbf{k}) \cdot \mathbf{n}ds$, where S is that part of the surface $z = x^2 + y^2$ cut out by the planes $x = 1$, $x = -1$, $y = 1$, $y = -1$, and $\mathbf{n}$ is the unit lower normal to S.
   (c) $\iint_S (x\mathbf{i} + y\mathbf{j}) \cdot \mathbf{n}ds$, where S is that part of the surface $z = \sqrt{x^2 + y^2}$ below $z = 1$, and $\mathbf{n}$ is the unit normal to S with negative z-component.
   (d) $\iint_S (x^2 y\mathbf{i} + xy\mathbf{j} + z\mathbf{k}) \cdot \mathbf{n}ds$, where S is defined by $z = 2 - x^2 - y^2$, $0 \le z$, and $\mathbf{n}$ is the unit normal to S with negative z-component.
   (e) $\iint_S (x\mathbf{i} + y\mathbf{j}) \cdot \mathbf{n}ds$, where S is the surface $x^2 + y^2 + z^2 = 4, 1 \le z$, and $\mathbf{n}$ is the unit upper normal to S.
   (f) $\iint_S (y\mathbf{i} - x\mathbf{j} + \mathbf{k}) \cdot \mathbf{n}ds$, where S is the smaller surface cut by the plane $y + z = 1$, from the sphere $x^2 + y^2 + z^2 = 1$ and $\mathbf{n}$ is the unit upper normal to S.

18. Evaluate $\iint_S (y\mathbf{i} - x\mathbf{j} + z\mathbf{k}) \cdot ds$ where
   (a) S is that part of $z = 9 - x^2 - y^2$ cut out by $z = 2y$.
   (b) S is that part of $z = 2y$ cut out by $z = 9 - x^2 - y^2$, and $\mathbf{n}$ is the unit upper normal to S in each case. [*Hint:* Use polar coordinates with pole at $(0, -1)$].

19. A circular tube $S: x^2 + z^2 = 1, 0 \le y \le 2$ is a model for a part of an artery. Blood flows through the artery and the force over unit area at any point on the arterial wall is given by

$$F = e^{-y}n + \frac{1}{y^2 + 1}j,$$

where n is the unit outer normal to the arterial wall. Blood diffuses through the wall in such a way that if dS is a small area on S, the amount of diffusion through dS in one second is $F \cdot ndS$. Find the total amount of blood leaving the entire wall per second.

20. Evaluate the triple integral over the region given
    (a) $\iiint_V xy \, dV$, where V is enclosed by $z = \sqrt{1 - x^2 - y^2}$, $z = 0$.
    (b) $\iiint_V (x + y + z) \, dV$, where V is bounded by $x = 0$, $x = 1$, $z = 0$, $y + z = 2$, $y = z$.
    (c) $\iiint_V x^2 y \, dV$, where V is the first octant volume bounded by $z = 1$, $z = x^2/4 + y^2/9$.
    (d) $\iiint_V (x^2 + 2z) \, dV$, where V is bounded by $z = 0$, $y + z = 4$, $y = x^2$.

21. Find the volume bounded by the given surfaces
    (a) $z = \sqrt{2 - x^2 - y^2}$, $z = x^2 + y^2$.
    (b) $z = x^2 + y^2$, $z = 4 - x^2 - y^2$.

22. Find the volume inside the sphere $x + y^2 + z^2 = 4$ but outside the cylinder $x^2 + y^2 = 1$.

23. Find the volume described.
    (a) Bounded by $z = \sqrt{x^2 + y^2}$, $z = \sqrt{1 - x^2 - y^2}$.
    (b) Bounded by $x^2 + y^2 + z^2 = 1$, $y = x$, $y = 2x$, $z = 0$ (in first octant).
    (c) Bounded by $z = 2\sqrt{x^2 + y^2}$, $x^2 + y^2 = 4$, $z = 0$.

24. Determine the integral of the function $x^2 z$ taken over the entire surface of the right circular cylinder of height h which stands on the circle $x^2 + y^2 = a^2$. What is the integral of the given function taken throughout the volume of the cylinder?

25. Repeat the calculation of the above exercise for the case of the upper half of the sphere $x^2 + y^2 + z^2 = a^2$.

26. Show that Green's theorem can be written in the form $\oint_C F \cdot T ds = \iint_D \nabla \times F \cdot k dx dy$, where $F = iF_1 + jF_2$ and T is the unit tangent vector to C. [Hint: $T ds = i dx + j dy$.]

27. Show that Green's theorem can be written in the form $\oint_C F \cdot N ds = \iint_D \nabla \cdot F dx dy$, where $F = iF_2 - jF_1$ and N is the outer unit normal to the curve C. [Hint: $N ds = i dy - j dx$].

28. Evaluate $\iint_S (x + y + z) dS$, where S is the part of the surface of the sphere $x^2 + y^2 + z^2 = a^2$ which lies in the first octant. [Hint: Use spherical coordinates. Answer $= 3\pi a^3/2$.]

Chapter 6

Integral theorems

6.1 Introduction

In this chapter we examine the integral theorems of vector analysis, which can be extremely useful in the evaluation of various integrals such as those discussed in Chapter 5. There are mainly three integral theorems that will be examined due to their importance in vector analysis applications: Green's theorem, Stokes' theorem and the Divergence theorem (Gauss' theorem). These theorems are very useful in developing numerical or analytical solution techniques arising in different branches of applied science and engineering.

6.2 Green's theorem

Double integrals over a plane region may be transformed into line integrals over the boundary of the region. Conversely, line integrals over the boundary of a region may be transformed into double integrals over the plane region. This transformation is of practical interest because it may help make the evaluation of an integral easier. It also helps in the theory whenever one wants to switch from one type of integral to another. The transformation can be done by the use of Green's theorem in the plane.

George Green(1793–1841) was a self-educated English mathematician who started out as a baker and at his death was a fellow of Caius College, Cambridge

University. His work concerned potential theory in connection with electricity and magnetism, vibrations, waves and elasticity. The work he accomplished remained almost unknown, even in England, until after his death.

Line integrals in the xy-plane are of the form $\int_C f(x, y)\, ds$, and in the special case that $f(x, y)$ is the tangential component of some vector field $\mathbf{F}(x, y) = P(x, y)\mathbf{i} + Q(x, y)\mathbf{j}$ along C the integrals take the form

$$\int_C \mathbf{F} \cdot d\mathbf{R} = \int_C P(x, y)\, dx + Q(x, y)\, dy. \tag{6.1}$$

We now show that when C is a closed curve, the line integral in (6.1) can usually be replaced by a double integral. The precise result is contained in the following theorem which is sometimes known as Green's lemma.

Theorem 6.1   *If C is a plane region bounded by a finite number of simple closed curves, and if $P(x, y)$ and $Q(x, y)$ have continuous first partial derivatives in a domain D containing C and R, where dA is the elementary area of the domain D then*

$$\oint_C P\, dx + Q\, dy = \iint_R \left(\frac{\partial Q}{\partial x} - \frac{\partial P}{\partial y} \right) dA, \tag{6.2}$$

provided the line integral is taken in the positive direction around C.

Comment: Formula (6.2) can be written in vector form as

$$\iint_R (\text{curl } \mathbf{F}) \cdot \mathbf{k}\, dx\, dy = \oint_C \mathbf{F} \cdot d\mathbf{R}.$$

This follows from the fact that the third component of the curl of $\mathbf{F}$ is $\partial F_2/\partial x - \partial F_1/\partial y$ (where $\mathbf{F} = F_1\mathbf{i} + F_2\mathbf{j}$).

Proof   We first consider a simple region R for which every line parallel to the x- and y-axis that intersects C does so in at most two points (Fig. 6.1). Then C can be subdivided into an upper and a lower part,

$$C_2: y = h(x) \quad \text{and} \quad C_1: y = g(x).$$

If we consider the second term on the right of (6.2), we have

$$\iint_R -\frac{\partial P}{\partial y}\, dA = \int_a^b \int_{g(x)}^{h(x)} -\frac{\partial P}{\partial y}\, dy\, dx$$

$$= \int_a^b \{-P\}_{g(x)}^{h(x)}\, dx$$

$$= \int_a^b \{P(x, g(x)) - P(x, h(x))\}\, dx.$$

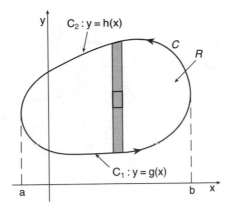

Figure 6.1   A simple region R.

On the other hand, the first term on the left-hand side of (6.2) is

$$\oint_C P \, dx = \int_{C_1} P \, dx + \int_{C_2} P \, dx = \int_{C_1} P \, dx - \int_{-C_2} P \, dx,$$

and if we use x as a parameter along C_1 and $-C_2$, then we have

$$\oint_C P \, dx = \int_a^b P(x, g(x)) \, dx - \int_a^b P(x, h(x)) \, dx$$

$$= \int_a^b \{P(x, g(x)) - P(x, h(x))\} \, dx.$$

Thus, we have shown that

$$\oint_C P \, dx = \iint_R -\frac{\partial P}{\partial y} \, dA.$$

By subdividing C into two parts of the type $x = g(y)$ and $x = h(y)$(with $g(y) \leq h(y)$), we can also show that

$$\oint_C Q \, dy = \iint_R \frac{\partial Q}{\partial x} \, dA$$

$$= \int_a^b \int_{g(y)}^{h(y)} \frac{\partial Q}{\partial x} \, dx \, dy$$

$$= \int_a^b \{Q(h(y), y) - Q(g(y), y)\} \, dy.$$

Addition of these results gives Green's theorem for this C and R.

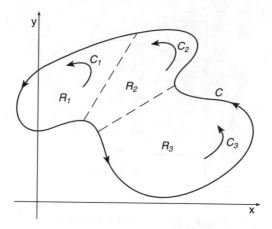

Figure 6.2   General region R decomposed into n subregions R_i.

Consider a more general region R such as that in Fig. 6.2; it can be decomposed into n subregions R_i, each of which satisfies the condition that lines parallel to the coordinate axes intersect its boundary in at most two points. For each subregion R_i, Green's theorem gives

$$\oint_{C_i} P \, dx + Q \, dy = \iint_{R_i} \left(\frac{\partial Q}{\partial x} - \frac{\partial P}{\partial y} \right) dA.$$

If these results are added, we get

$$\sum_{i=1}^{n} \left\{ \oint_{C_i} P \, dx + Q \, dy \right\} = \sum_{i=1}^{n} \iint_{R_i} \left(\frac{\partial Q}{\partial x} - \frac{\partial P}{\partial y} \right) dA.$$

Now, R is composed of the subregions R_i; therefore, the right-hand side of this equation is the double integral over R. Figure 6.2 illustrates that when line integrals over C_i are added, contributions from interior curves cancel in pairs, leaving the line integral around C. This completes the proof. ∎

The proof for even more general regions which cannot be divided into a finite number of subregions is left for more advanced texts. We continue with further applications of Green's theorem. Later in this chapter we shall use Green's theorem as the basic tool in deriving two very important theorems (by Gauss and Stokes) for transforming integrals.

Example 6.1   Evaluate the line integral $\oint_C y^2 \, dx + x^2 \, dy$.   □

Solution By Green's theorem (see Fig. 6.3), we have

$$\oint_C y^2 \, dx + x^2 \, dy = \iint_R (2x - 2y) \, dA$$

$$= 2 \int_0^1 \int_0^x (x - y) \, dy \, dx$$

$$= 2 \int_0^1 \left\{ xy - \frac{y^2}{2} \right\}_0^x dx = \int_0^1 x^2 \, dx$$

$$= \left\{ \frac{x^3}{3} \right\}_0^1 = \frac{1}{3}.$$

■

Example 6.2 Evaluate the integral

$$\oint_C (5x^2 + ye^x + y) \, dx + (e^x + e^y) \, dy,$$

where C is the circle $(x - 1)^2 + y^2 = 1$. □

Solution By Green's theorem (see Fig. 6.4), we again have

$$\oint_C (5x^2 + ye^x + y) \, dx + (e^x + e^y) \, dy = \iint_R (e^x - e^y - 1) \, dA$$

$$= -\iint_R dA$$

$$= -(\text{area of } R) = -\pi (1)^2$$

$$= -\pi.$$

■

Example 6.3 Find the work done by the force $\mathbf{F} = (-16y + \sin x^2)\mathbf{i} + (4e^y + 3x^2)\mathbf{j}$ acting along the simple closed curve C shown in Fig. 6.5. □

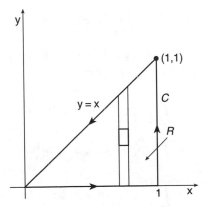

Figure 6.3 Line integral for Example 6.1.

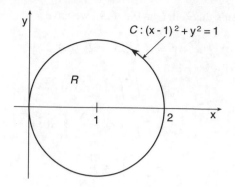

Figure 6.4   Circular region C for Example 6.2.

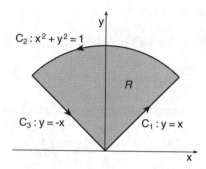

Figure 6.5   Simple closed curve for Example 6.3.

Solution   From Chapter 5, the work done by $\mathbf{F}$ is given by

$$W = \oint_C \mathbf{F} \cdot d\mathbf{R} = \oint_C (-16y + \sin x^2)\, dx + (4e^y + 3x^2)\, dy$$

and so by Green's theorem

$$W = \iint_R (6x + 16)\, dA.$$

In view of the region R, the last integral is best handled in polar coordinates. Since R is defined by $0 \le r \le 1$, $\pi/4 \le \theta \le 3\pi/4$,

$$\begin{aligned}
W &= \int_{\pi/4}^{3\pi/4} \int_0^1 (6r \cos\theta + 16)r\, dr\, d\theta \\
&= \int_{\pi/4}^{3\pi/4} (2r^3 \cos\theta + 8r^2)\big|_0^1\, d\theta \\
&= \int_{\pi/4}^{3\pi/4} (2\cos\theta + 8)\, d\theta = 4\pi.
\end{aligned}$$

■

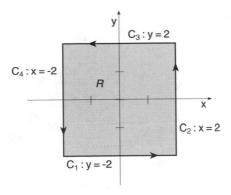

Figure 6.6   Four straight line segments which make up C.

Green's theorem cannot be used to evaluate a line integral around a closed curve that contains a point at which either $P(x, y)$ or $Q(x, y)$ fails to have continuous first partial derivatives. For example, let C be the closed curve consisting of four straight line segments C_1, C_2, C_3, C_4 shown in Fig. 6.6. Green's theorem is *not* applicable to the line integral

$$\oint \frac{-y}{x^2 + y^2}\,dx + \frac{x}{x^2 + y^2}\,dy$$

since p, Q, $\partial P/\partial y$, and $\partial Q/\partial x$ are not continuous at the origin.

6.3 Region with holes

Green's theorem can be extended to a region R with 'holes', i.e. a region bounded between two or more piecewise smooth simple closed curves. Figure 6.7a shows a region R bounded by a curve C that consists of two simple closed curves C_1 and C_2, that is, $C = C_1 \cup C_2$. The curve C is positively oriented, since if we traverse C_1 in a counterclockwise direction and C_2 in a clockwise direction, the region R is always to the left. If crosscuts are introduced as shown in Fig. 6.7b, the region R is divided into two subregions, R_1 and R_2. By applying Green's theorem to R_1 and R_2, we obtain

$$\iint_R \left(\frac{\partial Q}{\partial x} - \frac{\partial P}{\partial y} \right) dA = \iint_{R_1} \left(\frac{\partial Q}{\partial x} - \frac{\partial P}{\partial y} \right) dA + \iint_{R_2} \left(\frac{\partial Q}{\partial x} - \frac{\partial P}{\partial y} \right) dA$$

$$= \oint_{C_1} P\,dx + Q\,dy + \oint_{C_2} P\,dx + Q\,dy$$

$$= \oint_C P\,dx + Q\,dy. \tag{6.3}$$

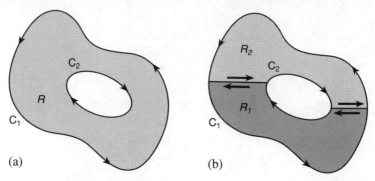

Figure 6.7   Regions R bounded by $C = C_1 \cup C_2$.

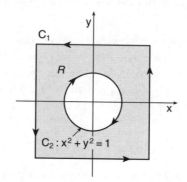

Figure 6.8   Region R for Example 6.4.

The last result follows from the fact that the line integrals on the crosscuts (paths with opposite orientations) cancel each other.

Example 6.4   Evaluate

$$\oint_C \frac{-y}{x^2 + y^2}\, dx + \frac{y}{x^2 + y^2}\, dy,$$

where $C = C_1 \cup C_2$ is the boundary of the shaded region R shown in Fig. 6.8. □

Solution   Because

$$P(x, y) = -\frac{y}{x^2 + y^2} \quad \text{and} \quad Q(x, y) = \frac{x}{x^2 + y^2},$$

$$\frac{\partial P}{\partial y} = \frac{y^2 - x^2}{(x^2 + y^2)^2} \quad \text{and} \quad \frac{\partial Q}{\partial x} = \frac{y^2 - x^2}{(x^2 + y^2)^2}.$$

are continuous on the region R bounded by C, it follows from the preceding discussion that

$$\oint_C \frac{-y}{x^2 + y^2} \, dx + \frac{x}{x^2 + y^2} \, dy = \iint_R \left\{ \frac{y^2 - x^2}{(x^2 + y^2)^2} - \frac{y^2 - x^2}{(x^2 + y^2)^2} \right\} dA = 0.$$

■

As a consequence of the discussion preceding Example 6.4, we can establish a result for line integrals that enables us, under certain circumstances, to replace a complicated closed path with a path that is simpler.

Suppose, as shown in Fig. 6.9, that C_1 and C_2 are two non-intersecting piecewise smooth simple closed paths that have the same counterclockwise orientation. Suppose further that P and Q have continuous first partial derivatives such that

$$\frac{\partial P}{\partial y} = \frac{\partial Q}{\partial x}$$

in the region R bounded between C_1 and C_2. Then from (6.3) we have

$$\oint_{C_1} P \, dx + Q \, dy + \oint_{-C_2} P \, dx + Q \, dy = 0$$

or

$$\oint_{C_1} P \, dx + Q \, dy = \oint_{C_2} P \, dx + Q \, dy. \tag{6.4}$$

Example 6.5   Evaluate the line integral

$$\oint_C \frac{-y}{x^2 + y^2} \, dx + \frac{x}{x^2 + y^2} \, dy.$$

□

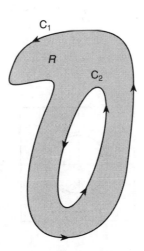

Figure 6.9   Two piecewise smooth simple closed paths.

Solution   One method of evaluating the line integral is to write

$$\oint_C = \int_{C_1} + \int_{C_2} + \int_{C_3} + \int_{C_4}$$

and then evaluate the four integrals on the line segments C_1, C_2, C_3 and C_4.

Alternatively, if we note that the circle C': $x^2 + y^2 = 1$ lies entirely within C (see Fig. 6.10), then from Example 6.4 it is apparent that $P = -y/(x^2 + y^2)$ and $Q = x/(x^2 + y^2)$ have continuous first partial derivatives in the region R bounded between C and C'. In addition

$$\frac{\partial P}{\partial y} = \frac{y^2 - x^2}{(x^2 + y^2)^2} = \frac{\partial Q}{\partial x}$$

in R. Therefore, it follows from (6.5) that

$$\oint_C \frac{-y}{x^2 + y^2}\, dx + \frac{x}{x^2 + y^2}\, dy = \oint_{C'} \frac{-y}{x^2 + y^2}\, dx + \frac{x}{x^2 + y^2}\, dy.$$

Using the parameterization for C':

$$x = \cos t, \quad y = \sin t, 0 \le t \le 2\pi$$

we obtain

$$\oint_C \frac{-y}{x^2 + y^2}\, dx + \frac{x}{x^2 + y^2}\, dy = \int_0^{2\pi} (-\sin t(-\sin t) + \cos t(\cos t))\, dt$$

$$= \int_0^{2\pi} (\sin^2 t + \cos^2 t)\, dt$$

$$= \int_0^{2\pi} dt = 2\pi. \tag{6.5}$$

■

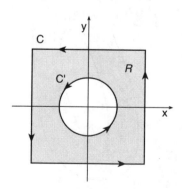

Figure 6.10   Region R containing the circle $x^2 + y^2 = 1$.

It is interesting to note that the result in (6.6):

$$\oint_C \frac{-y}{x^2 + y^2}\, dx + \frac{x}{x^2 + y^2}\, dy = 2\pi$$

is true for every piecewise smooth simple closed curve C with the origin in its interior. We need only choose C' to be $x^2 + y^2 = a^2$, where a is small enough so that the circle lies entirely within C.

6.4 Integrals over vector fields

If $\mathbf{F}(x, y, z) = P(x, y, z)\mathbf{i} + Q(x, y, z)\mathbf{j} + R(x, y, z)\mathbf{k}$ is the velocity field of a fluid element, then the volume of the fluid flowing through an element of surface area ΔS per unit time is approximated by

$$(\text{Height})(\text{Area of base}) = (\mathbf{F} \cdot \mathbf{n})\Delta S,$$

where $\mathbf{n}$ is a unit normal to the surface (see Fig. 6.11). The total volume of a fluid passing through S per unit time is called the *flux of* $\mathbf{F}$ *through* S and is given by

$$\iint_S f(x, y, z)\, dS = \iint_S (\mathbf{F} \cdot \mathbf{n})\, dS. \tag{6.6}$$

In the case of a closed surface S, if $\mathbf{n}$ is the outer(inner) normal, then (6.7) gives the volume of fluid flowing out(in) through S per unit time. Note in the integral given by (6.7) that $f(x, y, z)$ is the component of $\mathbf{F}$ in the direction of $\mathbf{n}$.

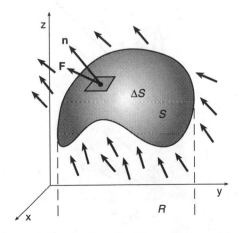

Figure 6.11   Surface S with a unit normal to it.

Example 6.6 Let $\mathbf{F}(x, y, z) = z\mathbf{j} + z\mathbf{k}$ represent the flow of a liquid. Find the flux of $\mathbf{F}$ through the surface S given by that portion of the plane $z = 6 - 3x - 2y$ in the first octant oriented upward.   ☐

Solution   The vector field and the surface are illustrated in Fig. 6.12. By defining the plane $g(x, y, z) = 3x + 2y + z - 6 = 0$, we see that a unit normal with a positive $\mathbf{k}$ component is

$$\mathbf{n} = \frac{\nabla g}{|\nabla g|} = \frac{3}{\sqrt{14}}\mathbf{i} + \frac{2}{\sqrt{14}}\mathbf{j} + \frac{1}{\sqrt{14}}\mathbf{k}.$$

Hence,

$$\text{Flux} \ = \ \iint_S f(x, y, z)\, \mathrm{d}S = \iint (\mathbf{F} \cdot \mathbf{n})\, \mathrm{d}S = \frac{1}{\sqrt{14}} \iint_S 3z\, \mathrm{d}S,$$

with R the projection of the surface onto the xy-plane, we find from (6.7) that

$$\iint_S (\mathbf{F} \cdot \mathbf{n})\, \mathrm{d}S = \frac{1}{\sqrt{14}} \iint_R 3(6 - 3x - 2y)(\sqrt{14}\, \mathrm{d}A)$$

$$= 3 \int_0^2 \int_0^{3-3x/2} (6 - 3x - 2y)\mathrm{d}y\, \mathrm{d}x = 18.$$

 ∎

Depending on the nature of the vector field, the integral (6.7) can represent other kinds of flux. For example, (6.7) could also give electric flux, magnetic flux, flux of heat and so on.

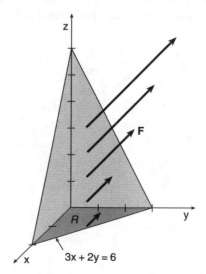

Figure 6.12   Surface S with flux $\mathbf{F}$ for Example 6.6.

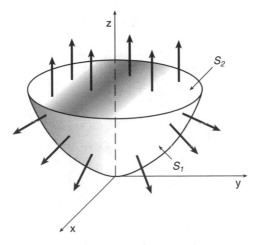

Figure 6.13   Orientable, piecewise smooth, closed surface S.

Remark 6.1   If the surface S is piecewise defined, we express a surface integral over S as the sum of the surface integrals over the various pieces of the surface. For example, suppose S is the orientable piecewise smooth closed surface bounded by the paraboloid $z = x^2 + y^2$ (S_1) and the plane $z = 1$ (S_2), then the flux of a vector field $\mathbf{F}$ out of the surface S is

$$\iint_S \mathbf{F} \cdot \mathbf{n}\, dS = \iint_{S_1} \mathbf{F} \cdot \mathbf{n}\, dS + \iint_{S_2} \mathbf{F} \cdot \mathbf{n}\, dS$$

where we take S_1 oriented upward and S_2 oriented downward (see Fig. 6.13)   ■

Example 6.7   Evaluate

$$\iint_S \mathbf{F} \cdot \mathbf{n}\, dS,$$

where $\mathbf{F} = x\mathbf{i} + y\mathbf{j} + z\mathbf{k}$ and $\mathbf{n}$ is the unit outward-pointing normal to the surface S enclosing the volume bounded by $x^2 + y^2 = 4$, $z = 0$, $z = 2$.   □

Solution   We divide S into four parts (Fig. 6.14):

$$S_1: \quad z = 0, \quad x^2 + y^2 \le 4;$$
$$S_2: \quad z = 2, \quad x^2 + y^2 \le 4;$$
$$S_3: \quad y = \sqrt{4 - x^2}, \quad 0 \le z \le 2;$$
$$S_4: \quad y = -\sqrt{4 - x^2}, \quad 0 \le z \le 2.$$

On S_1, $\mathbf{n} = -\mathbf{k}$; on S_2, $\mathbf{n} = \mathbf{k}$; and on S_3 and S_4

$$\mathbf{n} = \frac{\nabla(x^2 + y^2 - 4)}{|\nabla(x^2 + y^2 - 4)|} = \frac{(2x, 2y, 0)}{\sqrt{4x^2 + 4y^2}} = \frac{(x, y, 0)}{2}.$$

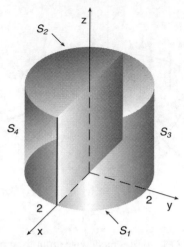

Figure 6.14   Surface S divided into four regions for Example 6.7.

We know that S_1 and S_2 project onto $S_{xy}: x^2 + y^2 \leq 4$ in the xy-plane, and S_3 and S_4 project onto the rectangle $S_{xz}: -2 \leq x \leq 2, 0 \leq z \leq 2$ in the xz-plane. Consequently,

$$\iint_S \mathbf{F} \cdot \mathbf{n} \, dS = \iint_{S_1} -z \, dS + \iint_{S_2} d \, dS + \iint_{S_3} \left(\frac{x^2 + y^2}{2} \right) dS$$

$$+ \iint_{S_4} \left(\frac{x^2 + y^2}{2} \right) dS$$

$$= 0 + \iint_{S_{xy}} 2 \, dA$$

$$+ \frac{1}{2} \iint_{S_{xz}} (x^2 + 4 - x^2) \sqrt{1 + \left(\frac{-x}{\sqrt{4 - x^2}} \right)^2} \, dA$$

$$+ \frac{1}{2} \iint_{S_{xz}} (x^2 + 4 - x^2) \sqrt{1 + \left(\frac{x}{\sqrt{4 - x^2}} \right)^2} \, dA$$

$$= 2 \iint_{S_{xy}} dA + 4 \iint_{S_{xz}} \frac{2}{\sqrt{4 - x^2}} \, dA$$

$$= 2(\text{area of } S_{xy}) + 8 \int_{-2}^{2} \int_{0}^{2} \frac{1}{\sqrt{4 - x^2}} \, dz \, dx$$

$$= 2(4\pi) + 16 \int_{-2}^{2} \frac{1}{\sqrt{4 - x^2}} \, dx.$$

If we set $x = 2 \sin \theta$, then $dx = 2 \cos \theta \, d\theta$ and

$$\iint_S \mathbf{F} \cdot \mathbf{n} \, dS = 8\pi + 16 \int_{-\pi/2}^{\pi/2} \frac{1}{2 \cos \theta} 2 \cos \theta \, d\theta$$

$$= 8\pi + 16 \{\theta\}_{-\pi/2}^{\pi/2} = 24\pi.$$

■

Example 6.8 If a spherical object is submerged in a fluid of density ρ, it experiences a buoyant force due to fluid pressure. Show that the magnitude of this force is exactly the weight of fluid displaced by the object.   □

Solution   Suppose the object is represented by the sphere $x^2 + y^2 + z^2 = R^2$ and the surface of the fluid by the plane $z = h$ ($h > R$) (see Fig. 6.15). If dS is a small area on the surface, then the force due to fluid pressure on dS has magnitude $P \, dS$, where P is pressure, and this force acts normal to the surface of the sphere. If $\mathbf{n}$ is the unit inward-pointing normal to the sphere, then the force on dS is

$$(P \, dS)\mathbf{n}.$$

Clearly, the resultant force on the sphere will be in the z-direction, the y-components cancelling because of the symmetry of the sphere. The z-component of the force on dS is $(P \, dS)\mathbf{n} \cdot \mathbf{k}$; thus the magnitude of the resultant force on the sphere is

$$\iint_S P \mathbf{k} \cdot \mathbf{n} \, dS.$$

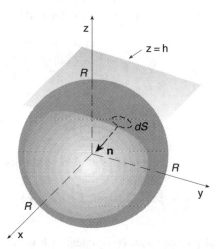

Figure 6.15   Object represented by the sphere and surface of fluid by a plane.

A normal to the surface is $\nabla(x^2 + y^2 + z^2 - R^2) = (2x, 2y, 2z)$ and, therefore, the unit inward-pointing normal is

$$\mathbf{n} = -\frac{(x, y, z)}{\sqrt{x^2 + y^2 + z^2}} = -\frac{1}{R}(x, y, z).$$

Using spherical coordinates (see Chapter 5) for an element on the sphere, and $P = 9.81\rho(h - z)$,

$$\iint_S P\mathbf{k} \cdot \mathbf{n} \, dS = \iint_S 9.81\rho(h - z)\left(-\frac{z}{R}\right) dS$$

$$= -\frac{9.81\rho}{R} \int_0^{2\pi} \int_0^\pi (h - R\cos\phi)(R\cos\phi)R^2 \sin\phi \, d\phi \, d\theta$$

$$= -9.81\rho R^2 \int_0^{2\pi} \left\{ -\frac{h}{2}\cos^2\phi + \frac{R}{3}\cos^3\phi \right\}_0^\pi d\theta$$

$$= \frac{2(9.81)\rho R^3}{3} \{\theta\}_0^{2\pi}$$

$$= \frac{4}{3}\pi R^3 (9.81\rho).$$

■

6.5 Green's theorem in vector form

Green's theorem of the preceding section has two vector forms. In this and the next section we shall generalize these forms to three dimensions.

If $\mathbf{F}(x, y) = P(x, y)\mathbf{i} + Q(x, y)\mathbf{j}$ is a two-dimensional vector field, then

$$\text{curl } \mathbf{F} = \nabla \times \mathbf{F} = \begin{vmatrix} \mathbf{i} & \mathbf{j} & \mathbf{k} \\ \dfrac{\partial}{\partial x} & \dfrac{\partial}{\partial y} & \dfrac{\partial}{\partial z} \\ P & Q & 0 \end{vmatrix}$$

$$= \left(\frac{\partial Q}{\partial x} - \frac{\partial P}{\partial y}\right)\mathbf{k}.$$

From Chapter 4, we can write Green's theorem in vector notation as

$$\oint_C \mathbf{F} \cdot d\mathbf{R} = \oint_C \mathbf{F} \cdot \mathbf{T} \, ds = \iint_R (\nabla \times \mathbf{F}) \cdot \mathbf{k} \, dA, \tag{6.7}$$

i.e., the line integral of the tangential component of $\mathbf{F}$ is the double integral of the normal component of the curl of $\mathbf{F}$ ($\nabla \times \mathbf{F}$).

6.6 Stokes' theorem in 3D

The vector form of Green's theorem given in (6.7) relates a line integral around a piecewise smooth simple closed curve C forming the boundary of a plane region R to a double integral over R. Green's theorem in 3-space (Fig. 6.16) relates a line integral around a piecewise smooth simple closed curve C forming the boundary of a surface S with a surface integral over S. Suppose $z = f(x, y)$ is a continuous function whose graph is a piecewise smooth orientable surface over a region R on the xy-plane.

Let C form the boundary of S and let the projection of C onto the xy-plane form the boundary of R. The positive direction on C is induced by the orientation of the surface S; the positive direction on C corresponds to the direction a person would have to walk on C to have his or her head pointing in the direction of the orientation of the surface while keeping the surface to the left (see Fig. 6.17). More precisely, the positive orientation of C is in accordance with the right-hand rule: If the thumb of the right hand points in the direction of the orientation of the surface, then roughly the fingers of the right hand wrap around the surface in the positive direction. Finally, let $\mathbf{T}$ be a unit tangent vector to C that points in the positive direction.

The three-dimensional form of Green's theorem is known as Stokes' theorem after the Irish mathematician and physicist George G. Stokes (1819-1903). Like George Green, George Stokes was a don at Cambridge University. In 1854, Stokes posed this theorem as a problem on a prize examination for Cambridge students.

Theorem 6.2 *Let C be a closed, piecewise-smooth curve that does not intersect itself and let S be a piecewise-smooth, orientable surface with C as boundary*

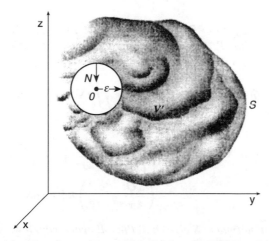

Figure 6.16 Singular point excluded from three-dimensional region.

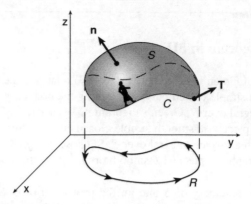

Figure 6.17 Region R showing orientation of surface.

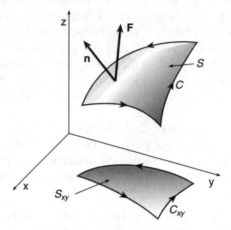

Figure 6.18 Piecewise smooth curve C that does not intersect itself.

(Fig. 6.18). Let $\mathbf{F}(x, y, z) = P(x, y, z)\mathbf{i} + Q(x, y, z)\mathbf{j} + R(x, y, z)\mathbf{k}$ be a vector field whose components P, Q, and R have continuous first partial derivatives in a domain that contains S and C. Then,

$$\oint_C \mathbf{F} \cdot d\mathbf{R} = \oint_C \mathbf{F} \cdot \mathbf{T}\, ds = \iint_S (\nabla \times \mathbf{F}) \cdot \mathbf{n}\, dS \qquad (6.8)$$

or

$$\oint_C P\, dx + Q\, dy + R\, dz = \iint_S \left\{ \left(\frac{\partial R}{\partial y} - \frac{\partial Q}{\partial z} \right)\mathbf{i} + \left(\frac{\partial P}{\partial z} - \frac{\partial R}{\partial x} \right)\mathbf{j} \right.$$
$$\left. + \left(\frac{\partial Q}{\partial x} - \frac{\partial P}{\partial y} \right)\mathbf{k} \right\} \cdot \mathbf{n}\, dS, \qquad (6.9)$$

where $\mathbf{n}$ is the unit normal to S chosen in the following way: If when moving along C the surface S is on the left, then $\mathbf{n}$ must be chosen as the unit normal on that side *of S. On the other hand if, when moving along C, the surface is on the right, then* **n** *must be chosen on the opposite side of S.*

Proof We first consider a surface S that projects in a one-to-one fashion onto each of the three coordinate planes. Because S projects one-to-one onto some region S_{xy} in the xy-plane, we can take the equation for S in the form $z = f(x, y)$. If the direction along C is as indicated in Fig. 6.18, and C_{xy} is the projection of C on the xy-plane, then

$$\oint_C P\,dx = \oint_{C_{xy}} P\,[x, y, f(x, y)]\,dx.$$

If we use Green's theorem on this line integral around C_{xy}, we have

$$\oint_C P\,dx = \iint_{S_{xy}} -\left\{\frac{\partial P}{\partial y} + \frac{\partial P}{\partial z}\frac{\partial z}{\partial y}\right\}dA.$$

On the other hand, since a unit normal to S is

$$\mathbf{n} = \frac{\left(-\dfrac{\partial z}{\partial x}, -\dfrac{\partial z}{\partial y}, 1\right)}{\sqrt{1 + \left(\dfrac{\partial z}{\partial x}\right)^2 + \left(\dfrac{\partial z}{\partial y}\right)^2}},$$

it also follows that

$$\iint_S \left(\frac{\partial P}{\partial z}\mathbf{j} - \frac{\partial P}{\partial y}\mathbf{k}\right)\cdot \mathbf{n}\,dS = \iint_S \frac{\left(-\dfrac{\partial P}{\partial z}\dfrac{\partial z}{\partial y} - \dfrac{\partial P}{\partial y}\right)}{\sqrt{1 + \left(\dfrac{\partial z}{\partial x}\right)^2 + \left(\dfrac{\partial z}{\partial y}\right)^2}}\,dS$$

$$= \iint_{S_{xy}} \frac{\left(-\dfrac{\partial P}{\partial z}\dfrac{\partial z}{\partial y} - \dfrac{\partial P}{\partial y}\right)}{\sqrt{1 + \left(\dfrac{\partial z}{\partial x}\right)^2 + \left(\dfrac{\partial z}{\partial y}\right)^2}}$$

$$\times \sqrt{1 + \left(\dfrac{\partial z}{\partial x}\right)^2 + \left(\dfrac{\partial z}{\partial y}\right)^2}\,dA$$

$$= \iint_{S_{xy}} -\left\{\frac{\partial P}{\partial y} + \frac{\partial P}{\partial z}\frac{\partial z}{\partial y}\right\}dA.$$

We have shown, then, that

$$\oint_C P\,dx = \iint_S \left(\frac{\partial P}{\partial z}\mathbf{j} - \frac{\partial P}{\partial y}\mathbf{k}\right)\cdot \mathbf{n}\,dS.$$

By projecting C and S onto the xz- and yz-planes, we can show similarly that

$$\oint_C R \, dz = \iint_S \left(\frac{\partial P}{\partial y} \mathbf{i} - \frac{\partial R}{\partial x} \mathbf{j} \right) \cdot \mathbf{n} \, dS$$

and

$$\oint_C Q \, dy = \iint_S \left(\frac{\partial Q}{\partial x} \mathbf{k} - \frac{\partial Q}{\partial z} \mathbf{i} \right) \cdot \mathbf{n} \, dS.$$

Addition of these results gives Stokes' theorem for $\mathbf{F}$ and S and completes the proof. ∎

The proof can be extended to more general curves and surfaces that do not project in a one-to-one fashion onto all three coordinate planes. Most surfaces S with bounding curves C can be subdivided into n subsurfaces S_i with bounding curves C_i that do satisfy this condition (Fig. 6.19). For each such subsurface, Stokes' theorem applies

$$\oint_{C_i} \mathbf{F} \cdot d\mathbf{R} = \iint_{S_i} (\nabla \times \mathbf{F}) \cdot \mathbf{n} \, dS, \quad i = 1, \ldots, n.$$

If these n equations are now added together, we have

$$\sum_{i=1}^{n} \oint_{C_i} \mathbf{F} \cdot d\mathbf{R} = \sum_{i=1}^{n} \iint_{S_i} (\nabla \times \mathbf{F}) \cdot \mathbf{n} \, dS.$$

Since the S_i constitute S, the right-hand side of this equation is the surface integral of $(\nabla \times \mathbf{F}) \cdot \mathbf{n}$ over S. Figure 6.19 illustrates that when line integrals over C_i are added, contributions from interior curves cancel in pairs, and the remaining line integrals give the line integral $\mathbf{F} \, d\mathbf{r}$ along C. For general surfaces that cannot be

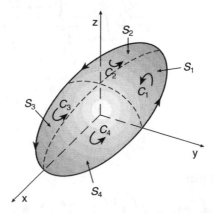

Figure 6.19   Line integrals over C_i cancelling out in pairs.

divided into a finite number of subsurfaces of this type, the reader should consult a more advanced text.

Green's theorem is a special case of Stokes' theorem. If $\mathbf{F}(x, y) = P(x, y)\mathbf{i} + Q(x, y)\mathbf{j}$, and C is a closed curve in the xy-plane, then by Stokes' theorem

$$\oint_C P\,\mathrm{d}x + Q\,\mathrm{d}y = \iint_S \left(\frac{\partial Q}{\partial x} - \frac{\partial P}{\partial y}\right)\mathbf{k}\cdot\mathbf{n}\,\mathrm{d}S,$$

where S is any surface for which C is the boundary. If we choose S as that part of the xy-plane bounded by C, then $\mathbf{n} = \mathbf{k}$ and

$$\oint_C P\,\mathrm{d}x + Q\,\mathrm{d}y = \iint_S \left(\frac{\partial Q}{\partial x} - \frac{\partial P}{\partial y}\right)\mathrm{d}A.$$

Suppose that the curl of a vector field $\mathbf{F}$ vanishes in a simply connected domain D. If C is any piecewise-smooth, closed curve in D, then there exists a piecewise-smooth surface S in D with C as boundary. By Stokes' theorem

$$\oint_C \mathbf{F}\cdot\mathrm{d}\mathbf{R} = \iint (\nabla \times \mathbf{F})\cdot\mathbf{n}\,\mathrm{d}S = 0.$$

This implies that the line integral is independent of path in D, and that there exists a function $f(x, y, z)$ such that $\nabla f = \mathbf{F}$.

Example 6.9 Verify Stokes' theorem if $\mathbf{F} = x^2\mathbf{i} + x\mathbf{j} + xyz\mathbf{k}$, and S is that part of the sphere $x^2 + y^2 + z^2 = 4$ above the plane $z = 1$.   □

Solution If we choose $\mathbf{n}$ as the upper normal to s, then C, the boundary of S, must be traversed in the direction shown in Fig. 6.20. (If $\mathbf{n}$ is chosen as the lower normal,

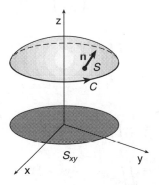

Figure 6.20 Surface S with upper normal $\mathbf{n}$ for Example 6.9.

then C must be traversed in the opposite direction). Since parametric equations for C are

$$x\sqrt{3}\cos t, \quad y = \sqrt{3}\sin t, \quad z = 1, \quad \text{and} \quad 0 \leq t \leq 2\pi,$$

we have

$$\oint_C \mathbf{F} \cdot d\mathbf{r} = \oint_C x^2\,dx + x\,dy + xyz\,dz$$

$$= \int_0^{2\pi} \left\{ 3\cos^2 t(-\sqrt{3}\sin t\,dt) + \sqrt{3}\cos t(\sqrt{3}\cos t\,dt) \right\}$$

$$= \int_0^{2\pi} \left\{ -3\sqrt{3}\cos^2 t\,\sin t + \frac{3}{2}(1 + \cos 2t) \right\}\,dt$$

$$= \left\{ \sqrt{3}\cos^2 t + \frac{3t}{2} + \frac{3\sin 2t}{4} \right\}_0^{2\pi}$$

$$= 3\pi.$$

Alternatively,

$$\iint_S (\nabla \times \mathbf{F}) \cdot \mathbf{n}\,dS = \iint_S (xz\mathbf{i} - yz\mathbf{j} + \mathbf{k}) \cdot \frac{(2x, 2y, 2z)}{\sqrt{4x^2 + 4y^2 + 4z^2}}\,dS$$

$$= \iint_S \frac{x^2 - y^2 z + z}{\sqrt{x^2 + y^2 + z^2}}\,dS$$

$$= \iint_{S_{xy}} \frac{z(x^2 - y^2 + 1)}{2}$$

$$\times \sqrt{1 + \left(\frac{\partial z}{\partial x}\right)^2 + \left(\frac{\partial z}{\partial y}\right)^2}\,dA$$

$$= \frac{1}{2}\iint_{S_{xy}} z(x^2 - y^2 + 1)$$

$$\times \sqrt{1 + (-x/z)^2 + (-y/z)^2}\,dA$$

$$= \frac{1}{2}\iint_{S_{xy}} z(x^2 - y^2 + 1)\frac{\sqrt{x^2 + y^2 + z^2}}{z}\,dA$$

$$= \iint_{S_{xy}} (x^2 - y^2 + 1)\,dA.$$

If we use polar coordinates to evaluate this double integral over $S_{xy}: x^2 + y^2 \le 3$, we have

$$\iint_S (\nabla \times \mathbf{F}) \cdot \mathbf{n} \, dS = \int_{-\pi}^{\pi} \int_0^{\sqrt{3}} (r^2 \cos^2 \theta - r^2 \sin^2 \theta + 1) r \, dr \, d\theta$$

$$= \int_{-\pi}^{\pi} \left\{ \frac{r^4}{4} (\cos^2 \theta - \sin^2 \theta) + \frac{r^2}{2} \right\}_0^{\sqrt{3}} d\theta$$

$$= \int_{-\pi}^{\pi} \left\{ \frac{9}{4} \cos 2\theta + \frac{3}{2} \right\} d\theta$$

$$= \left\{ \frac{9}{8} \sin 2\theta + \frac{3\theta}{2} \right\}_{-\pi}^{\pi} = 3\pi.$$

∎

Example 6.10   Let S be the part of the cylinder $z = 1 - x^2$ for $0 \le x \le 1$, $-2 \le y \le 2$. Verify Stokes' theorem if $\mathbf{F} = xy\mathbf{i} + yz\mathbf{j} + xz\mathbf{k}$.   □

Solution   The surface S, the curve C (which is composed of the union of C_1, C_2, C_3, and C_4), and the region R are shown in Fig. 6.21.

The surface integral: From $\mathbf{F} = xy\mathbf{i} + yz\mathbf{j} + xz\mathbf{k}$, we find

$$\nabla \times \mathbf{F} = \begin{vmatrix} \mathbf{i} & \mathbf{j} & \mathbf{k} \\ \dfrac{\partial}{\partial x} & \dfrac{\partial}{\partial y} & \dfrac{\partial}{\partial z} \\ xy & yz & xz \end{vmatrix} = -y\mathbf{i} - z\mathbf{j} - x\mathbf{k}.$$

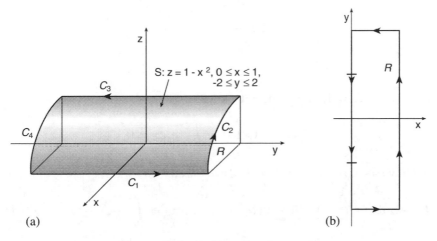

(a)   (b)

Figure 6.21   Surface S and curve C.

Now, if $g(x, y, z) = z + x^2 - 1 = 0$ defines the cylinder, then the upper normal is

$$\mathbf{n} = \frac{\nabla g}{|\nabla g|} = \frac{2x\mathbf{i} + \mathbf{k}}{\sqrt{4x^2 + 1}}.$$

Therefore,

$$\iint_S (\nabla \times \mathbf{F}) \cdot \mathbf{n} \, dS = \iint_S \frac{(-2xy - x)}{\sqrt{4x^2 + 1}} \, dS.$$

To evaluate the latter surface integral, we have the following (see Chapter 5):

$$\iint_S \frac{(-2xy - x)}{\sqrt{4x^2 + 1}} \, dS = \iint_R (-2xy - x) \, dA$$

$$= \int_0^1 \int_{-2}^2 (-2xy - x) \, dy \, dx$$

$$= \int_0^1 \left(-xy^2 - xy \right)\Big|_{-2}^2 \, dx$$

$$= \int_0^1 (-4x) \, dx = -2. \qquad (6.10)$$

The line integral: We write $\oint_C = \int_{C_1} + \int_{C_2} + \int_{C_3} + \int_{C_4}$. On $C_1: x = 0, z = 0$, $dx = 0, dz = 0$, so

$$\int_{C_1} y(0) + y(0)dy + 0 = 0.$$

On $C_2: y = 2, z = 1 - x^2, dy = 0, dz = -2xdx$, so

$$\int_{C_2} 2x \, dx + 2(1 - x^2)(0) + x(1 - x^2)(-2x \, dx)$$

$$= \int_1^0 (2x - 2x^2 + 2x^4) \, dx$$

$$= -\frac{11}{15}.$$

On $C_3: x = 0, z = 1, dx = 0, dz = 0$, so

$$\int_{C_3} 0 + y \, dy + 0 = \int_2^{-2} y \, dy = 0.$$

On $C_4: y = -2, z = 1 - x^2, dy = 0, dz = -2x \, dx$, so

$$\int_{C_4} -2x \, dx - 2(1 - x^2)(0) + x(1 - x^2)(-2x \, dx) = \int_0^1 (-2x - 2x^2 + 2x^4) \, dx$$

$$= -\frac{19}{15}.$$

Hence,

$$\oint_C xy \, dx + yz \, dy + xz \, dz = 0 - \frac{11}{15} + 0 - \frac{19}{15} = -2,$$

which of course, agrees with (6.11).   ■

Example 6.11   Evaluate the integral

$$\oint_C 2xy^3 \, dx + 3x^2 y^2 \, dy + (2z + z) \, dz,$$

where C consists of line segments joining $A(2, 0, 0)$ to $B(0, 1, 0)$ to $D(0, 0, 1)$
to A.   □

Solution   By Stokes' theorem,

$$\oint_C 2xy^3 \, dx + 3x^2 y^2 \, dy + (2z + x) \, dz = \iint_S \nabla \times (2xy^3, 3x^2 y^2, 2z + x) \cdot \mathbf{n} \, dS,$$

where S is any surface with C as boundary. If we choose S as the flat triangle
bounded by C (Fig. 6.22), then a normal vector to S is

$$\mathbf{BD} \times \mathbf{BA} = \begin{vmatrix} \mathbf{i} & \mathbf{j} & \mathbf{k} \\ 0 & -1 & 1 \\ 2 & -1 & 0 \end{vmatrix} = (1, 2, 2)$$

and, therefore,

$$\mathbf{n} = \frac{(1, 2, 2)}{3}.$$

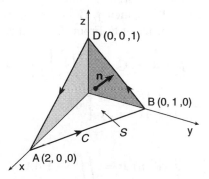

Figure 6.22   Flat triangle S bounded by C.

Since

$$\nabla \times (2xy^3, 3x^2y^2, 2z + x) = \begin{vmatrix} \mathbf{i} & \mathbf{j} & \mathbf{k} \\ \dfrac{\partial}{\partial x} & \dfrac{\partial}{\partial y} & \dfrac{\partial}{\partial z} \\ 2xy^3 & 3x^2y^2 & 2z + x \end{vmatrix} = -\mathbf{j},$$

it follows that

$$\nabla \times (2xy^3, 3x^2y^2, 2z + x) \cdot \mathbf{n} = -\mathbf{j} \cdot \frac{(1, 2, 2)}{3} = -\frac{2}{3}$$

and

$$\oint_C 2xy^3 \, dx + 3x^2y^2 \, dy + (2z + x) \, dz = \iint_S -\frac{2}{3} dS = -\frac{2}{3}(\text{area of } S).$$

The area of triangle S is given by

$$\frac{1}{2} |\mathbf{BD} \times \mathbf{BA}| = \frac{1}{2} |(1, 2, 2)| = \frac{3}{2}.$$

Finally,

$$\oint_C 2xy^3 \, dx + 3x^2y^2 \, dy + (2z + x) \, dz = -\frac{2}{3} \left(\frac{3}{2} \right) = -1.$$

■

6.7 The divergence theorem

6.7.1 Another vector form of Green's theorem

Let $\mathbf{F} = P\mathbf{i} + Q\mathbf{j}$ and let $\mathbf{T} = (dx/ds)\mathbf{i} + (dy/ds)\mathbf{j}$ be a unit tangent to a simple closed plane curve C. In the last section we saw that $\oint_C (\mathbf{F} \cdot \mathbf{T}) dS$ can be evaluated by a double integral involving the curl of $\mathbf{F}$. Similarly, if $\mathbf{n} = (dy/ds)\mathbf{i} - (dx/ds)\mathbf{j}$ is a unit normal to C (check $\mathbf{T} \cdot \mathbf{n}$), then $\oint_C (\mathbf{F} \cdot \mathbf{n}) dS$ can be expressed in terms of a double integral of the divergence of $\mathbf{F}$. From Green's theorem,

$$\oint_C (\mathbf{F} \cdot \mathbf{n}) \, ds = \oint_C P \, dy - Q \, dx = \iint_R \left\{ \frac{\partial P}{\partial x} - \left(-\frac{\partial Q}{\partial y} \right) \right\} dA$$

$$= \iint_R \left\{ \frac{\partial P}{\partial x} + \frac{\partial Q}{\partial y} \right\} dA,$$

i.e.

$$\oint_C (\mathbf{F} \cdot \mathbf{n}) \, ds = \iint_R \nabla \cdot \mathbf{F} \, dA. \qquad (6.11)$$

The result in (6.11) is a special case of the *divergence* or *Gauss' theorem*.

The divergence theorem relates certain surface integrals that enclose volumes to triple integrals over the enclosed volume. More precisely, we have the following.

Theorem 6.3   *Let S be a piecewise-smooth surface enclosing a region V (Fig. 6.23). Let* $\mathbf{F}(x, y, z) = L(x, y, z)\mathbf{i} + M(x.y.z)\mathbf{j} + N(x, y, z)\mathbf{k}$ *be a vector field whose components L, M and N have continuous first partial derivatives in a domain containing S and V. If* $\mathbf{n}$ *is the unit outer normal to S, then*

$$\iint_S \mathbf{F} \cdot \mathbf{n} \, dS = \iiint_V \nabla \cdot \mathbf{F} \, dV. \tag{6.12}$$

or writing the above formula explicitly, we obtain

$$\iint_S (L\mathbf{i} + M\mathbf{j} + N\mathbf{k}) \cdot \mathbf{n} \, dS = \iiint_V \left(\frac{\partial L}{\partial x} + \frac{\partial M}{\partial y} + \frac{\partial N}{\partial z} \right) dV. \tag{6.13}$$

Proof   We first consider a surface S for which any line parallel to any coordinate axis intersects S in at most two points (Fig. 6.24). We can then divide S into an upper and a lower portion, $S_2: z = f_2(x, y)$ and $S_1: z = f_1(x, y)$, both of which have the same projection S_{xy} in the xy-plane. We consider the third term in the surface integral on the left-hand side of (6.13):

$$\iint_R N\mathbf{k} \cdot \mathbf{n} \, dS = \iint_{S_1} N\mathbf{k} \cdot \mathbf{n} \, dS + \iint_{S_2} N\mathbf{k} \cdot \mathbf{n} \, dS. \tag{6.14}$$

On S_1,

$$\mathbf{n} = \frac{\left(\dfrac{\partial f_1}{\partial x}, \dfrac{\partial f_1}{\partial y}, -1 \right)}{\sqrt{1 + \left(\dfrac{\partial f_1}{\partial x} \right)^2 + \left(\dfrac{\partial f_1}{\partial y} \right)^2}}.$$

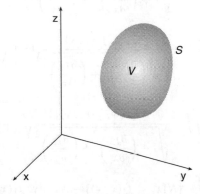

Figure 6.23   Piecewise smooth surface V and surface S.

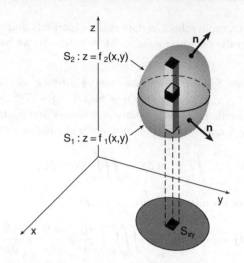

Figure 6.24 Addition of surface integrals S_i.

On S_2,

$$\mathbf{n} = \frac{\left(\dfrac{\partial f_2}{\partial x}, \dfrac{\partial f_2}{\partial y}, -1\right)}{\sqrt{1 + \left(\dfrac{\partial f_2}{\partial x}\right)^2 + \left(\dfrac{\partial f_2}{\partial y}\right)^2}},$$

consequently,

$$
\iint_S N\mathbf{k} \cdot \mathbf{n}\, dS = \iint_{S_1} \frac{-N}{\sqrt{1 + \left(\dfrac{\partial f_1}{\partial x}\right)^2 + \left(\dfrac{\partial f_1}{\partial y}\right)^2}}\, dS
$$

$$
+ \iint_{S_2} \frac{N}{\sqrt{1 + \left(\dfrac{\partial f_2}{\partial x}\right)^2 + \left(\dfrac{\partial f_2}{\partial y}\right)^2}}\, dS
$$

$$
= \iint_{S_{xy}} \frac{-N\{x, y, f_1(x, y)\}}{\sqrt{1 + \left(\dfrac{\partial f_1}{\partial x}\right)^2 + \left(\dfrac{\partial f_1}{\partial y}\right)^2}} \sqrt{1 + \left(\dfrac{\partial f_1}{\partial x}\right)^2 + \left(\dfrac{\partial f_1}{\partial y}\right)^2}\, dA
$$

$$
+ \iint_{S_{xy}} \frac{N\{x, y, f_2(x, y)\}}{\sqrt{1 + \left(\dfrac{\partial f_2}{\partial x}\right)^2 + \left(\dfrac{\partial f_2}{\partial y}\right)^2}} \sqrt{1 + \left(\dfrac{\partial f_2}{\partial x}\right)^2 + \left(\dfrac{\partial f_2}{\partial y}\right)^2}\, dA
$$

$$
= \iint_{S_{xy}} \{N[x, y, f_2(x, y)] - N[x, y, f_1(x, y)]\}\, dA.
$$

On the other hand,

$$\iiint_V \frac{\partial N}{\partial z}\, dV = \iint_{S_{xy}} \left\{ \int_{f_1(x,y)}^{f_2(x,y)} \frac{\partial N}{\partial z}\, dz \right\} dA$$

$$= \iint_{S_{xy}} \{N\}_{f_1(x,y)}^{f_2(x,y)}\, dA$$

$$= \iint_{S_{xy}} \{N[x, y, f_2(x, y)] - N[x, y, f_1(x, y)]\}\, dA.$$

We have just shown that

$$\iint_S N\mathbf{k} \cdot \mathbf{n}\, dS = \iiint_V \frac{\partial N}{\partial z}\, dV.$$

Projections of S onto the xz- and yz-planes lead in a similar way to

$$\iint_S M\mathbf{j} \cdot \mathbf{n}\, dS = \iiint_V \frac{\partial M}{\partial y}\, dV$$

$$\text{and } \iint_S L\mathbf{i} \cdot \mathbf{n}\, dS = \iiint_V \frac{\partial L}{\partial x}\, dV.$$

By adding these three results, we obtain the divergence theorem for $\mathbf{F}$ and S.   ■

We can extend the proof to general surfaces for which lines parallel to the coordinate axes intersect the surfaces in more than two points. Most volumes V bounded by surfaces S can be subdivided into n sub-volumes V_i whose bounding surfaces S_i do satisfy this condition (Fig. 6.24).

For each such sub-volume, the divergence theorem is now known to apply:

$$\iint_{S_i} \mathbf{F} \cdot \mathbf{n}\, dS = \iiint_{V_i} \nabla \cdot \mathbf{F} dV, \quad i = 1, \ldots, n.$$

If these n equations are then added together, we have

$$\sum_{i=1}^n \iint_{S_i} \mathbf{F} \cdot \mathbf{n}\, dS = \sum_{i=1}^n \iiint_{V_i} \nabla \cdot \mathbf{F} dV.$$

The right-hand side is the triple integral of $\nabla \cdot \mathbf{F}$ over V since the V_i constitute V.

Figure 6.24 illustrates that when surface integrals over the S_i are added, contributions from interior surfaces cancel out in pairs, and the remaining surface integrals add to give the surface integral of $\mathbf{F} \cdot \mathbf{n}$ over S. The proof for more general surfaces that cannot be subdivided into a finite number of subsurfaces of this type is left for more advanced texts.

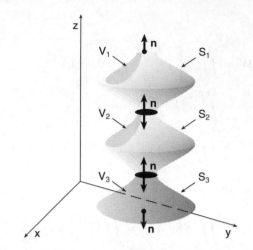

Figure 6.25 The closed region for Example 6.12.

Example 6.12 Let D be the region bounded by the hemisphere $x^2 + y^2 + (z-1)^2 = 9$, $1 \le z \le 4$, and the plane $z = 1$. Verify the divergence theorem if $\mathbf{F} = x\mathbf{i} + y\mathbf{j} + (z-1)\mathbf{k}$. □

Solution The closed region is shown in Fig. 6.25.

The triple integral: Since $\mathbf{F} = x\mathbf{i} + y\mathbf{j} + (z-1)\mathbf{k}$, we see that $\nabla \cdot \mathbf{F} = 3$. Hence,

$$\iiint_D \nabla \cdot \mathbf{F} \, dV = \iiint_D 3 \, dV = 3 \iiint_D dV$$

$$= 3\left(\frac{2}{3}\pi (3)^3\right) = 54\pi. \qquad (6.15)$$

In the last calculation, we have used the fact that $\iiint_D dV$ gives the volume of the hemisphere.

The surface integral: We write $\iint_S = \iint_{S_1} + \iint_{S_2}$, where S_1 is the hemisphere and S_2 is the plane $z = 1$. If S_1 is a level surface of $g(x, y, z) = x^2 + y^2 + (z-1)^2$, then a unit outer normal is

$$\mathbf{n} = \frac{\nabla g}{|\nabla g|} = \frac{x\mathbf{i} + y\mathbf{j} + (z-1)\mathbf{k}}{\sqrt{x^2 + y^2 + z^2}} = \frac{x}{3}\mathbf{i} + \frac{y}{3}\mathbf{j} + \frac{(z-1)}{3}\mathbf{k}.$$

Now

$$\mathbf{F} \cdot \mathbf{n} = \frac{x^2}{3} + \frac{y^2}{3} + \frac{(z-1)^2}{3} = 3$$

and so

$$\iint_{S_1} (\mathbf{F} \cdot \mathbf{n}) \, dS = \iint_R (3) \left(\frac{3}{\sqrt{9 - x^2 - y^2}} \, dA \right)$$

and using polar coordinates,

$$\iint_R (3) \left(\frac{3}{\sqrt{9 - x^2 - y^2}} \, dA \right) = 9 \int_0^{2\pi} \int_0^3 (9 - r^2)^{-1/2} \, r \, dr \, d\theta$$
$$= 54\pi.$$

On S_2, we take $\mathbf{n} = -\mathbf{k}$ so that $\mathbf{F} \cdot \mathbf{n} = -z + 1$. But since $z = 1$,

$$\iint_{S_2} (-z + 1) \, dS = 0.$$

Hence, we see that

$$\iint_R (\mathbf{F} \cdot \mathbf{n}) \, dS = 54\pi + 0 = 54\pi$$

which agrees with (6.16). ∎

Example 6.13   If $\mathbf{F} = xy\mathbf{i} + y^2 z\mathbf{j} + z^3\mathbf{k}$, evaluate $\iint_S (\mathbf{F} \cdot \mathbf{n}) \, dS$, where S is the unit cube defined by $0 \le x \le 1, 0 \le y \le 1, 0 \le z \le 1$ (see Fig. 6.26)   □

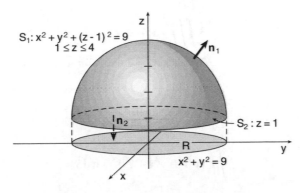

Figure 6.26   The unit cube S for Example 6.13.

Solution Rather than evaluate six surface integrals, we apply the divergence theorem. Then we need $\nabla \cdot \mathbf{F} = y + 2xy + 3z^2$, so that

$$\iint_S (\mathbf{F} \cdot \mathbf{n}) \, dS = \iiint_D (y + 2yz + 3z^2) \, dV$$

$$= \int_0^1 \int_0^1 \int_0^1 (y + 2yz + 3z^2) \, dx \, dy \, dz$$

$$= \int_0^1 \int_0^1 (y + 2yz + 3z^2) \, dy \, dz$$

$$= \int_0^1 \left(\frac{y^2}{2} + y^2 z + 3yz^2 \right) \Big|_0^1 \, dz$$

$$= \int_0^1 \left(\frac{1}{2} + z + 3z^2 \right) \, dz$$

$$= \left(\frac{1}{2} z + \frac{z^2}{2} + z^3 \right) \Big|_0^1 = 2.$$

■

Example 6.14 Use the divergence theorem to evaluate the surface integral of Example 6.7. □

Solution By the divergence theorem (see Fig. 6.14), we have

$$\iint_S \mathbf{F} \cdot \mathbf{n} \, dS = \iiint_V \nabla \cdot \mathbf{F} \, dV$$

$$= \iiint_V (1 + 1 + 1) \, dV$$

$$= 3 \iiint_V dV$$

$$= 3(\text{Volume of } V) = 3(4\pi)(2) = 24\pi.$$

■

Example 6.15 Evaluate

$$\iint_S (x^3 \mathbf{i} + y^3 \mathbf{j} + z^3 \mathbf{k}) \cdot \mathbf{n} \, dS$$

where $\mathbf{n}$ is the unit inner normal to the surface bounding the volume defined by the surfaces $y = x$, $z = 0$, $z = 2$, $y = 0$ and $x = 4$. □

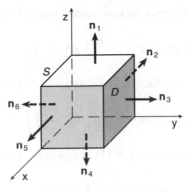

Figure 6.27   Volume defined by surfaces given in Example 6.15.

Solution   By the divergence theorem (see Fig. 6.27), we have

$$\iint_S (x^3\mathbf{i} + y^3\mathbf{j} + z^3\mathbf{k}) \cdot \mathbf{n}\, dS = -\iiint_V \nabla \cdot (x^3\mathbf{i} + y^3\mathbf{j} + z^3\mathbf{k})\, dV$$

$$= -3\int_0^4 \int_0^x \int_0^2 (x^2 + y^2 + z^2)\, dz\, dy\, dx$$

$$= -3\int_0^4 \int_0^x \left\{ (x^2 + y^2)z + \frac{z^3}{3} \right\}_0^2 dy\, dx$$

$$= -2\int_2^4 \int_0^x \left\{ 3(x^2 + y^2) + 4 \right\} dy\, dx$$

$$= -2\int_0^4 \left\{ y^3 + (3x^2 + 4)y \right\}_0^x dx$$

$$= -2\int_0^4 (4x^3 + 4x)\, dx = -2 \left(x^4 + 2x^2 \right)\Big|_0^4$$

$$= -576.$$

∎

Remark 6.2   Various important theorems stem from the divergence theorem. For instance, if u and v are two sufficiently differentiable scalar point functions and if we let

$$\mathbf{F} = u\nabla v,$$

then by application of the vector formulae in Chapter 4, we have

$$\nabla \cdot \mathbf{F} = \nabla \cdot (u\nabla v)$$

$$= u\nabla \cdot \nabla v + \nabla u \cdot \nabla v$$

$$= \nabla u \cdot \nabla v + u\nabla^2 v.$$

Hence, applying the divergence theorem to the vector $\mathbf{F} = u\nabla v$, we have

$$\iiint_V \left(\nabla u \cdot \nabla v + u\nabla^2 v \right) \, dV = \iint_S \mathbf{n} \cdot u\nabla v \, dS. \tag{6.16}$$

Similarly, if we interchange the roles of u and v in (6.16), we obtain

$$\iiint_V \left(\nabla v \cdot \nabla u + v\nabla^2 u \right) \, dV = \iint_S \mathbf{n} \cdot v\nabla u \, dS. \tag{6.17}$$

Finally, if we subtract (6.16) from (6.17), we obtain Green's Theorem

$$\iiint_V \left(u\nabla^2 v - v\nabla^2 u \right) \, dV = \iint_S \mathbf{n} \cdot (u\nabla v - v\nabla u) \, dS, \tag{6.18}$$

where $\mathbf{n}$ is the outward normal to the surface S which bounds the volume V. Green's Theorem in the form (6.18), is of great importance in many areas of research. Some specific areas are in the development of the boundary element method, diffraction of water waves by offshore structures and many other physical problems. ■

6.8 Summary

- *Definite integrals in one dimension*: The basic definite integral of calculus is: $\int_a^b (df/dx)dx = f(b) - f(a)$, where $f(x)$ is a function of x alone.
- *Line integrals in space*: If the vector field $\mathbf{F}$ is conservative, that is, $\nabla \times \mathbf{F} = 0$ such that $\mathbf{F} = \nabla\Phi$ where Φ is a scalar function, then the scalar line integral of this conservative field vector is $\int_P^Q \mathbf{F} \cdot d\mathbf{R} = \Phi(Q) - \Phi(P)$. This result can be extended to a **vector line integral**, defined by

$$\int_P^Q \frac{d\mathbf{F}}{ds} ds = \mathbf{F}(Q) - \mathbf{F}(P).$$

- *Green's theorem in two dimensions*: This theorem states as follows: Let C be a closed contour that surrounds the two dimensional region S. Let $\mathbf{F}(x, y) = \mathbf{i}F_1(x, y) + \mathbf{j}F_2(x, y)$ be a vector field whose components F_1 and F_2 have continuous first partial derivatives in a domain that contains S and C. Then Green's theorem states that

$$\oint_C \mathbf{F} \cdot d\mathbf{R} = \oint_C \mathbf{F} \cdot \mathbf{T} ds = \iint_S \nabla \times \mathbf{F} \cdot \mathbf{k} dS.$$

This integral theorem can be explicitly written as

$$\oint_C F_1 dx + F_2 dy = \iint_S \left\{ \frac{\partial F_2}{\partial x} - \frac{\partial F_1}{\partial y} \right\} dx dy.$$

- *Stokes' theorem in three dimensions*: Let C be a closed piecewise-smooth curve that does not intersect itself and let S be a piecewise-smooth surface with C as the boundary. Let $\mathbf{F}(x, y, z) = \mathbf{i}F_1(x, y, z) + \mathbf{j}F_2(x, y, z) + \mathbf{k}F_3(x, y, z)$ be a vector field whose components F_1, F_2, and F_3 have continuous first partial derivatives in a domain that contains S and C. Then Stokes theorem states that

$$\oint_C \mathbf{F} \cdot d\mathbf{R} = \oint_C \mathbf{F} \cdot \mathbf{T} ds = \iint_S \nabla \times \mathbf{F} \cdot \mathbf{n} dS.$$

Here $\mathbf{n}$ is the outward unit normal from the surface S.

- *Divergence theorem*: Let S be a piecewise-smooth surface enclosing a region V. Let $\mathbf{F}(x, y, z) = \mathbf{i}F_1 + \mathbf{j}F_2 + \mathbf{k}F_3$ be a vector field whose components are F_1, F_2 and F_3 have continuous first partial derivatives in a domain containing S and V. If $\mathbf{n}$ is the outward unit normal vector to S, then the divergence theorem states that

$$\iint_S \mathbf{F} \cdot \mathbf{n} dS = \iiint_V \nabla \cdot \mathbf{F} dV.$$

Remark 6.3 If u and v are two sufficiently differentiable scalar point functions, and if we let $\mathbf{F} = u\nabla v$, then, by applying the vector formulae, we obtain $\nabla \cdot \mathbf{F} = \nabla u \cdot \nabla v + u\nabla^2 v$. Then the divergence theorem can be obtained in another important form

$$\iiint_V (u\nabla^2 v - v\nabla^2 u)dV = \iint_S (u\nabla v - v\nabla u) \cdot \mathbf{n} dS.$$

This theorem has enormous applications in many practical problems. ∎

Exercises

1. Evaluate the line integral by two methods: (i) directly and (ii) using Green's theorem.
   (a) $\oint_C x^2 y dx + xy^3 dy$, C is the square with vertices $(0, 0)$, $(1, 0)$, and $(0, 1)$.
   (b) $\oint_C x dx - x^2 y^2 dy$, C is the triangle with vertices $(0, 0)$, $(1, 1)$, and $(0, 1)$.
   (c) $\oint_C (x + 2y)dx + (x - 2y)dy$, C consists of the arc of the parabola $y = x^2$ from $(0, 0)$ to $(1, 1)$ followed by the line segment from $(1, 1)$ to $(0, 0)$.
   (d) $\oint_C (x^2 + y^2)dx + 2xy dy$, C consists of the arc of the parabola $y = x^2$ from $(0, 0)$ to $(2, 4)$ and the line segments from $(2, 4)$ to $(0, 4)$ and from $(0, 4)$. to $(0, 0)$

2. Use Green's theorem to evaluate the line integral along the given positively oriented curve.
   (a) $\oint_C xy dx + y^5 dy$, C is the triangle with vertices $(0, 0)$, $(2, 0)$, and $(2, 1)$.
   (b) $\oint_C (y + e^{\sqrt{x}})dx + (2x + \cos y^2)dy$, C is the boundary of the region enclosed by the parabolas $y = x^2$ and $x = y^2$.

(c) $\oint_C (y - \tan^{-1} x)dx + (3x + \sin y)dy$, C is the boundary of the region enclosed by the parabola $y = x^2$ and the line $y = 4$.

(d) $\oint_C x^2 dx + y^2 dy$, C is the curve $x^6 + y^6 = 1$.

(e) $\oint_C 2xy dx + x^2 dy$, C is the cardioid $r = 1 + \cos\theta$.

(f) $\oint_C (x^3 - y^3)dx + (x^3 + y^3)dy$, C is the boundary of the region between the circles $x^2 + y^2 = 1$ and $x^2 + y^2 = 9$.

(g) $\int_C \mathbf{F} \cdot d\mathbf{r}$, where $\mathbf{F}(x, y) = (y^2 - x^2 y)\mathbf{i} + xy\mathbf{j}$ and C consists of the circle $x^2 + y^2 = 4$ from $(2, 0)$ to $(\sqrt{2}, \sqrt{2})$ to $(0, 0)$ and from $(0, 0)$ to $(2, 0)$.

(h) $\int_C \mathbf{F} \cdot d\mathbf{r}$, where $\mathbf{F}(x, y) = x^3 y\mathbf{i} + x^4\mathbf{j}$ and C is the curve $x^4 + y^4 = 1$.

3. Use Green's theorem to find the work done by the force $\mathbf{F}(x, y) = x(x + y)\mathbf{i} + xy^2\mathbf{j}$ in moving a particle from the origin along the x-axis to $(1, 0)$, then along the line segment to $(0, 1)$, and then back to the origin along the y-axis.

4. A particle starts at the point $(-2, 0)$, moves along the x-axis to $(2, 0)$, and then along the semi-circle $y = \sqrt{4 - x^2}$ to the starting point. Use Green's theorem to find the work done on this particle by the force field $\mathbf{F}(x, y) = (x, x^3 + 3xy^2)$.

5. Use Stokes' theorem to evaluate $\iint_S \nabla \times \mathbf{F} \cdot d\mathbf{S}$.

(a) $\mathbf{F}(x, y, z) = xyz\mathbf{i} + xj + e^{xy}\cos z\mathbf{k}$, S is the hemisphere $x^2 + y^2 + z^2 = 1, 0 \le z$, oriented upward.

(b) $\mathbf{F}(x, y, z) = y^2 z\mathbf{i} + xz\mathbf{j} + x^2 y^2\mathbf{k}$, S is the part of the paraboloid $z = x^2 + y^2$ that lies inside the cylinder $x^2 + y^2 = 1$, oriented upward.

(c) $\mathbf{F}(x, y, z) = yz^3\mathbf{i} + \sin(xyz)\mathbf{j} + x^3\mathbf{k}$, S is the part of the paraboloid $y = 1 - x^2 - z^2$ that lies to the right of the xz-plane, oriented towards the xz-plane.

(d) $\mathbf{F}(x, y, z) = (x + \tan^{-1} yz)\mathbf{i} + y^2 z\mathbf{j} + z\mathbf{k}$, S is the part of the hemisphere $x = \sqrt{9 - y^2 - z^2}$ that lies inside the cylinder $y^2 + z^2 = 4$, oriented in the direction of the positive x-axis.

In the next four questions, use Stokes' theorem to evaluate $\int_C \mathbf{F} \cdot d\mathbf{r}$. In each case C is oriented counterclockwise as viewed from above.

6. $\mathbf{F}(x, y, z) = xz\mathbf{i} + 2xy\mathbf{j} + 3xy\mathbf{k}$, C is the boundary of the part of the plane $3x + y + z = 3$ in the first octant.

7. $\mathbf{F}(x, y, z) = z^2\mathbf{i} + y^2\mathbf{j} + xy\mathbf{k}$, C is the triangle with vertices $(1, 0, 0)$, $(0, 1, 0)$, and $(0, 0, 2)$.

8. $\mathbf{F}(x, y, z) = 2z\mathbf{i} + 4x\mathbf{j} + 5y\mathbf{k}$, C is the curve of intersection of the plane $z = x + 4$ and the cylinder $x^2 + y^2 = 4$.

9. $\mathbf{F}(x, y, z) = x\mathbf{i} + y\mathbf{j} + (x^2 + y^2)\mathbf{k}$, C is the part of the paraboloid $z = 1 - x^2 - y^2$ in the first octant.

10. Use Stokes' theorem to evaluate $\int_C \mathbf{F} \cdot d\mathbf{r}$, where

$$\mathbf{F}(x, y, z) = x^2 z\mathbf{i} + xy^2\mathbf{j} + z^2\mathbf{k}$$

and C is the curve of intersection of the plane $x + y + z = 1$ and the cylinder $x^2 + y^2 = 9$ oriented counterclockwise as viewed from above.

11.   Use Stoke's theorem to evaluate $\int_C \mathbf{F} \cdot d\mathbf{r}$, where $\mathbf{F}(x, y, z) = x^2 y \mathbf{i} + \frac{1}{3} x^3 \mathbf{j} + xy\mathbf{k}$ and C is the curve of intersection of the hyperbolic paraboloid $z = y^2 - x^2$ and the cylinder $x^2 + y^2 = 1$ oriented counterclockwise as viewed from above.

12.   Verify that Stoke's theorem is true for the given vector field $\mathbf{F}$ and surface S.

(a) $\mathbf{F}(x, y, z) = 3y\mathbf{i} + 4z\mathbf{j} - 6x\mathbf{k}$, S is the part of the paraboloid

$$z = 9 - x^2 - y^2,$$

that lies above the xy-plane, oriented upward.

(b) $\mathbf{F}(x, y, z) = xy\mathbf{i} + yz\mathbf{j} + xz\mathbf{k}$, S is the hemisphere

$$z = \sqrt{a^2 - x^2 - y^2},$$

oriented upward.

(c) $\mathbf{F}(x, y, z) = y\mathbf{i} + z\mathbf{j} + x\mathbf{k}$, S is the part of the plane $x + y + z = 1$ that lies in the first octant, oriented upward.

13.   Calculate the work done by the force field

$$\mathbf{F}(x, y, z) = (x^x + z^2)\mathbf{i} + (y^y + x^2)\mathbf{j} + (z^z + y^2)\mathbf{k}$$

when a particle moves under its influence around the edge of the part of the sphere $x^2 + y^2 + z^2 = 4$ that lies in the first octant, in a counterclockwise direction as viewed from above.

14.   Evaluate $\int_C (y + \sin x)dx + (z^2 + \cos y)dy + x^3 dz$, where C is the curve $\mathbf{r}(t) = (\sin t, \cos t, \sin 2t)$, $0 \leq t \leq 2\pi$. [*Hint:* Observe that C lies on the surface $z = 2xy$.]

In the next seven questions, use the Divergence theorem to calculate the surface integral $\iint_S \mathbf{F} \cdot d\mathbf{S}$, i.e. calculate the flux of $\mathbf{F}$ across S.

15. $\mathbf{F}(x, y, z) = -xz\mathbf{i} - yz\mathbf{j} + z^2\mathbf{k}$, S is the ellipsoid $x^2/a^2 + y^2/b^2 + z^2/c^2 = 1$.

16. $\mathbf{F}(x, y, z) = z \cos y\mathbf{i} + x \sin z\mathbf{j} + xz\mathbf{k}$, S is the surface of the tetrahedron bounded by the planes $x = 0$, $y = 0$, $z = 0$, and $2x + y + z = 2$.

17. $\mathbf{F}(x, y, z) = (x + e^{y \tan z})\mathbf{i} + 3xe^{xz}\mathbf{j} + (\cos y - z)\mathbf{k}$, S is the surface with equation $x^4 + y^4 + z^4 = 1$.

18. $\mathbf{F}(x, y, z) = x^3\mathbf{i} + y^3\mathbf{j} + z^3\mathbf{k}$, S is the sphere $x^2 + y^2 + z^2 = 1$.

19. $\mathbf{F}(x, y, z) = (x^3 + y \sin z)\mathbf{i} + (y^3 + z \sin x)\mathbf{j} + 3z\mathbf{k}$, S is the surface of the solid bounded by the hemispheres $z = \sqrt{4 - x^2 - y^2}, z = \sqrt{1 - x^2 - y^2}$ and the plane $z = 0$.

20. $\mathbf{F}(x, y, z) = xy^2\mathbf{i} + yz\mathbf{j} + zx^2\mathbf{k}$, S is the surface of the solid that lies between the cylinders $x^2 + y^2 = 1$ and $x^2 + y^2 = 4$ and between the planes $z = 1$ and $z = 3$.

21. $F(x, y, z) = (x^3 + yz)i + x^2yj + xy^2k$, S is the surface of the solid bounded by the spheres $x^2 + y^2 + z^2 = 4$ and $x^2 + y^2 + z^2 = 9$.

22. Use the Divergence theorem to evaluate $\iint_S F \cdot dS$, where

$$F(x, y, z) = z^2xi + \left(\frac{1}{3}y^3 + \tan z\right)j + (x^2z + y^2)k$$

and S is the top half of the sphere $x^2 + y^2 + z^2 = 1$. [*Hint:* Note that S is not a closed surface. First compute integrals over S_1 where S_1 is the disk $x^2 + y^2 \leq 1$ oriented downward and $S_2 = S \cup S_1$.]

23. Let $F(x, y, z) = z\tan^{-1}(y^2)i + z^2\ln(x^2 + 1)j + zk$. Find the flux of F across the part of the paraboloid $x^2 + y^2 + z = 2$ that lies above the plane $z = 1$ and is oriented upward.

24. Verify that div $E = 0$ for the electric field $E(R) = \varepsilon QR/|R|^3$.

25. Use the Divergence theorem to evaluate

$$\iint_S (2x + 2y + z^2)dS,$$

where S is the sphere $x^2 + y^2 + z^2 = 1$.

26. Verify that Stokes' theorem is true for the vector field $F(x, y, z) = x^2i + y^2j + z^2k$, where S is the part of the paraboloid $z = 1 - x^2 - y^2$ that lies above the xy-plane, and S has upward orientation.

27. Use Stokes' theorem to evaluate $\iint_S \nabla \times F \cdot dS$, where $F(x, y, z) = x^2yzi + yz^2j + z^3e^{xy}k$, S is the part of the sphere $x^2 + y^2 + z^2 = 5$ that lies above the plane $z = 1$, and S is oriented upward.

28. Use Stokes' theorem to evaluate $\int_C F \cdot dr$, where $F(x, y, z) = xyi + yzj + zxk$ and C is the triangle with vertices $(1, 0, 0)$, $(0, 1, 0)$, and $(0, 0, 1)$, oriented counterclockwise as viewed from above.

29. Use the Divergence theorem to calculate the surface integral $\iint_S F \cdot dS$, where $F(x, y, z) = x^3i + y^3j + z^3k$ and S is the surface of the solid bounded by the cylinder $x^2 + y^2 = 1$ and the planes $z = 0$ and $z = 2$.

30. Verify that the Divergence theorem is true for the vector field

$$F(x, y, z) = xi + yj + zk,$$

where S is the surface of a unit ball $x^2 + y^2 + z^2 \leq 1$.

31. Compute the outward flux of

$$F(x, y, z) = \frac{xi + yj + zk}{(x^2 + y^2 + z^2)^{3/2}}$$

through the ellipsoid $4x + 9y^2 + 6z^2 = 36$.

Chapter 7

Orthogonal curvilinear coordinate systems

7.1 Introduction

In this chapter we discuss, in a systematic way, how we can represent scalar fields, vector fields and, in particular, the gradient, divergence, curl and Laplacian operators in any convenient coordinate system in which the base vectors are *orthogonal*. Knowledge of these representations enables us to readily apply vector field concepts to many other situations where Cartesian coordinates would no longer be applicable.

So far, in this text, we have chosen to describe quantities of physical interest in terms of the rectangular Cartesian coordinate system. However, where the situation warranted, we have used two other coordinate systems, namely the cylindrical and spherical polars using elementary knowledge. In some particular problems such as problems related to the cylindrical structures or spherical structures, we have a simple relationship between the Cartesian and polar coordinates. And in that situation, we would normally work using cylindrical or spherical coordinates. We are well versed with the use of the Cartesian system for vector quantities such as *gradient, divergence* and *curl* but not with any other coordinate systems. It is our aim in this chapter to define these quantities in the orthogonal curvilinear systems. We shall also develop the expressions for the basic *elements of space* — the line element, the surface element and the volume element.

7.2 Derivation of curvilinear coordinates

The analysis of general orthogonal curvilinear coordinates begins with the representation of vectors. It is worth mentioning here that the concept of a vector is independent of the coordinate system. A particular coordinate system is used only when we want to develop an expression for the components of a vector.

Let us consider the position vector $\mathbf{R}$ of a point with curvilinear coordinates (u_1, u_2, u_3). We should first note that we cannot express this position vector as

$$\mathbf{R} = \mathbf{i}u_1 + \mathbf{j}u_2 + \mathbf{k}u_3,$$

since this is only true if (u_1, u_2, u_3) are the Cartesian coordinates of the endpoint of the vector $\mathbf{R}$.

We know that $d\mathbf{R}/ds = \mathbf{T}$ is a unit tangent vector where s describes a curve. On a u_1 coordinate curve, where u_2 and u_3 are held constant, we can regard u_1 as a parameter where u_2 and u_3 are held fixed, so that $\partial\mathbf{R}/\partial u_1$ must be a tangent vector. Similarly, $\partial\mathbf{R}/\partial u_2$ and $\partial\mathbf{R}/\partial u_3$ are tangent vectors to the u_2 and u_3 coordinate curves. Hence, we can obtain three unit tangent vectors say $\mathbf{e}_1$, $\mathbf{e}_2$ and $\mathbf{e}_3$, using the relations

$$\mathbf{e}_1 = \frac{\partial\mathbf{R}}{\partial u_1} \Big/ \left|\frac{\partial\mathbf{R}}{\partial u_1}\right|,$$

$$\mathbf{e}_2 = \frac{\partial\mathbf{R}}{\partial u_2} \Big/ \left|\frac{\partial\mathbf{R}}{\partial u_2}\right|,$$

$$\mathbf{e}_3 = \frac{\partial\mathbf{R}}{\partial u_3} \Big/ \left|\frac{\partial\mathbf{R}}{\partial u_3}\right|. \tag{7.1}$$

Now let us define

$$\left|\frac{\partial\mathbf{R}}{\partial u_1}\right| = h_1,$$

$$\left|\frac{\partial\mathbf{R}}{\partial u_2}\right| = h_2,$$

$$\left|\frac{\partial\mathbf{R}}{\partial u_3}\right| = h_3. \tag{7.2}$$

Here h_1, h_2 and h_3 are defined as the *scale factors* and, hence, (7.1) can be rewritten as

$$\frac{\partial\mathbf{R}}{\partial u_1} = h_1\mathbf{e}_1,$$

$$\frac{\partial\mathbf{R}}{\partial u_2} = h_2\mathbf{e}_2,$$

$$\frac{\partial\mathbf{R}}{\partial u_3} = h_3\mathbf{e}_3. \tag{7.3}$$

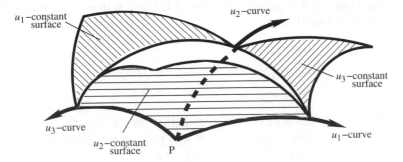

Figure 7.1   Orthogonal curvilinear coordinates.

At each point of space, then, we can define a triad of unit vectors $\mathbf{e}_1$, $\mathbf{e}_2$ and $\mathbf{e}_3$ as shown in Fig. 7.1, which corresponds directly to the introduction of $\mathbf{i}$, $\mathbf{j}$ and $\mathbf{k}$ in the Cartesian description of vectors. It is worth noting that the curvilinear system of coordinates (u_1, u_2, u_3) will be an orthogonal system if and only if (iff), $\mathbf{e}_1$, $\mathbf{e}_2$ and $\mathbf{e}_3$ are mutually perpendicular to each other at every point.

It is also clear that an arbitrary vector $\mathbf{A}$ located at a particular point P in space may, by the parallelogram law of combination, be written in curvilinear coordinate form as

$$\mathbf{A} = \mathbf{e}_1 a_1 + \mathbf{e}_2 a_2 + \mathbf{e}_3 a_3 \tag{7.4}$$

where, using the orthogonality of the unit vectors,

$$|\mathbf{A}|^2 = \mathbf{A} \cdot \mathbf{A} = a_1^2 + a_2^2 + a_3^2. \tag{7.5}$$

7.3 The gradient in generalized curvilinear coordinates

We have already defined the gradient of a scalar field G in the Cartesian coordinate system. In this section we shall define it for the generalized coordinate system. Let G be a scalar function of x, y and z such that

$$x = x(u_1, u_2, u_3),$$

$$y = y(u_1, u_2, u_3),$$

$$z = z(u_1, u_2, u_3),$$

where u_1, u_2 and u_3 are curvilinear coordinates. The gradient of scalar field G in Cartesian coordinates is

$$\nabla G = \mathbf{i}\frac{\partial G}{\partial x} + \mathbf{j}\frac{\partial G}{\partial y} + \mathbf{k}\frac{\partial G}{\partial z}, \tag{7.6}$$

where $G = G(x, y, z)$.

In order to extend this definition to other orthogonal systems, we express G in generalized coordinates

$$
\begin{aligned}
G &= G(x, y, z) \\
&= G(x(u_1, u_2, u_3), y(u_1, u_2, u_3), z(u_1, u_2, u_3)) \\
&= G(u_1, u_2, u_3),
\end{aligned}
$$

so that any level surface of G can be described by the equation of the form $G(u_1, u_2, u_3) = $ constant. Suppose that P is a point in the scalar field and that S_1 is the level surface at P. Let C is any arbitrary curve passing through P and that we move from P to a neighbouring point Q such that the generalized coordinates change by du_1, du_2 and du_3 as depicted in Fig. 7.2. The change in ϕ as we move from P to Q is

$$
dG \approx \frac{\partial G}{\partial u_1} du_1 + \frac{\partial G}{\partial u_2} du_2 + \frac{\partial G}{\partial u_3} du_3. \tag{7.7}
$$

If C is parameterized using the arc-length parameter s, then the rate of change of G along C is

$$
\frac{dG}{ds} = \frac{\partial G}{\partial u_1} \frac{du_1}{ds} + \frac{\partial G}{\partial u_2} \frac{du_2}{ds} + \frac{\partial G}{\partial u_3} \frac{du_3}{ds}. \tag{7.8}
$$

But if $\mathbf{R}$ is the position vector of the point P, then the unit tangent vector at P is $\mathbf{T} = d\mathbf{R}/ds$, or, using generalized coordinates, $\mathbf{R} = \mathbf{R}(u_1, u_2, u_3)$,

$$
\begin{aligned}
\mathbf{T} &= \frac{d\mathbf{R}}{ds} \\
&= \frac{\partial \mathbf{R}}{\partial u_1} \frac{du_1}{ds} + \frac{\partial \mathbf{R}}{\partial u_2} \frac{du_2}{ds} + \frac{\partial \mathbf{R}}{\partial u_3} \frac{du_3}{ds} \\
&= h_1 \mathbf{e}_1 \frac{du_1}{ds} + h_2 \mathbf{e}_2 \frac{du_2}{ds} + h_3 \mathbf{e}_3 \frac{du_3}{ds}. \tag{7.9}
\end{aligned}
$$

For an orthogonal system, we can express (7.8) in the form of a scalar product:

$$
\frac{dG}{ds} = \left\{ \frac{1}{h_1} \frac{\partial G}{\partial u_1} \mathbf{e}_1 + \frac{1}{h_2} \frac{\partial G}{\partial u_2} \mathbf{e}_2 + \frac{1}{h_3} \frac{\partial G}{\partial u_3} \mathbf{e}_3 \right\} \cdot \left\{ h_1 \mathbf{e}_1 \frac{du_1}{ds} + h_2 \mathbf{e}_2 \frac{du_2}{ds} + h_3 \mathbf{e}_3 \frac{du_3}{ds} \right\}.
$$

$$
\tag{7.10}
$$

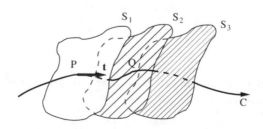

Figure 7.2   The gradient in generalized curvilinear coordinates.

The second vector here is **T** and we define the first vector as the gradient of G, such that

$$\nabla G = \frac{\mathbf{e}_1}{h_1}\frac{\partial G}{\partial u_1} + \frac{\mathbf{e}_2}{h_2}\frac{\partial G}{\partial u_2} + \frac{\mathbf{e}_3}{h_3}\frac{\partial G}{\partial u_3}. \tag{7.11}$$

Therefore, we can write (7.10) in compact form as

$$\frac{dG}{ds} = \nabla G \cdot \mathbf{T}. \tag{7.12}$$

If $G = G(u_1, u_2, u_3) = $ constant then $dG/ds = 0$ which implies that $\nabla G \cdot \mathbf{T} = 0$ and, hence, ∇G must be a normal vector to the surface $G = $ constant.

We can now use equation (7.11) very readily to obtain the form ∇G for the Cartesian, cylindrical and spherical polar coordinate systems. It can be easily verified that the scale factors for these three systems are, respectively, $(1, 1, 1), (1, r, 1)$ and $(1, r, r\sin\theta)$ with proper geometrical definitions of these three sets of axes. It is worth noting here that the unit tangent vectors corresponding to these three sets of axes can be determined very easily and they are given by

Cartesian: $\mathbf{e}_1 = \mathbf{x} = \mathbf{i},$

$\mathbf{e}_2 = \mathbf{y} = \mathbf{j},$

$\mathbf{e}_3 = \mathbf{z} = \mathbf{k},$

Cylindrical: $\mathbf{e}_1 = \mathbf{r} = \mathbf{i}\cos\theta + \mathbf{j}\sin\theta,$

$\mathbf{e}_2 = \boldsymbol{\theta} = -\mathbf{i}\sin\theta + \mathbf{j}\cos\theta,$

$\mathbf{e}_3 = \mathbf{z} = \mathbf{k},$

Spherical: $\mathbf{e}_1 = \mathbf{r} = \mathbf{i}\sin\theta\cos\phi + \mathbf{j}\sin\theta\sin\phi + \mathbf{k}\cos\theta,$

$\mathbf{e}_2 = \boldsymbol{\theta} = \mathbf{i}\cos\theta\cos\phi + \mathbf{j}\cos\theta\sin\phi - \mathbf{k}\sin\theta,$

$\mathbf{e}_3 = \boldsymbol{\phi} = -\mathbf{i}\sin\phi + \mathbf{j}\cos\phi.$

and it can be easily verified that these unit tangent vectors are orthogonal to each other for each set of axes and, hence, the ∇G are obtained as follows:

Cartesian: $\nabla G = \mathbf{i}\dfrac{\partial G}{\partial x} + \mathbf{j}\dfrac{\partial G}{\partial y} + \mathbf{k}\dfrac{\partial G}{\partial z},$

Cylindrical: $\nabla G = \mathbf{r}\dfrac{\partial G}{\partial r} + \boldsymbol{\theta}\dfrac{1}{r}\dfrac{\partial G}{\partial \theta} + \mathbf{k}\dfrac{\partial G}{\partial z},$

Spherical: $\nabla G = \mathbf{r}\dfrac{\partial G}{\partial r} + \boldsymbol{\theta}\dfrac{1}{r}\dfrac{\partial G}{\partial \theta} + \boldsymbol{\phi}\dfrac{1}{r\sin\theta}\dfrac{\partial G}{\partial \phi}.$

Now it can be easily verified that the elementary surface areas (surface elements) of the three major coordinate systems are:

Cartesian: $dS = [(dxdy), (dydz), (dzdx)],$

Cylindrical: $dS = [(rd\theta dr), (rd\theta dz), (drdz)],$

Spherical: $dS = [(rd\theta dr), (r^2\sin\theta d\phi)d\theta, (r\sin\theta d\phi dr)].$

Similarly, the elementary volumes dV for these three major coordinate systems, respectively, are as follows:

$$dV = [(dx\,dy\,dz), (dr)(r\,d\theta)(dz), (dr)(r\,d\theta)(r\sin\theta\,d\phi)].$$

Example 7.1   Convert the vector $\mathbf{A} = \mathbf{r}A_r + \boldsymbol{\theta}A_\theta + \boldsymbol{\phi}A_\phi$ in spherical coordinates into Cartesian coordinates.   □

Solution   The given vector in spherical coordinates can be written in Cartesian coordinates as follows:

$$\mathbf{A} = \mathbf{r}A_r + \boldsymbol{\theta}A_\theta + \boldsymbol{\phi}A_\phi \quad \text{in spherical,}$$

$$= \mathbf{i}A_x + \mathbf{j}A_y + \mathbf{k}A_z \quad \text{in Cartesian.}$$

To determine the values of A_x, A_y and A_z we take the scalar product of $\mathbf{A}$ with $\mathbf{i}, \mathbf{j}, \mathbf{k}$ in turn. Thus we get

$$A_x = \mathbf{i} \cdot \mathbf{A} = \mathbf{i} \cdot (\mathbf{r}A_r + \boldsymbol{\theta}A_\theta + \boldsymbol{\phi}A_\phi)$$

$$= \sin\theta\cos\phi A_r + \cos\theta\cos\phi A_\theta - \sin\phi A_\phi$$

$$= \frac{xA_r}{\sqrt{x^2+y^2+z^2}} + \frac{xzA_\theta}{\sqrt{x^2+y^2}\sqrt{x^2+y^2+z^2}} - \frac{yA_\phi}{\sqrt{x^2+y^2}}.$$

Similarly, we can obtain A_y and A_z:

$$A_y = \mathbf{j} \cdot \mathbf{A} = \mathbf{j} \cdot (\mathbf{r}A_r + \boldsymbol{\theta}A_\theta + \boldsymbol{\phi}A_\phi)$$

$$= \sin\theta\sin\phi A_r + \cos\theta\sin\phi A_\theta + \cos\phi A_\phi$$

$$= \frac{yA_r}{\sqrt{x^2+y^2+z^2}} + \frac{yzA_\theta}{\sqrt{x^2+y^2}\sqrt{x^2+y^2+z^2}} + \frac{xA_\phi}{\sqrt{x^2+y^2}},$$

$$A_z = \mathbf{k} \cdot \mathbf{A} = \mathbf{k} \cdot (\mathbf{r}A_r + \boldsymbol{\theta}A_\theta + \boldsymbol{\phi}A_\phi)$$

$$= \cos\theta A_r - \sin\theta A_\theta$$

$$= \frac{zA_r}{\sqrt{x^2+y^2+z^2}} - \sqrt{\frac{x^2+y^2}{x^2+y^2+z^2}}A_\theta.$$

If A_r, A_θ and A_ϕ are themselves functions of r, θ and ϕ, they too need to be converted into functions of x, y and z by the relationship, $x = r\sin\theta\cos\phi$, $y = r\sin\theta\sin\phi$ and $z = r\cos\theta$.   ■

7.4 The divergence in orthogonal curvilinear coordinates

In this section we derive expressions for the divergence of a vector field in orthogonal curvilinear coordinates. There are two ways in which this can be accomplished.

Firstly, we could choose to use the coordinate-free integral definition of this quantity and work consistently in curvilinear coordinates to deduce the required results. Secondly, we could regard ∇ as a vector differential operator and simply determine the scalar vector product of ∇ with a vector field $\mathbf{A}$, expressing all quantities in curvilinear coordinates.

Method 1: The coordinate-free definition of the divergence of a vector field $\mathbf{A}$ is

$$\nabla \cdot \mathbf{A} = \lim_{\delta V \to 0} \left\{ \frac{1}{\delta V} \iint_S \mathbf{A} \cdot \mathbf{n} dS \right\}, \tag{7.13}$$

where δV is an elementary volume bounded by a closed surface S.

Let P be a point with curvilinear coordinates (u_1, u_2, u_3) and δV a small parallelepiped with edges parallel to the unit vectors $\mathbf{e}_1, \mathbf{e}_2, \mathbf{e}_3$ at P as shown in Fig. 7.3. We calculate the flux integral $\iint_S \mathbf{A} \cdot \mathbf{n} dS$ over the six faces of the box. We shall assume that the dimension of the box are so small that the vector field $\mathbf{A}$ may be regarded as a constant over each bounding face.

On the face *PEHD* we assume that $\mathbf{A}$ is a constant vector equal to the actual vector at $P, \mathbf{A}^{(P)}$ say. On this face, $\mathbf{n} = -\mathbf{e}_1$ (outward pointing unit normal vector), hence,

$$\iint_{PEHD} \mathbf{A} \cdot \mathbf{n} dS \approx -\mathbf{A}^{(P)} \cdot \mathbf{e}_1 \iint_{PEHD} dS_1$$

$$= -\mathbf{A}^{(P)} \cdot \mathbf{e}_1 (h_2 h_3)^{(P)} du_2 du_3$$

$$= -a_1^{(P)} (h_2 h_3)^{(P)} du_2 du_3,$$

where we have evaluated $h_2 h_3$ at the point P.

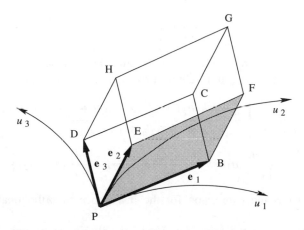

Figure 7.3   The divergence in orthogonal curvilinear coordinates.

On the face *BFGC* we assume that $\mathbf{A}$ is a constant vector equal to the actual vector at B, $\mathbf{A}^{(B)}$ say. On this face $\mathbf{n} = +\mathbf{e}_1$, hence,

$$\iint_{BFGC} \mathbf{A} \cdot \mathbf{n} dS \approx a_1^{(B)} (h_2 h_3)^{(B)} du_2 du_3,$$

where $h_2 h_3$ is evaluated at point B. But to a good approximation, it can be obtained by using Taylor's expansion about the point P and keeping only the first two terms

$$(a_1 h_2 h_3)^{(B)} \approx (a_1 h_2 h_3)^{(P)} + \frac{\partial}{\partial u_1}(a_1 h_2 h_3)^{(P)} du_1.$$

Adding these two expressions, we obtain the approximate value

$$\frac{\partial}{\partial u_1}(a_1 h_2 h_3)^{(P)} du_1 du_2 du_3,$$

for the flux across these two faces.

A similar calculation of the surface integral over the other two pairs of faces (*PBCD* and *EFGH*, and *PEFB* and *DHGC*) give the further contributions,

$$\frac{\partial}{\partial u_2}(a_2 h_1 h_3)^{(P)} du_1 du_2 du_3$$

and

$$\frac{\partial}{\partial u_3}(a_3 h_1 h_2)^{(P)} du_1 du_2 du_3,$$

respectively. Adding all the contributions yields

$$\iint_S \mathbf{A} \cdot \mathbf{n} dS \approx \left[\frac{\partial}{\partial u_1}(a_1 h_2 h_3)^{(P)} + \frac{\partial}{\partial u_2}(a_2 h_1 h_3)^{(P)} \right.$$
$$\left. + \frac{\partial}{\partial u_3}(a_3 h_1 h_2)^{(P)} \right] du_1 du_2 du_3.$$

Since P is an arbitrary point, we may drop the superscript. Then, from the definition (7.13) and the volume element $\delta V = h_1 h_2 h_3 du_1 du_2 du_3$, we obtain

$$\nabla \cdot \mathbf{A} = \lim_{du_1 du_2 du_3 \to 0} \left\{ \frac{\frac{\partial}{\partial u_1}(a_1 h_2 h_3) + \frac{\partial}{\partial u_2}(a_2 h_1 h_3) + \frac{\partial}{\partial u_3}(a_3 h_1 h_2)}{h_1 h_2 h_3 du_1 du_2 du_3} \right\} du_1 du_2 du_3$$

$$= \frac{1}{h_1 h_2 h_3} \left\{ \frac{\partial}{\partial u_1}(a_1 h_2 h_3) + \frac{\partial}{\partial u_2}(a_2 h_1 h_3) + \frac{\partial}{\partial u_3}(a_3 h_1 h_2) \right\}. \qquad (7.14)$$

Equation (7.6) is the expression for the divergence in orthogonal curvilinear coordinates.

The expression for the Laplacian ∇^2 in curvilinear coordinates can now be easily obtained if we use the definition $\nabla^2 G = \nabla \cdot (\nabla G)$ and simply use the expression for div and grad in equation (7.6). We find, after a straightforward calculation,

$$\nabla^2 G = \frac{1}{h_1 h_2 h_3} \left\{ \frac{\partial}{\partial u_1} \left(\frac{h_2 h_3}{h_1} \frac{\partial G}{\partial u_1} \right) + \frac{\partial}{\partial u_2} \left(\frac{h_1 h_3}{h_2} \frac{\partial G}{\partial u_2} \right) + \frac{\partial}{\partial u_3} \left(\frac{h_1 h_2}{h_3} \frac{\partial G}{\partial u_3} \right) \right\}.$$

(7.15)

The Laplacian in the three most important coordinate systems are:

- Cartesian:

$$\nabla^2 G = \frac{\partial^2 G}{\partial x^2} + \frac{\partial^2 G}{\partial y^2} + \frac{\partial^2 G}{\partial z^2},$$

- Cylindrical polars:

$$\nabla^2 G = \frac{1}{r} \frac{\partial}{\partial r} \left(r \frac{\partial G}{\partial r} \right) + \frac{1}{r^2} \frac{\partial^2 G}{\partial \theta^2} + \frac{\partial^2 G}{\partial z^2},$$

- Spherical polars:

$$\nabla^2 G = \frac{1}{r^2} \frac{\partial}{\partial r} \left(r^2 \frac{\partial G}{\partial r} \right) + \frac{1}{r^2 \sin \theta} \frac{\partial}{\partial \theta} \left(\sin \theta \frac{\partial G}{\partial \theta} \right) + \frac{1}{r^2 \sin^2 \theta} \frac{\partial^2 G}{\partial \phi^2}.$$

Method 2: We note that by choosing the scalar field G in equation (7.6) to be each of the generalized coordinates u_1, u_2 and u_3 in turn, we obtain

$$\nabla u_1 = \frac{\mathbf{e}_1}{h_1},$$

$$\nabla u_2 = \frac{\mathbf{e}_2}{h_2},$$

$$\nabla u_3 = \frac{\mathbf{e}_3}{h_3}. \qquad (7.16)$$

Hence, if $\mathbf{A} = \mathbf{e}_1 a_1 + \mathbf{e}_2 a_2 + \mathbf{e}_3 a_3$ is a vector field expressed in orthogonal coordinate system,

$$\mathbf{A} = h_1 a_1 \nabla u_1 + h_2 a_2 \nabla u_2 + h_3 a_3 \nabla u_3$$

$$= \sum_{i=1}^{3} h_i a_i \nabla u_i. \qquad (7.17)$$

Therefore, the expression for $\nabla \cdot \mathbf{A}$ is given by

$$
\begin{aligned}
\nabla \cdot \mathbf{A} &= \sum_{i=1}^{3} \nabla \cdot (h_i a_i \nabla u_i) \\
&= \sum_{i=1}^{3} [\nabla(h_i a_i) \cdot \nabla u_i + h_i a_i \nabla \cdot \nabla u_i] \\
&= \sum_{i=1}^{3} \left[\nabla(h_i a_i) \cdot \frac{\mathbf{e}_i}{h_i} + h_i a_i \nabla^2 u_i \right] \\
&= \frac{1}{h_1^2} \frac{\partial}{\partial u_1}(h_1 a_1) + \frac{1}{h_2} \frac{\partial}{\partial u_2}(h_2 a_2) + \frac{1}{h_3^2} \frac{\partial}{\partial u_3}(h_3 a_3) \\
&\quad + \frac{1}{h_1 h_2 h_3} \left[(h_1 a_1) \frac{\partial}{\partial u_1} \left(\frac{h_2 h_3}{h_1} \right) + (h_2 a_2) \frac{\partial}{\partial u_2} \left(\frac{h_1 h_3}{h_2} \right) \right. \\
&\quad \left. + (h_3 a_3) \frac{\partial}{\partial u_3} \left(\frac{h_1 h_2}{h_3} \right) \right] \\
&= \frac{1}{h_1 h_2 h_3} \left[\left(\frac{h_2 h_3}{h_1} \right) \frac{\partial}{\partial u_1}(h_1 a_1) + \left(\frac{h_1 h_3}{h_2} \right) \frac{\partial}{\partial u_2}(h_2 a_2) \right. \\
&\quad + \left(\frac{h_1 h_2}{h_3} \right) \frac{\partial}{\partial u_3}(h_3 a_3) + (h_1 a_1) \frac{\partial}{\partial u_1} \left(\frac{h_2 h_3}{h_1} \right) \\
&\quad \left. + (h_2 a_2) \frac{\partial}{\partial u_2} \left(\frac{h_1 h_3}{h_2} \right) + (h_3 a_3) \frac{\partial}{\partial u_3} \left(\frac{h_1 h_2}{h_3} \right) \right] \\
&= \frac{1}{h_1 h_2 h_3} \left[\frac{\partial}{\partial u_1}(a_1 h_2 h_3) + \frac{\partial}{\partial u_2}(a_2 h_1 h_3) + \frac{\partial}{\partial u_3}(a_3 h_1 h_2) \right].
\end{aligned}
$$

This is the direct second method to determine $\nabla \cdot \mathbf{A}$. Application of divergence is given below for the three major coordinate systems.

7.4.1 The application of divergence to Cartesian, cylindrical and spherical systems

1. *Cartesian system:* In this case, the scale factors are: $h_1 = 1, h_2 = 1$ and $h_3 = 1$. Therefore,

$$
\nabla \cdot \mathbf{A} = \frac{\partial a_1}{\partial x} + \frac{\partial a_2}{\partial y} + \frac{\partial a_3}{\partial z}. \tag{7.18}
$$

2. *Cylindrical polar system:* In this case, the scale factors are: $h_1 = 1, h_2 = r$ and $h_3 = 1$. Therefore,

$$
\begin{aligned}
\nabla \cdot \mathbf{A} &= \frac{1}{r} \left\{ \frac{\partial}{\partial r}(r a_1) + \frac{\partial}{\partial \theta}(a_2) + \frac{\partial}{\partial z}(r a_3) \right\} \\
&= \frac{1}{r} \frac{\partial}{\partial r}(r a_1) + \frac{1}{r} \frac{\partial}{\partial \theta}(a_2) + \frac{\partial}{\partial z}(a_3). \tag{7.19}
\end{aligned}
$$

3. *Spherical polar system:* In this case, the scale factors are: $h_1 = 1$, $h_2 = r$ and $h_3 = r \sin \theta$. Therefore,

$$\nabla \cdot \mathbf{A} = \frac{1}{r^2 \sin \theta} \left\{ \frac{\partial}{\partial r} (r^2 \sin \theta a_1) + \frac{\partial}{\partial \theta} (r \sin \theta a_2) + \frac{\partial}{\partial \phi} (r a_3) \right\}$$

$$= \frac{1}{r^2} \frac{\partial}{\partial r} (r^2 a_1) + \frac{1}{r \sin \theta} \frac{\partial}{\partial \theta} (\sin \theta a_2) + \frac{1}{r \sin \theta} \frac{\partial}{\partial \phi} (a_3). \quad (7.20)$$

7.5 The curl in orthogonal curvilinear coordinates

In this section we derive expressions for the curl of a vector field in orthogonal curvilinear coordinates. There are again two ways in which this can be done. Firstly, we could regard ∇ as a vector differential operator and simply determine the vector product of ∇ with a vector field $\mathbf{A}$, expressing all quantities in curvilinear coordinates. Secondly, we could choose to use the coordinate-free integral definition of this quantity and work consistently in curvilinear coordinates to deduce the required results.

Method 1: We note that by choosing the scalar field G in equation (7.6) to be each of the generalized coordinates u_1, u_2 and u_3 in turn, we obtain

$$\nabla u_1 = \frac{\mathbf{e}_1}{h_1},$$

$$\nabla u_2 = \frac{\mathbf{e}_2}{h_2},$$

$$\nabla u_3 = \frac{\mathbf{e}_3}{h_3}. \quad (7.21)$$

Hence, if $\mathbf{A} = \mathbf{e}_1 a_1 + \mathbf{e}_2 a_2 + \mathbf{e}_3 a_3$ is a vector field expressed in orthogonal coordinate system,

$$\mathbf{A} = h_1 a_1 \nabla u_1 + h_2 a_2 \nabla u_2 + h_3 a_3 \nabla u_3$$

$$= \sum_{i=1}^{3} h_i a_i \nabla u_i. \quad (7.22)$$

Therefore, the expression for $\nabla \times \mathbf{A}$ is given by

$$\nabla \times \mathbf{A} = \sum_{i=1}^{3} \nabla \times (h_i a_i \nabla u_i)$$

$$= \sum_{i=1}^{3} [\nabla (h_i a_i) \times \nabla u_i + h_i a_i \nabla \times \nabla u_i]$$

$$= \sum_{i=1}^{3} \nabla (h_i a_i) \times \nabla u_i.$$

In obtaining the last expression we have used the vector identity $\nabla \times \nabla u_i = 0$. Thus we have,

$$\nabla \times \mathbf{A} = \sum_{i=1}^{3} \nabla \times (h_i a_i \nabla u_i)$$

$$= \sum_{i=1}^{3} \nabla(h_i a_i) \times \nabla u_i$$

$$= \sum_{i=1}^{3} \nabla(h_i a_i) \times \frac{\mathbf{e}_i}{h_i}.$$

Expanding the three gradient terms on the right-hand side using (7.6),

$$\nabla \times \mathbf{A} = \sum_{i=1}^{3} \left[\frac{\mathbf{e}_1}{h_1} \frac{\partial}{\partial u_1}(h_i a_i) + \frac{\mathbf{e}_2}{h_2} \frac{\partial}{\partial u_2}(h_i a_i) + \frac{\mathbf{e}_3}{h_3} \frac{\partial}{\partial u_3}(h_i a_i) \right] \times \frac{\mathbf{e}_i}{h_i},$$

which is a total of nine terms. Also, note that $\mathbf{e}_1 \times \mathbf{e}_2 = \mathbf{e}_3$, and $\mathbf{e}_i \times \mathbf{e}_i = 0$, $i = 1, 2, 3$, etc., for orthogonal coordinate systems (see Fig. 7.4). Therefore, we get only six non zero terms, and, hence,

$$\nabla \times \mathbf{A} = -\frac{\mathbf{e}_3}{h_1 h_2} \frac{\partial}{\partial u_2}(h_1 a_1) + \frac{\mathbf{e}_2}{h_1 h_3} \frac{\partial}{\partial u_3}(h_1 a_1) + \frac{\mathbf{e}_3}{h_1 h_2} \frac{\partial}{\partial u_1}(h_2 a_2)$$

$$-\frac{\mathbf{e}_1}{h_2 h_3} \frac{\partial}{\partial u_3}(h_2 a_2) - \frac{\mathbf{e}_2}{h_1 h_3} \frac{\partial}{\partial u_1}(h_3 a_3) + \frac{\mathbf{e}_1}{h_2 h_3} \frac{\partial}{\partial u_2}(h_3 a_3).$$

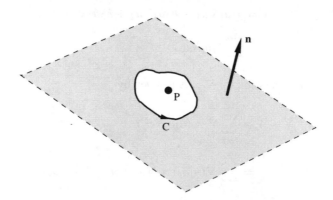

Figure 7.4   A general definition of curl.

Collecting like components,

$$\nabla \times \mathbf{A} = \frac{\mathbf{e}_1}{h_2 h_3} \left[\frac{\partial}{\partial u_2} (h_3 a_3) - \frac{\partial}{\partial u_3} (h_2 a_2) \right]$$

$$+ \frac{\mathbf{e}_2}{h_1 h_3} \left[\frac{\partial}{\partial u_3} (h_1 a_1) - \frac{\partial}{\partial u_1} (h_3 a_3) \right]$$

$$+ \frac{\mathbf{e}_3}{h_1 h_2} \left[\frac{\partial}{\partial u_1} (h_2 a_2) - \frac{\partial}{\partial u_2} (h_1 a_1) \right]. \tag{7.23}$$

Equation (7.23) is the general form of the curl of a vector field $\mathbf{A}$ in orthogonal curvilinear coordinates. We note that equation (7.23) may be expressed in the more concise determinant form

$$\nabla \times \mathbf{A} = \frac{1}{h_1 h_2 h_3} \begin{vmatrix} h_1 \mathbf{e}_1 & h_2 \mathbf{e}_2 & h_3 \mathbf{e}_3 \\ \dfrac{\partial}{\partial u_1} & \dfrac{\partial}{\partial u_2} & \dfrac{\partial}{\partial u_3} \\ h_1 a_1 & h_2 a_2 & h_3 a_3 \end{vmatrix}. \tag{7.24}$$

7.5.1 The application of curl to Cartesian, cylindrical and spherical systems

For the three major coordinate systems:

1. *Cartesian system:* In this case, the scale factors are: $h_1 = 1$, $h_2 = 1$ and $h_3 = 1$. Therefore,

$$\nabla \times \mathbf{A} = \begin{vmatrix} \mathbf{i} & \mathbf{j} & \mathbf{k} \\ \dfrac{\partial}{\partial x} & \dfrac{\partial}{\partial y} & \dfrac{\partial}{\partial z} \\ a_1 & a_2 & a_3 \end{vmatrix}. \tag{7.25}$$

2. *Cylindrical polar system:* In this case, the scale factors are: $h = 1$, $h_2 = r$ and $h_3 = 1$. Therefore,

$$\nabla \times \mathbf{A} = \frac{1}{r} \begin{vmatrix} \mathbf{r} & r\boldsymbol{\theta} & \mathbf{k} \\ \dfrac{\partial}{\partial r} & \dfrac{\partial}{\partial \theta} & \dfrac{\partial}{\partial z} \\ a_1 & r a_2 & a_3 \end{vmatrix}. \tag{7.26}$$

3. *Spherical polar system:* In this case, the scale factors are: $h_1 = 1$, $h_2 = r$ and $h_3 = r \sin \theta$. Therefore,

$$\nabla \times \mathbf{A} = \frac{1}{r^2 \sin \theta} \begin{vmatrix} \mathbf{r} & r\boldsymbol{\theta} & r \sin \theta \boldsymbol{\phi} \\ \dfrac{\partial}{\partial r} & \dfrac{\partial}{\partial \theta} & \dfrac{\partial}{\partial \phi} \\ a_1 & r a_2 & r \sin \theta a_3 \end{vmatrix}. \tag{7.27}$$

Method 2 – A general definition of curl: In Method 1 we have given the explicit description of how to evaluate $\nabla \times \mathbf{A}$. Here we shall give more general definition of curl of a vector field. Consider a point P in a vector field $\mathbf{A}$ and define a direction by means of a unit vector $\mathbf{n}$ (see Fig. 7.5). If we consider an elementary closed path C lying in a plane through P with normal $\mathbf{n}$, we can apply Stokes' theorem,

$$\oint_C \mathbf{A} \cdot d\mathbf{R} = \iint_{\Delta S} \nabla \times \mathbf{A} \cdot \mathbf{n} dS,$$

where ΔS is any elementary surface bounded by C. However, if ΔS is small enough for $\nabla \times \mathbf{A}$ not to vary significantly over it, the flux of $\nabla \times \mathbf{A}$ can be approximated by using its value at P, that is,

$$\oint_C \mathbf{A} \cdot d\mathbf{R} \approx (\nabla \times \mathbf{A})_P \cdot \mathbf{n} \Delta S.$$

This suggests that

$$(\nabla \times \mathbf{A})_P \cdot \mathbf{n} \approx \frac{1}{\Delta S} \oint_C \mathbf{A} \cdot d\mathbf{R}.$$

If we now take limit in such a way that ΔS shrinks to the point P, we obtain a point definition of curl as follows:

$$(\nabla \times \mathbf{A})_P \cdot \mathbf{n} = \lim_{\Delta S \to 0} \left\{ \frac{1}{\Delta S} \oint_C \mathbf{A} \cdot d\mathbf{R} \right\}. \qquad (7.28)$$

This is the component of curl $\mathbf{A}$ along the direction $\mathbf{n}$ and it is defined as the limit of a circulation per unit area. The dimensions of $\nabla \times \mathbf{A}$ are, therefore, the dimensions

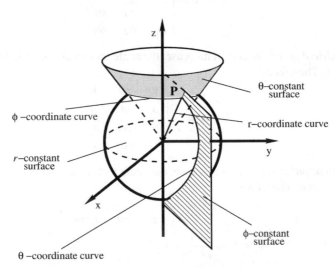

Figure 7.5   Scale factors for the spherical coordinates.

of **A** divided by length. Equation (7.28) is a coordinate-free definition of $\nabla \times \mathbf{A}$ at a point. We have, in fact, used this definition above to obtain the expression for $\nabla \times \mathbf{A}$ in orthogonal curvilinear coordinate systems.

Example 7.2 Determine the scale factors for the spherical coordinate system and express the unit vectors in this system in terms of **i**, **j** and **k**. □

Solution The position vector **R** in spherical coordinate system is given by

$$\mathbf{R} = \mathbf{i}x + \mathbf{j}y + \mathbf{k}z$$
$$= \mathbf{i}(r\sin\theta\cos\phi) + \mathbf{j}(r\sin\theta\sin\phi) + \mathbf{k}(r\cos\theta).$$

Therefore, the unit vectors $\mathbf{r}, \boldsymbol{\theta}, \boldsymbol{\phi}$ along the three directions can be obtained as follows:

$$\frac{\partial \mathbf{R}}{\partial r} = \mathbf{i}\sin\theta\cos\phi + \mathbf{j}\sin\theta\sin\phi + \mathbf{k}\cos\theta,$$

$$\frac{\partial \mathbf{R}}{\partial \theta} = \mathbf{i}r\cos\theta\cos\phi + \mathbf{j}r\cos\theta\sin\phi - \mathbf{k}r\sin\theta,$$

$$\frac{\partial \mathbf{R}}{\partial \phi} = -\mathbf{i}r\sin\theta\sin\phi + \mathbf{j}r\sin\theta\cos\phi + \mathbf{k}0.$$

It can be easily seen that the scale factors are:

$$h_r = \left|\frac{\partial \mathbf{R}}{\partial r}\right| = 1,$$

$$h_\theta = \left|\frac{\partial \mathbf{R}}{\partial \theta}\right| = r,$$

$$h_\phi = \left|\frac{\partial \mathbf{R}}{\partial \phi}\right| = r\sin\theta.$$

Hence, the three unit vectors are:

$$\mathbf{r} = \frac{1}{h_r}\frac{\partial \mathbf{R}}{\partial r} = \mathbf{i}\sin\theta\cos\phi + \mathbf{j}\sin\theta\sin\phi + \mathbf{k}\cos\theta,$$

$$\boldsymbol{\theta} = \frac{1}{h_\theta}\frac{\partial \mathbf{R}}{\partial \theta} = \mathbf{i}\cos\theta\cos\phi + \mathbf{j}\cos\theta\sin\phi - \mathbf{k}\sin\theta,$$

$$\boldsymbol{\phi} = \frac{1}{h_\phi}\frac{\partial \mathbf{R}}{\partial \phi} = -\mathbf{i}\sin\phi + \mathbf{j}\cos\phi.$$

■

Note: It can be easily shown that these coordinates are mutually orthogonal.

Example 7.3   Determine the volume of the elliptical cylinder

$$\frac{x^2}{a^2} + \frac{y^2}{b^2} = 1,$$

between $z = 0$ and $z = h$.   □

Solution   This problem deals with the determination of volume of an elliptical cylinder of height $z = h$ standing on the $z = 0$ plane (Fig. 7.6). There are a number of ways to obtain the volume of this cylinder. But here, we shall take the opportunity to introduce a less well-known orthogonal coordinate system — *elliptical cylindrical coordinates* (u, v, z). These are defined by the transformation of coordinates

$$x = \lambda \cosh u \cos v,$$

$$y = \lambda \sinh u \sin v,$$

$$z = z,$$

in which λ is a parameter to be specified later. The ranges of the new variables are $u \geq 0, 0 \leq v \leq 2\pi$ and $-\infty < z < \infty$.

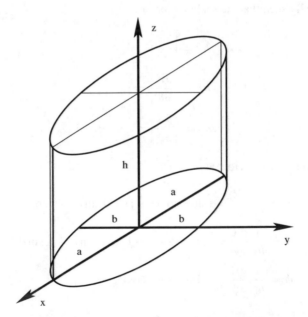

Figure 7.6   Volume of the elliptical cylinder.

The given transformation renders two important coordinate surfaces — hyperbola and ellipse. We study the properties of these coordinates:

- A v constant coordinate surface has representation

$$\frac{x^2}{\lambda^2 \cos^2 v} - \frac{y^2}{\lambda^2 \sin^2 v} = \cosh^2 u - \sinh^2 u = 1, \quad -\infty < z < \infty,$$

which represents a hyperbola drawn out along the $z-$ direction, there being one such hyperbola for each value of v.

- A u constant coordinate surface has representation

$$\frac{x^2}{\lambda^2 \cosh^2 v} + \frac{y^2}{\lambda^2 \sinh^2 v} = \cos^2 u + \sin^2 u = 1, \quad -\infty < z < \infty,$$

which is the equation of a cylinder with an elliptic cross-section, there being one such ellipse for each value of u.

For the particular case, $u = 0$, $x = \lambda \cos v$, $y = 0$ and $-\infty < z < \infty$, which is the equation of the strip $-\lambda < x < \lambda$. For the elliptical cylinder given in the problem, we must have $\lambda^2 \cosh^2 u = a^2$ and $\lambda^2 \sinh^2 u = b^2$. Therefore, subtracting, $\lambda = \sqrt{a^2 - b^2}$. The $u-$ value for this cylinder is clearly given by $u = \sinh^{-1}(b/\lambda)$.

In order to obtain the required volume, we must clearly determine the volume element for this coordinate system and then integrate. This implies that we must first determine the scale factors. It can be easily shown that this coordinate system is orthogonal. The scale factors are obtained using the position vector $\mathbf{R}$ given as

$$\mathbf{R} = \mathbf{i}x + \mathbf{j}y + \mathbf{k}z$$
$$= \mathbf{i}(\lambda \cosh u \cos v) + \mathbf{j}(\lambda \sinh u \sin v) + \mathbf{k}z.$$

We find that

$$h_u = \left| \frac{\partial \mathbf{R}}{\partial u} \right|$$
$$= |\mathbf{i}(\lambda \sinh u \cos v) + \mathbf{j}(\lambda \cosh u \sin v)|$$
$$= \lambda \sqrt{(\sinh^2 u \cos^2 v + \cosh^2 u \sin^2 v)}$$
$$= \lambda \sqrt{(\sinh^2 u + \sin^2 v)}$$
$$= \lambda \sqrt{(\cosh^2 u + \cos^2 v)}.$$

Similarly, we obtain

$$h_v = \left| \frac{\partial \mathbf{R}}{\partial v} \right|$$

$$= |\mathbf{i}(-\lambda \cosh u \sin v) + \mathbf{j}(\lambda \sinh u \cos v)|$$

$$= \lambda \sqrt{(\cosh^2 u \sin^2 v + \sinh^2 u \cos^2 v)}$$

$$= \lambda \sqrt{(\sinh^2 u + \sin^2 v)}$$

$$= \lambda \sqrt{(\cosh^2 u + \cos^2 v)}$$

and

$$h_z = \left| \frac{\partial \mathbf{R}}{\partial u} \right|$$

$$= |\mathbf{k}| = 1.$$

Thus, the volume element in elliptical cylindrical coordinates is given by

$$dV = h_u h_v h_z du dv dz$$

$$= \lambda^2 (\sinh^2 u + \sin^2 v) du dv dz,$$

for our problem. The total volume V of the given elliptical cylinder is then obtained by integrating this elementary volume over the cylinder as z ranges from 0 to h, u ranges from 0 to $\sinh^{-1}(b/\lambda)$ and v ranges from 0 to 2π. Therefore, we have

$$V = \lambda^2 \int_{z=0}^{h} \int_{u=0}^{\sinh^{-1}(\frac{b}{\lambda})} \int_{v=0}^{2\pi} [\sinh^2 u + \sin^2 v] du dv dz$$

$$= \lambda^2 h \pi \int_{u=0}^{\sinh^{-1}(\frac{b}{\lambda})} (2 \sinh^2 u + 1) du$$

$$= \pi h \lambda^2 \int_{u=0}^{\sinh^{-1}(\frac{b}{\lambda})} (\cosh 2u) du$$

$$= \frac{\pi}{2} h \lambda^2 \sinh 2u \Big|_0^{\sinh^{-1}(\frac{b}{\lambda})}$$

$$= \pi h a b.$$

Note: Therefore, the required volume of the elliptic cylinder is $V = \pi h a b$. If $a = b$ for the circular cylinder the volume is $V = \pi a^2 h$. It can be easily noticed that $\mathbf{u} \cdot \mathbf{v} = 0$ and $\mathbf{u} \cdot \mathbf{u} = 1$. A similar trend can be observed for the other unit vectors confirming that these elliptical coordinates are mutually orthogonal. ■

Example 7.4 Verify Stokes' theorem for the vector field $\mathbf{A} = \mathbf{i}(x - y) + \mathbf{j}2z + \mathbf{k}x^2$ using the cone $z = +\sqrt{x^2 + y^2}$, $0 < z < 2$ (Fig. 7.7). $\qquad\square$

Solution Stokes' theorem is as follows:

$$\oint_C \mathbf{A} \cdot d\mathbf{R} = \iint_S \nabla \times \mathbf{A} \cdot \mathbf{n}dS. \tag{7.29}$$

The position vector $\mathbf{R} = \mathbf{i}x + \mathbf{j}y + \mathbf{k}z$ at any point on the cone. But along $C, z = 2$ and the circle is given by $x^2 + y^2 = 4$. Hence, the line integral is

$$\oint_C \mathbf{A} \cdot d\mathbf{R} = \oint_C \{(x - y)dx + 4dy\} = 4\pi.$$

This integral can be readily evaluated to give a value 4π.

The flux integral is more difficult. The unit normal vector to the conical surface is given by

$$\mathbf{n} = \frac{\nabla f}{|\nabla f|},$$

where $f = z - \sqrt{x^2 + y^2}$.

The gradient can be evaluated to yield

$$\nabla f = -\mathbf{i}\frac{x}{z} - \mathbf{j}\frac{y}{z} + \mathbf{k},$$

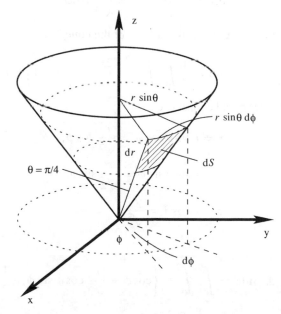

Figure 7.7 Verification of Stokes' theorem using a cone.

so that

$$|\nabla f| = \sqrt{\frac{x^2 + y^2}{z^2} + 1} = \sqrt{2}.$$

Therefore, the unit normal vector is

$$\mathbf{n} = \frac{1}{\sqrt{2}} \left\{ -\mathbf{i}\frac{x}{z} - \mathbf{j}\frac{y}{z} + \mathbf{k} \right\}.$$

Also,

$$\nabla \times \mathbf{A} = \begin{vmatrix} \mathbf{i} & \mathbf{j} & \mathbf{k} \\ \dfrac{\partial}{\partial x} & \dfrac{\partial}{\partial y} & \dfrac{\partial}{\partial z} \\ x - y & 2z & x^2 \end{vmatrix} = -2\mathbf{i} - 2x\mathbf{j} + \mathbf{k}.$$

Thus we have,

$$\nabla \times \mathbf{A} \cdot \mathbf{n} = \frac{1}{\sqrt{2}} \left[\frac{2x}{z} + \frac{2xy}{z} + 1 \right].$$

On the surface of the cone S, the spherical coordinate θ has the value $\pi/4$, so the coordinate r $(= z/\cos\theta)$ varies from 0 to $2\sqrt{2}$ as z varies from 0 to 2.

The surface element on the cone is

$$dS = r\sin\theta d\phi dr = \frac{r}{\sqrt{2}} dr d\phi.$$

Hence, the flux of $\nabla \times \mathbf{A}$ across the surface of the cone is

$$\iint_S \nabla \times \mathbf{A} \cdot \mathbf{n} dS = \frac{1}{2} \int_0^{2\pi} \int_0^{2\sqrt{2}} \left\{ \frac{2x}{z} + \frac{2xy}{z} + 1 \right\} r dr d\phi.$$

Finally, using the spherical coordinates

$$x = r\sin\theta\cos\phi = \frac{r}{\sqrt{2}}\cos\phi,$$

$$y = r\sin\theta\sin\phi = \frac{r}{\sqrt{2}}\sin\phi,$$

$$z = r\cos\theta = \frac{r}{\sqrt{2}},$$

we obtain

$$\iint_S \nabla \times \mathbf{A} \cdot \mathbf{n} dS = \int_0^{2\pi} \int_0^{2\sqrt{2}} \left\{ \cos\phi + \frac{r}{\sqrt{2}}\cos\phi\sin\phi + \frac{1}{2} \right\} r dr d\phi$$

$$= \frac{1}{2} 2\pi \left[\frac{r^2}{2} \right]_0^{2\sqrt{2}} = 4\pi.$$

Thus, this has the same value as the line integral, and, hence, Stokes' theorem is completely verified. ∎

Example 7.5   Verify the divergence theorem for the vector field $\mathbf{F} = 2xz\mathbf{i} + yz\mathbf{j} + z^2\mathbf{k}$ over the upper half of the sphere $x^2 + y^2 + z^2 = a^2$. □

Solution   The divergence theorem states that

$$\oiint_S \mathbf{F} \cdot \mathbf{n} dS = \iiint_V \nabla \cdot \mathbf{F} dV,$$

where S is the surface of the upper hemisphere, V is its interior volume and $\mathbf{n}$ is the outward unit normal vector. We shall evaluate the surface integral first and then the volume integral. Thus we need to determine the unit normal vector $\mathbf{n}$ for the curved surface S_1 and the base of the hemisphere $S_2 = z = 0$. The outward unit normal vector for the base is obviously $\mathbf{n} = -\mathbf{k}$ but for the curved surface we need to evaluate this normal vector using the familiar formula

$$\mathbf{n} = \frac{\nabla f}{|\nabla f|},$$

where $f = x^2 + y^2 + z^2 - a^2$. With this information the divergence formula can be rewritten as

$$\oiint_S \mathbf{F} \cdot \mathbf{n} dS = \iint_{S_1} \mathbf{F} \cdot \mathbf{n} dS_1 + \iint_{S_2=(z=0)} \mathbf{F} \cdot (-\mathbf{k}) dS_2.$$

Now it is an easy matter to find $\mathbf{n} = \nabla f / |\nabla f| = (x/a)\mathbf{i} + (y/a)\mathbf{j} + (z/a)\mathbf{k}$. It can be easily obtained that $\mathbf{F} \cdot \mathbf{n} = 2x^2 z/a + y^2 z/a + z^3/a$ for the S_1 integral and $\mathbf{F} \cdot (-\mathbf{k}) = -z^2$ for the S_2 integral. It is obvious that the integrand of S_2 is zero and so it contributes nothing. Now the problem reduces to evaluate

$$\oiint_S \mathbf{F} \cdot \mathbf{n} dS = \iint_{S_1} \mathbf{F} \cdot \mathbf{n} dS_1$$

$$= \iint_{S_1} \left\{ \frac{2x^2 z}{a} + \frac{y^2 z}{a} + \frac{z^3}{a} \right\} dS_1.$$

To evaluate the right-hand side integral on S_1 we need to transform the variables into spherical coordinates giving

$$x = a \sin\theta \cos\phi, \quad y = a \sin\theta \sin\phi \quad \text{and} \quad z = a\cos\theta,$$

and the $dS_1 = (a d\theta)(a \sin\theta d\phi)$. The limits of θ and ϕ are given by $0 \le \theta \le \pi/2$ and $0 \le \phi \le 2\pi$. Substituting these values into the integral on the right-hand side, we obtain

$$\oiint_S \mathbf{F} \cdot \mathbf{n} dS = \int_{\theta=0}^{\pi/2} \int_{\phi=0}^{2\pi} [2a^2 \sin^2\theta \cos^2\phi \cos\theta + a^2 \sin^2\theta \sin^2\phi \cos\theta$$

$$+ a^2 \cos^3\theta](a^2 \sin\theta d\theta d\phi)$$

$$= 2a^4 \int_0^{\pi/2} \int_0^{2\pi} \sin^3\theta \cos\theta \cos^2\phi d\theta d\phi$$

$$+ a^4 \int_0^{\pi/2} \int_0^{2\pi} \sin^3\theta \cos\theta \sin^2\phi d\theta d\phi$$

$$+ a^4 \int_0^{\pi/2} \int_0^{2\pi} \cos^3\theta \sin\theta d\theta d\phi$$

$$= \frac{2a^4\pi}{4} + \frac{a^4\pi}{4} + \frac{2\pi a^4}{4} = \frac{5a^4\pi}{4}.$$

The volume integral on the right-hand side is simply

$$\iiint_V \nabla \cdot \mathbf{F} dV = \iiint_V 5z dV$$

$$= \int_{r=0}^a \int_{\theta=0}^{\pi/2} \int_{\phi=0}^{2\pi} 5r \cos\theta (r^2 \sin\theta dr d\theta d\phi) = \frac{5a^4\pi}{4}.$$

These two results are identical confirming that the divergence is verified.   ∎

Example 7.6   Verify the divergence theorem for the vector field $\mathbf{F} = y\mathbf{i} + x\mathbf{j} + z^2\mathbf{k}$ over the cylindrical region bounded by $x^2 + y^2 = a^2$, $z = 0$ and $z = h$.   □

Solution   The divergence theorem states that

$$\oiint_S \mathbf{F} \cdot \mathbf{n} dS = \iiint_V \nabla \cdot \mathbf{F} dV,$$

where S is the entire surface area of the cylinder and V is its interior volume. In the cylinder case, there are three surfaces: S_1 is the curved surface, $S_2 = (z = 0)$ base surface and $S_3 = (z = h)$ the top surface. The outward unit normal vectors in these faces are, respectively, $\mathbf{n} = \nabla f/|\nabla f|$ where $f = x^2 + y^2 - a^2$ is the curved surface; $\mathbf{n} = (-\mathbf{k})$ for $S_2 = (z = 0)$ the bottom surface; and $\mathbf{n} = \mathbf{k}$ for $S_3 = (z = h)$ the top surface. The unit normal for the curved surface is obtained as

$$\mathbf{n} = \frac{\nabla f}{|\nabla f|} = \frac{x}{a}\mathbf{i} + \frac{y}{a}\mathbf{j} + 0\mathbf{k}.$$

Thus the left-hand side of the surface integral can be written as

$$\oiint_S \mathbf{F} \cdot \mathbf{n}\,dS = \iint_{S_1} \mathbf{F} \cdot \mathbf{n}\,dS_1 + \iint_{S_2} \mathbf{F} \cdot \mathbf{n}\,dS_2 + \iint_{S_3} \mathbf{F} \cdot \mathbf{n}\,dS_3$$

$$= \iint_{S_1} \frac{2xy}{a}\,dS_1 + \iint_{S_2} (-z^2)\,dS_2 + \iint_{S_3} z^2\,dS_3$$

$$= 2a \int_{\theta=0}^{2\pi}\int_{z=0}^{h} (\cos\theta \sin\theta)(a d\theta dz) + 0 + h^2 \int_{r=0}^{a}\int_{\theta=0}^{2\pi} r\,dr\,d\theta$$

$$= \pi h^2 a^2.$$

The volume integral on the right-hand side is given by

$$\iiint_V \nabla \cdot \mathbf{F}\,dV = \int_{r=0}^{a}\int_{z=0}^{h}\int_{\theta=0}^{2\pi} 2z(r dr d\theta dz).$$

$$- \pi h^2 a^2.$$

These two results are identical and so the divergence theorem is verified.   ■

7.6 Potentials

A particle of unit mass in a gravitational field experiences a force which may be represented by a vector field $\mathbf{F}(\mathbf{R})$. Gravitational field $\mathbf{F}$ must be irrotational; otherwise energy could be extracted from the field by taking the particle around a loop for which the circulation of $\mathbf{F}$ is positive, and so a perpetual motion machine could be devised. For similar reasons electrostatic fields are irrotational. It is conventional to express irrotational force fields $\mathbf{F} = -\nabla\Phi$, where $\Phi = \Phi(\mathbf{R})$ is known as a *potential* for $\mathbf{F}$. Similarly $\Phi(\mathbf{R}) - \Phi(\mathbf{R}_0)$ is a potential, for any choice of reference point $\mathbf{R}_0$, and measures the energy given up by the field when a particle is moved from $\mathbf{R}$ to $\mathbf{R}_0$. Consequently, $\Phi(r)$ $(r = |\mathbf{R}|)$ measures the ability of the field to supply energy. Irrotational force fields are often called *conservative* since the total work required in moving a particle around any loop (closed curve) is always zero. We manifest this concept by a simple example.

Example 7.7   Find the potential $\Phi(r)$ for the electrostatic force field $\mathbf{E} = Q\mathbf{R}/4\pi\varepsilon_0 r^3$ due to an electric charge Q situated at the origin, where $r = |\mathbf{R}| = \sqrt{x^2 + y^2 + z^2}$, and ε_0 is the permittivity of the free space.   □

Solution   We know that the force field $\mathbf{E}$ is conservative and hence $\nabla \times \mathbf{E} = 0$ which implies that $\mathbf{E} = -\nabla\Phi$ such that $\Phi = \Phi(r)$. It is to be noted that $\nabla\Phi$ is everywhere normal to the level surfaces of $\Phi = $ constant. Now the given expression for $\mathbf{E}$ can be written as $\nabla\Phi(r) = -Q\mathbf{R}/4\pi\varepsilon_0 r^3$. This equation yields

$$\frac{d\Phi}{dr}(\nabla r) = -\frac{Q\mathbf{R}}{4\pi\varepsilon_0 r^3},$$

which reduces to

$$\frac{\mathbf{R}}{r}\frac{d\Phi}{dr} = -\frac{Q\mathbf{R}}{4\pi\varepsilon_0 r^3}.$$

Now equating the coefficient of $\mathbf{R}/r$ we obtain the differential equation

$$\frac{d\Phi}{dr} = -\frac{Q}{4\pi\varepsilon_0 r^2}.$$

Note that if we consider the unit radial vector $\mathbf{R}/r = \mathbf{r}$ then using the curvilinear coordinates for $\nabla\Phi$, we will have the same differential equation cited above. Integration with respect to r yields

$$\Phi(r) = \frac{Q}{4\pi\varepsilon_0 r} + C,$$

where C is a constant. It is convenient to set $C = 0$. Then $\Phi(r)$ measures the work stored when a unit charge is brought to $\mathbf{R}$ from infinitely large distance. ∎

7.7 Summary

- The curvilinear coordinates (u_1, u_2, u_3) are defined by the *transformation equations*

$$x = x(u_1, u_2, u_3),$$
$$y = y(u_1, u_2, u_3),$$
$$z = z(u_1, u_2, u_3),$$

together with their inverses

$$u_1 = u_1(x, y, z),$$
$$u_2 = u_2(x, y, z),$$
$$u_3 = u_3(x, y, z).$$

These two set of coordinates are related to the position vector $\mathbf{R} = \mathbf{R}(x, y, z) = \mathbf{R}(u_1, u_2, u_3)$.

- The basic vectors of a curvilinear system are unit vectors e_1, e_2 and e_3 defined by

$$e_1 = \frac{\partial R}{\partial u_1} \bigg/ \left| \frac{\partial R}{\partial u_1} \right|,$$

$$e_2 = \frac{\partial R}{\partial u_2} \bigg/ \left| \frac{\partial R}{\partial u_2} \right|,$$

$$e_3 = \frac{\partial R}{\partial u_3} \bigg/ \left| \frac{\partial R}{\partial u_3} \right|.$$

- The system (u_1, u_2, u_3) is *orthogonal* if e_1, e_2 and e_3 are mutually perpendicular. This implies that

$$e_1 \times e_2 = e_3 = -e_2 \times e_1,$$

$$e_2 \times e_3 = e_1 = -e_3 \times e_2,$$

$$e_3 \times e_1 = e_2 = -e_1 \times e_3,$$

$$e_1 \cdot e_1 = 1,$$

$$e_2 \cdot e_2 = 1,$$

$$e_3 \cdot e_3 = 1,$$

$$e_1 \cdot e_2 = e_2 \cdot e_3 = e_3 \cdot e_1 = 0.$$

- The *scale factors* are defined by

$$h_1 = \left| \frac{\partial R}{\partial u_1} \right|,$$

$$h_2 = \left| \frac{\partial R}{\partial u_2} \right|,$$

$$h_3 = \left| \frac{\partial R}{\partial u_3} \right|.$$

- The *differential element of arc lengths* along u_1, u_2, u_3 coordinate curves, respectively, are given by $h_1 du_1, h_2 du_2, h_3 du_3$.
- The *differential elements of area* on the coordinate surfaces $u_1 = $ constant, $u_2 = $ constant and $u_3 = $ constant are given by $h_2 h_3 du_2 du_3$, $h_1 h_3 du_1 du_3$, $h_1 h_2 du_1 du_2$, respectively.
- The *differential element of volume* is $h_1 h_2 h_3 du_1 du_2 du_3$.
- The three major orthogonal coordinate systems are:
  (a) Cartesian $(u_1, u_2, u_3) \rightarrow (x, y, z)$; $h_1 = 1, h_2 = 1, h_3 = 1$; $e_1 = i$, $e_2 = j$, $e_3 = k$.
  (b) Cylindrical $(u_1, u_2, u_3) \rightarrow (r, \theta, z)$; $h_1 = 1, h_2 = r, h_3 = 1$; $e_1 = r$, $e_2 = \theta$, $e_3 = k$.
  (c) Spherical $(u_1, u_2, u_3) \rightarrow (r, \theta, \phi)$; $h_1 = 1, h_2 = r, h_3 = r \sin \theta$; $e_1 = r$, $e_2 = \theta$, $e_3 = \phi$.

- If G is a scalar function and $\mathbf{A} = a_1\mathbf{e_1} + a_2\mathbf{e_2} + a_3\mathbf{e_3}$ is a vector function of orthogonal curvilinear coordinates, then

$$\nabla G = \frac{\mathbf{e_1}}{h_1}\frac{\partial G}{\partial u_1} + \frac{\mathbf{e_2}}{h_2}\frac{\partial G}{\partial u_2} + \frac{\mathbf{e_3}}{h_3}\frac{\partial G}{\partial u_3},$$

$$\nabla \cdot \mathbf{A} = \frac{1}{h_1 h_2 h_3}\left\{ \frac{\partial}{\partial u_1}(a_1 h_2 h_3) + \frac{\partial}{\partial u_2}(a_2 h_1 h_3) + \frac{\partial}{\partial u_3}(a_3 h_1 h_2) \right\},$$

$$\nabla \times \mathbf{A} = \frac{1}{h_1 h_2 h_3}\begin{vmatrix} h_1\mathbf{e_1} & h_2\mathbf{e_2} & h_3\mathbf{e_3} \\ \dfrac{\partial}{\partial u_1} & \dfrac{\partial}{\partial u_2} & \dfrac{\partial}{\partial u_3} \\ h_1 a_1 & h_2 a_2 & h_3 a_3 \end{vmatrix},$$

$$\nabla^2 G = \frac{1}{h_1 h_2 h_3}\left\{ \frac{\partial}{\partial u_1}\left(\frac{h_2 h_3}{h_1}\frac{\partial G}{\partial u_1} \right) + \frac{\partial}{\partial u_2}\left(\frac{h_1 h_3}{h_2}\frac{\partial G}{\partial u_2} \right) \right.$$
$$\left. + \frac{\partial}{\partial u_3}\left(\frac{h_1 h_2}{h_3}\frac{\partial G}{\partial u_3} \right) \right\}.$$

Exercises

1. Determine the scale factors for the Cartesian coordinate system and express the unit vectors in this system in terms of $\mathbf{i}$, $\mathbf{j}$ and $\mathbf{k}$.
2. Determine the scale factors for the elliptic cylindrical coordinate system and express the unit vectors in this system in terms of $\mathbf{i}$, $\mathbf{j}$ and $\mathbf{k}$.
3. Determine the scale factors for the spherical coordinate system and express the unit vectors in this system in terms of $\mathbf{i}$, $\mathbf{j}$ and $\mathbf{k}$.
4. Verify the relation $\mathbf{r} \times \boldsymbol{\theta} = \boldsymbol{\phi}, \boldsymbol{\theta} \times \boldsymbol{\phi} = \mathbf{r}$ and $\boldsymbol{\phi} \times \mathbf{r}$ in spherical polar, and the relation $\mathbf{r} \times \boldsymbol{\phi} = \mathbf{k}, \boldsymbol{\phi} \times \mathbf{k} = \mathbf{r}$ and $\mathbf{k} \times \mathbf{r} = \boldsymbol{\theta}$ in cylindrical polar.
5. Determine the gradients of the following scalar fields: (a) r^2; (b) $1/r$.
6. Verify the following equations for general orthogonal curvilinear coordinates:
   (a) $\mathbf{e_1} = h_2 h_3 \nabla u_2 \times \nabla u_3$.
   (b) $\mathbf{e_2} = h_3 h_1 \nabla u_3 \times \nabla u_1$.
   (c) $\mathbf{e_3} = h_1 h_2 \nabla u_1 \times \nabla u_2$.
7. Determine the volume of the elliptical cylinder

$$\frac{x^2}{9} + \frac{y^2}{4} = 1,$$

between $z = 0$ and $z = 4$.
8. Use appropriate coordinates to determine the Laplacian $\nabla^2 G$ if
   (a) $G = \ln \sqrt{(x^2 + y^2)}$.
   (b) $G = \ln \sqrt{(x^2 + y^2 + z^2)}$.
9. If $\phi = \sqrt{x^2 + y^2 + z^2}$ and $\psi = xy/z^2$, calculate $\nabla\phi$ and $\nabla\psi$. Show that the scalar product $(\nabla\phi) \cdot (\nabla\psi)$ is zero. Can you think of any way by which this

result could have been predicted in advance without separately calculating $\nabla\phi$ and $\nabla\psi$?

10. Verify the divergence theorem by evaluating the double and triple integrals as shown below:

$$\iiint_V \nabla \cdot \mathbf{F} dV = \iint_S \mathbf{F} \cdot \mathbf{n} dS,$$

where V is the sphere $x^2 + y^2 + z^2 \le a^2$ and $\mathbf{F} = \mathbf{i}(2xy^2 + 2xz^2) + \mathbf{j}x^2 y + \mathbf{k}x^2 z$.

11. If $\mathbf{F} = \mathbf{i}2x^2 y + \mathbf{j}(xz^2 - y^3) + \mathbf{k}xyz$, calculate div $\mathbf{F}$, curl $\mathbf{F}$ and div curl $\mathbf{F}$.

12. If $\mathbf{U}$ and $\mathbf{V}$ are any two vector fields, show that the divergence of their vector product takes the form

$$\nabla \cdot (\mathbf{U} \times \mathbf{V}) = \mathbf{V} \cdot \nabla \times \mathbf{U} - \mathbf{U} \cdot \nabla \times \mathbf{V}.$$

By applying the divergence theorem to $\mathbf{U} \times \mathbf{V}$ where $\mathbf{V}$ is any *constant vector*, show that

$$\iiint_V \nabla \times \mathbf{U} dV = \iint_S \mathbf{n} \times \mathbf{U} dS.$$

13. For the vector field $\mathbf{F}$ defined as

$$\mathbf{F} = \mathbf{i}(z - 2xr^{-1}) + \mathbf{j}(2y - 3z - 2yr^{-1}) + \mathbf{k}(x - 3y - 2zr^{-1}),$$

where $r = \sqrt{x^2 + y^2 + z^2}$, prove that $\nabla \times \mathbf{F} = 0$ and find the scalar potential ϕ such that $\mathbf{F} = \nabla\phi$.

14. Show that the vector $\mathbf{A} = \mathbf{r}A_r + \boldsymbol{\theta} A_\theta + \mathbf{k}A_z$ in cylindrical coordinates can be converted into Cartesian coordinates:

$$\mathbf{A} = \mathbf{i}\left\{ \frac{xA_r}{\sqrt{x^2 + y^2}} - \frac{yA_\theta}{\sqrt{x^2 + y^2}} \right\} + \mathbf{j}\left\{ \frac{yA_r}{\sqrt{x^2 + y^2}} + \frac{xA_\theta}{\sqrt{x^2 + y^2}} \right\} + \mathbf{k}A_z.$$

Chapter 8

Applications

8.1 Introduction

In this chapter we will consider a few applications concerning fluid flow in compressible and incompressible fluids, namely Poisson's equation, electromagnetic theory, Maxwell's wave equations and nonlinear ocean wave equations. A small section on the graphical simulation of a vector field has also been included. It is well known that one of the most important uses of vector analysis is in the concise formulation of physical laws and the derivation of other results from those laws.

We develop the mathematical formulae governing these problems using vector analysis in a concise form, and then we illustrate these theories with the practical problems cited above. We have selected elementary problems to demonstrate the usefulness of vector analysis.

Sections 8.15 and 8.16 deal with nonlinear ocean waves with resonance conditions. This material is very advanced and requires an advanced knowledge of the physical situation of the problem. The section will be useful to those working on the extraction of energy from nonlinear ocean waves. The last section demonstrates the computer simulation of a vector field with the help of *The Mathematica*.

8.2 Acceleration vector

The characteristics of a fluid as it flows past any fixed point in space are a function of the position vector $\mathbf{R} = x\mathbf{i} + y\mathbf{j} + z\mathbf{k}$ and the time variable t. As a particular instance, the velocity vector $\mathbf{V}$, at a point will be a function of the independent variables $\mathbf{R}$ and t. When obtaining the general equations of motion, the general acceleration vector of a moving fluid particle corresponding to a fixed set of orthogonal axes is established and then applied to a moving set of orthogonal axes.

Suppose that we let the fluid particle in space occupy the position P at time t. This is equivalent to letting the position vector P be given by the vector $\mathbf{R}$. We can also let the resultant velocity at P be given by $\mathbf{V} = \mathbf{V}(r, t)$. If after an interval of time δt, the fluid particle moves to the new position P', we can write the new position vector as $\mathbf{R} + \mathbf{V}\delta t$. The velocity vector of this new position vector at P' can be written as $\mathbf{V} + \delta\mathbf{V} = \mathbf{V}(\mathbf{R} + \mathbf{V}\delta t, t + \delta t)$ (see Fig. 8.1). This leads us to the following expression:

$$\delta\mathbf{V} = \mathbf{V}(\mathbf{R} + \mathbf{V}\delta t, t + \delta t) - \mathbf{V}(\mathbf{R}, t)$$
$$= \mathbf{V}(\mathbf{R} + \mathbf{V}\delta t, t + \delta t) - \mathbf{V}(\mathbf{R}, t + \delta t)$$
$$+ \mathbf{V}(\mathbf{R}, t + \delta) - \mathbf{V}(\mathbf{R}, t).$$

By Taylor's theorem, the preceding equation becomes

$$\mathbf{V}(\mathbf{R} + \mathbf{V}\delta t, t + \delta t) - \mathbf{V}(\mathbf{R}, t + \delta t)$$
$$= (\mathbf{V}\delta t \cdot \mathrm{grad})\mathbf{V}(\mathbf{R}, t + \delta t) + O(\delta t^2),$$

and

$$\mathbf{V}(\mathbf{R}, t + \delta t) - \mathbf{V}(\mathbf{R}, t) = \left(\frac{\partial\mathbf{V}(\mathbf{R}, t)}{\partial t}\right)\delta t + O(\delta t^2).$$

By addition, we get

$$\delta\mathbf{V} = \left(\frac{\partial\mathbf{V}(\mathbf{R}, t)}{\partial t} + (\mathbf{V} \cdot \mathrm{grad})\mathbf{V}(\mathbf{R}, t + \delta t)\right)\delta t + O(\delta t^2).$$

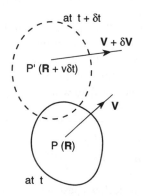

Figure 8.1   Acceleration vector.

Now if we divide throughout by δt and let $\delta t \to 0$,

$$\lim_{\delta t \to 0} \left(\frac{\delta \mathbf{V}}{\delta t}\right) = \left(\frac{\partial \mathbf{V}}{\partial t}\right) + (\mathbf{V} \cdot \text{grad})\mathbf{V}.$$

Then from the definition of differentiation, the total rate of change $D\mathbf{V}/Dt$ is defined as the rate of change of $\mathbf{V}$ following a particle of fluid

$$\frac{D\mathbf{V}}{Dt} = \frac{\partial \mathbf{V}}{\partial t} + (\mathbf{V} \cdot \text{grad})\mathbf{V}, \tag{8.1}$$

which is nothing but the sum of the local and convective rates of change of $\mathbf{V}$. In this expression, [equation (8.1)], the operator $(D/Dt) = (\partial/\partial t) + (\mathbf{V} \cdot \text{grad})$ is the total differentiation following the motion of the fluid. The interested reader is referred to the work by Lighthill (1986) which gives a very elegant description of this operator from the physical point of view.

The operator D/Dt is given below for the three commonly used orthogonal coordinate axes.

1. *Rectangular Cartesian coordinates*: The velocity vector is given by $\mathbf{V} = (u, v, w)$ (see Fig. 8.2). Elementary lengths are $(\delta x, \delta y, \delta z)$,

$$\text{grad} = \nabla = [(\partial/\partial x), (\partial/\partial y), (\partial/\partial z)].$$

Hence, we have the following:

$$\frac{D}{Dt} = \frac{\partial}{\partial t} + u\frac{\partial}{\partial x} + v\frac{\partial}{\partial y} + w\frac{\partial}{\partial z}.$$

2. *Cylindrical polar coordinates*: The well-known relationship between Cartesian and cylindrical polar coordinates is written as $x = r\cos\theta$, $y = r\sin\theta$ and $z = z$ (see Fig. 8.3). The elementary lengths in this case are given by $(\delta r, r\delta\theta, \delta z)$. The velocity vector $\mathbf{V} = (u_r, v_\theta, v_z)$ and

$$\nabla = [(\partial/\partial r), (\partial/r\partial\theta), (\partial/\partial z)].$$

This leads us to

$$\frac{D}{Dt} = \frac{\partial}{\partial t} + v_r\frac{\partial}{\partial r} + \frac{v_\theta}{r}\frac{\partial}{\partial\theta} + v_z\frac{\partial}{\partial z}.$$

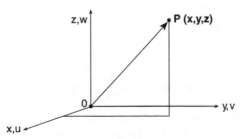

Figure 8.2   Cartesian coordinate axes.

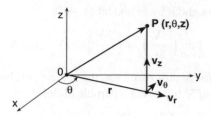

Figure 8.3   Cylindrical polar coordinate axes.

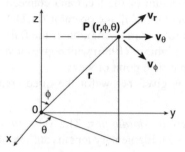

Figure 8.4   Spherical polar coordinate axes.

3. *Spherical polar coordinates*: The relationship between Cartesian and spherical polar coordinates can be written as $x = (r \sin \phi) \cos \theta$, $y = (r \sin \phi) \sin \theta$, and $z = r \cos \phi$ (see Fig. 8.4). The corresponding elementary lengths are given by $(\delta r, r\delta\phi, r \sin \phi \delta \theta)$, the velocity vector $\mathbf{V} = (v_r, v_\phi, v_\theta)$ and $\nabla = [(\partial/\partial r), (\partial/r\partial\phi), (1/r \sin \phi)(\partial/\partial\theta)]$. Therefore, we obtain

$$\frac{D}{Dt} = \frac{\partial}{\partial t} + v_r \frac{\partial}{\partial r} + \frac{v_\phi}{r} \frac{\partial}{\partial \phi} + \frac{v_\theta}{r \sin \phi} \frac{\partial}{\partial \theta}.$$

For a Cartesian coordinate system, the components of the velocity vector $\mathbf{V}$ would (at all times and at all points in space) be parallel to the Cartesian axes. However, when the velocity vector $\mathbf{V}$ is considered under a polar coordinate system, there is a rotation of the axes. At time t the position of P' will be $\mathbf{R}(r, \phi, \theta)$ where the velocity is given by $\mathbf{V}(v_r, v_\phi, v_\theta)$.

After the time interval δt, the fluid will have moved to position P''. At this position, the components of velocity, v_r, v_ϕ and v_θ will now be parallel to the axes r', ϕ' and θ'. Now if we refer to Curle and Davies (1968), we can define the acceleration of the fluid particle in the moving frame of axes as

$$\text{Acceleration} = \frac{D\mathbf{V}}{Dt} + [\mathbf{\Omega} \times \mathbf{V}],$$

where $\mathbf{\Omega}$ is the angular velocity vector of axes.

In the spherical polar coordinate system, the angular velocity vector is given by

$$\mathbf{\Omega} = [(v_\theta/r) \cot \phi, -(v_\theta/r), (v_\phi/r)]$$

and the linear velocity vector can be written as $\mathbf{V} = (v_r, v_\phi, v_\theta)$. This will then lead to

$$[\boldsymbol{\Omega} \times \mathbf{V}] = \begin{vmatrix} \mathbf{i} & \mathbf{j} & \mathbf{k} \\ \dfrac{v_\theta}{r} \cot\phi & -\dfrac{v_\theta}{r} & \dfrac{v_\phi}{r} \\ v_r & v_\phi & v_\theta \end{vmatrix}.$$

Expansion of this vector product leads to the component

$$\left(-\frac{v_\theta^2 + v_\phi^2}{r}, \quad \frac{v_r v_\phi}{r} - \frac{v_\theta^2}{r} \cot\phi, \quad \frac{v_\theta v_\phi}{r} \cot\phi + \frac{v_r v_\theta}{r} \right).$$

It follows that the components of the acceleration vector in spherical polar coordinates can be written in the following manner

$$\frac{Dv_r}{Dt} - \frac{v_\theta^2 + v_\phi^2}{r}, \quad \frac{Dv_\phi}{Dt} + \frac{v_r v_\phi}{r} - \frac{v_\theta^2}{r} \cot\phi, \quad \frac{Dv_\theta}{Dt} + \frac{v_r v_\theta}{r} + \frac{v_\theta v_\phi}{r} \cot\phi.$$

In the cylindrical polar coordinate axes, the angular velocity components are $\boldsymbol{\Omega} = [0, 0, (v_\theta/r)]$.

Note that this result can be obtained from the angular velocity vector $\boldsymbol{\Omega}$ with respect to the spherical polar coordinate system by changing ϕ to θ and then letting $v_\phi = 0$. The linear velocity vector is $\mathbf{V} = (v_r, v_\theta, v_z)$. This leads to

$$[\boldsymbol{\Omega} \times \mathbf{V}] = \begin{vmatrix} \mathbf{i} & \mathbf{j} & \mathbf{k} \\ 0 & 0 & \dfrac{v_\theta}{r} \\ v_r & v_\theta & v_z \end{vmatrix} = \left(-\frac{v_\theta^2}{r}, \quad \frac{v_r v_\theta}{r}, \quad 0 \right).$$

Thus the components of the acceleration vector in a cylindrical polar coordinate system can be written as

$$\left(\frac{Dv_r}{Dt} - \frac{v_\theta^2}{r}, \quad \frac{Dv_\theta}{Dt} + \frac{v_r v_\theta}{r}, \quad \frac{Dv_z}{Dt} \right).$$

8.3 Continuity equation of fluid flow

In many branches of engineering and physics we encounter the concepts of *flux* and *circulation*. We will explore the relationships between flux and divergence and between circulation and curl.

Consider a gas flowing through a region D of space. At time t and point $P(x, y, z)$ in D, gas flows through P with velocity $\mathbf{V}(x, y, z, t)$. If A is a unit area around P perpendicular to $\mathbf{V}$ (Fig. 8.5a) and $\rho(x, y, z, t)$ is the density of the gas at P, then

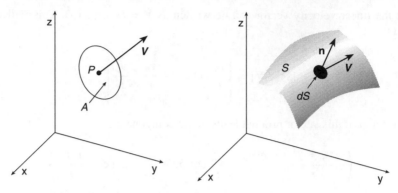

Figure 8.5   Gas flowing through (a) P and (b) surface S.

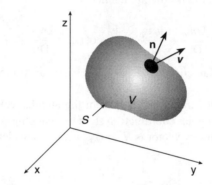

Figure 8.6   Closed surface S with unit outer normal **n**.

the amount of gas crossing A per unit area is $\rho\mathbf{V}$. At every point P in D, the vector $\rho\mathbf{V}$ is such that its direction $\mathbf{V}$ gives the velocity of gas flow, and its magnitude $\rho|\mathbf{V}|$ describes the mass of gas flowing in that direction per unit time.

Consider some surface S in D (Fig. 8.5b). If $\mathbf{n}$ is a unit normal to S, then $\rho\mathbf{V} \cdot \mathbf{n}$ is the component of $\rho\mathbf{V}$ normal to the surface S. If dS is an element of area on S, then $\rho\mathbf{V} \cdot \mathbf{n}\,dS$ describes the mass of gas flowing through dS per unit time. Consequently,

$$\iint_S \rho\mathbf{V} \cdot \mathbf{n}\,dS$$

is the mass of gas flowing through S per unit time. This quantity, as mentioned before, is known as the flux for surface S. If S is a closed surface (Fig. 8.6) and $\mathbf{n}$ is the unit outward normal to S then

$$\iint_S \rho\mathbf{V} \cdot \mathbf{n}\,dS$$

is the mass of gas flowing out of surface S per unit time. If this flux (for closed S) is positive then there is a net outward flow of gas through (i.e. more gas is leaving the volume bounded by S than is entering); if the flux is negative, the net flow is inward.

If we now apply the divergence theorem to the flux integral over the closed surface S, we have

$$\iint_S \rho \mathbf{V} \cdot \mathbf{n}\, dS = \iiint_V \nabla \cdot (\rho \mathbf{V})\, dV. \qquad (8.2)$$

Now the flux (the integral on the left of this equation) is the mass of gas per unit time leaving S. For the right-hand side to represent the same quantity, $\nabla \cdot (\rho \mathbf{V})$ must be interpreted as the mass of gas leaving unit volume per unit time, because then $\nabla \cdot (\rho \mathbf{V})\, dV$ represents the mass per unit time leaving dV and the triple integral is the mass per unit time leaving V. We have thus obtained an interpretation of the divergence of $\rho \mathbf{V}$. The divergence of $\rho \mathbf{V}$ is the flux per unit volume per unit time at a point: the mass of gas leaving unit volume in unit time. We can use this idea to derive the *equation of continuity* for fluid flow. The triple integral

$$\iiint_V \frac{\partial \rho}{\partial t}\, dV - \frac{\partial}{\partial t} \iiint_V \rho\, dV$$

measures the rate of change of mass in a volume V. If this triple integral is positive, then there is a net inward flow of mass; if it is negative, the net flow is outward. We conclude that this triple integral must be the negative of the flux for the volume V, i.e.

$$\iiint_V \frac{\partial \rho}{\partial t}\, dV = - \iiint_V \nabla \cdot (\rho \mathbf{V})\, dV,$$

or

$$\iiint_V \left(\nabla \cdot (\rho \mathbf{V}) + \frac{\partial \rho}{\partial t} \right) dV = 0.$$

If $\nabla \cdot (\rho \mathbf{V})$ and $\partial \rho / \partial t$ are continuous functions, then this equation can hold for arbitrary volume V only if

$$\nabla \cdot (\rho \mathbf{V}) + \frac{\partial \rho}{\partial t} = 0. \qquad (8.3)$$

This is called the *equation of continuity* and expresses conservation of mass. It is basic to all fluid flow. If C is a closed curve in the flow region D, then the *circulation* of the flow for the curve C is defined by

$$\Gamma = \oint_C \mathbf{V} \cdot d\mathbf{R}. \qquad (8.4)$$

To obtain an intuitive feeling for Γ, we consider two very simple two-dimensional flows. First, suppose $\mathbf{V} = x\mathbf{i} + y\mathbf{j}$, so that all particles of gas flow along radial lines directed away from the origin as shown in Fig. 8.7. In this case, the line integral defining Γ is independent of path and $\Gamma = 0$ for any curve whatsoever.

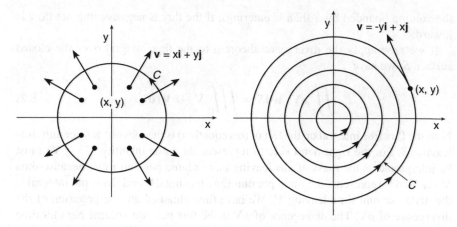

Figure 8.7   Two-dimensional gas flows.

Second, suppose $\mathbf{V} = -y\mathbf{i} + x\mathbf{j}$, so that all particles of the gas flow counter-clockwise around circles centred at the origin (Fig. 8.7). In this case Γ does not generally vanish. In particular, if C is the circle $x^2 + y^2 = r^2$, then $\mathbf{V}$ and $d\mathbf{R}$ are parallel, and

$$\Gamma = \oint_C \mathbf{V} \cdot d\mathbf{R}$$
$$= \oint_C \mathbf{V} \cdot \frac{d\mathbf{R}}{dS} dS,$$

which can be written as

$$\Gamma = \oint_C \mathbf{V} \cdot \frac{(d\mathbf{R}/dt)}{(dS/dt)} \, dS,$$

which in turn becomes

$$\Gamma = \oint_C \frac{\mathbf{V} \cdot \mathbf{V}}{|\mathbf{V}|} \, dS$$
$$= \oint_C |\mathbf{V}| \, dS,$$

which leads to

$$\Gamma = \oint_C \sqrt{x^2 + y^2} \, dS = \oint_C r \, dS$$
$$= 2\pi r^2.$$

These two flow patterns indicate perhaps that circulation is a measure of the tendency for the flow to be circulatory. If we apply Stokes' theorem to the circulation integral for the closed curve C, we have

$$\oint_C \mathbf{V} \cdot d\mathbf{R} = \iint_S (\nabla \times \mathbf{V}) \cdot d\mathbf{S} \tag{8.5}$$

where S is any surface in the flow with boundary C. If the right-hand side of this equation is to represent the circulation for C also, then $(\nabla \times \mathbf{V}) \cdot \mathbf{n} \, dS$ must be interpreted as the circulation for the curve bounding dS (or simply dS itself). Then the addition process of the surface integral (Fig. 8.8) gives the circulation around C, the circulation around all internal boundaries cancelling. But if $(\nabla \times \mathbf{V}) \cdot \mathbf{n} \, dS$ is the circulation for dS, then it follows that $(\nabla \times \mathbf{V}) \cdot$ must be the circulation for a unit area perpendicular to $\mathbf{n}$. Thus $\nabla \times \mathbf{V}$ describes the circulatory nature of the flow $\mathbf{V}$; its component $(\nabla \times \mathbf{V}) \cdot \mathbf{n}$ in any direction describes the circulation for unit area perpendicular to $\mathbf{n}$.

If we examine the continuity equation more closely, we can observe some additional physical applications

$$\nabla \cdot (\rho \mathbf{V}) + \frac{\partial \rho}{\partial t} = 0. \tag{8.6}$$

For the special case of an incompressible fluid, $\partial \rho / \partial t = 0$ so that

$$\nabla \cdot (\rho \mathbf{V}) = \rho \nabla \cdot \mathbf{V} = 0 \quad \text{or} \quad \nabla \cdot \mathbf{V} = 0.$$

In other words, the velocity vector of an incompressible fluid is one whose divergence vanishes everywhere within a region. The technical term for such vector fields is *solenoidal*.

If the fluid is irrotational, then the curl of the velocity vector must be zero, i.e.

$$\nabla \times \mathbf{V} = \mathbf{0},$$

which implies that $\mathbf{V}$ must be the gradient of a scalar function. If $\phi(x, y, z)$ is the scalar function which is known as the velocity potential, then

$$\mathbf{V} = \nabla \phi.$$

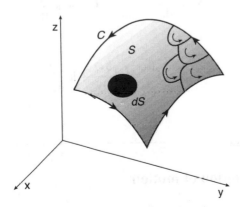

Figure 8.8   Surface integral giving circulation around C.

If the fluid is incompressible, then $\nabla \cdot \mathbf{V} = 0$ and, therefore,

$$\nabla \cdot \nabla \phi = 0$$

or

$$\nabla^2 \phi = 0.$$

This equation is known as Laplace's equation which is satisfied by the velocity potential ϕ.

Remark 8.1   Let m denote any general property of a particle of the fluid, such as its pressure, density, etc., by the function $Q(x, y, z, t)$. Thus by $\partial Q/\partial t$ is meant the variation of Q at a particular point in space as a function of time t. If we take the total differential $Q(x, y, z, t)$ with respect to time t, we obtain

$$\frac{\mathrm{d}Q}{\mathrm{d}t} = \frac{\partial Q}{\partial t} + \frac{\partial Q}{\partial x}\frac{\mathrm{d}x}{\mathrm{d}t} + \frac{\partial Q}{\partial y}\frac{\mathrm{d}y}{\mathrm{d}t} + \frac{\partial Q}{\partial z}\frac{\mathrm{d}z}{\mathrm{d}t}. \tag{8.7}$$

The quantity $\mathrm{d}Q/\mathrm{d}t$ is called the total derivative or material derivative of Q with respect to time when we fix our attention on the same particle of fluid. Equation (8.7) can very easily be put into vector form as follows

$$\frac{\mathrm{d}Q}{\mathrm{d}t} = \frac{\partial Q}{\partial t} + \nabla Q \cdot \frac{\mathrm{d}\mathbf{R}}{\mathrm{d}t}$$

$$= \frac{\partial Q}{\partial t} + \mathbf{V} \cdot \nabla Q, \tag{8.8}$$

where

$$\mathbf{V} = \frac{\mathrm{d}\mathbf{R}}{\mathrm{d}t} = \mathbf{i}\frac{\mathrm{d}x}{\mathrm{d}t} + \mathbf{j}\frac{\mathrm{d}y}{\mathrm{d}t} + \mathbf{k}\frac{\mathrm{d}z}{\mathrm{d}t}$$

and

$$\nabla Q = \mathbf{i}\frac{\partial Q}{\partial x} + \mathbf{j}\frac{\partial Q}{\partial y} + \mathbf{k}\frac{\partial Q}{\partial z}.$$

■

8.4 Euler's equation of motion

To develop the equation of motion of a non-viscous fluid, we consider the forces acting on an element of fluid with elementary volume $\mathrm{d}x\,\mathrm{d}y\,\mathrm{d}z$ such that if $\rho(x, y, z, t)$

is the density of fluid, then its mass is $\rho\,dx\,dy\,dz$. Thus if $P(x, y, z, t)$ is the pressure acting on the faces of the elementary volume of the fluid, then there will be a force in the x-direction on the element of the fluid under consideration of the magnitude

$$P\,dydz - \left(P + \frac{\partial P}{\partial x}dx\right)dydz = -\frac{\partial P}{\partial x}dxdydz. \qquad (8.9)$$

Let us also consider the action of an external force $\mathbf{F} = (F_x, F_y, F_z)$ acting on a unit mass of fluid. The external force acting on the element of fluid in the x-direction is $F_x(\rho\,dx\,dy\,dz)$. The acceleration of the element in the x-direction is du/dt, where u is the x component of the velocity vector $\mathbf{V} = (u, v, w)$. Hence, by Newton's second law of motion, we have

$$\left(\frac{Du}{Dt}\right)(\rho dxdydz) = F_x(\rho dxdydz) - \left(\frac{\partial P}{\partial x}\right)dxdydz.$$

Dividing throughout by $\rho\,dx\,dy\,dz$, we obtain

$$\frac{Du}{Dt} = F_x - \frac{1}{\rho}\frac{\partial P}{\partial x}.$$

By this reasoning, and considering the y and z direction, we obtain Euler's equations of motion:

$$\left.\begin{array}{l} \dfrac{Du}{dt} = F_x - \dfrac{1}{\rho}\dfrac{\partial P}{\partial x} \\[2mm] \dfrac{Dv}{dt} = F_y - \dfrac{1}{\rho}\dfrac{\partial P}{\partial y} \\[2mm] \dfrac{Dw}{dt} = F_z - \dfrac{1}{\rho}\dfrac{\partial P}{\partial z}. \end{array}\right\} \qquad (8.10)$$

The three scalar equations may be combined into the single vector equation

$$\frac{D\mathbf{V}}{Dt} = \mathbf{F} - \frac{1}{\rho}\nabla P. \qquad (8.11)$$

But

$$\frac{D\mathbf{V}}{Dt} = \frac{\partial \mathbf{V}}{\partial t} + (\mathbf{V}\cdot\nabla)\mathbf{V}. \qquad (8.12)$$

Therefore, (8.11) can be written as

$$\frac{\partial \mathbf{V}}{\partial t} + (\mathbf{V}\cdot\nabla)\mathbf{V} = \mathbf{F} - \frac{1}{\rho}\nabla P. \qquad (8.13)$$

This equation is usually known as Euler's equation of motion for inviscid fluids.

8.5 Bernoulli's equation

We can derive Bernoulli's equation from Euler's equation when the flow is irrotational with a conservative gravity force as an external force and if the fluid is incompressible. The following analysis is presented to derive this important equation of immense application in practical situations.

By the vector identity,

$$\nabla(\mathbf{U} \cdot \mathbf{V}) = (\mathbf{U} \cdot \nabla)\mathbf{V} + (\mathbf{V} \cdot \nabla)\mathbf{U} + \mathbf{U} \times (\nabla \times \mathbf{V})$$
$$+ \mathbf{V} \times (\nabla \times \mathbf{U}),$$

if we let $\mathbf{U} = \mathbf{V}$, we get

$$(\mathbf{V} \cdot \nabla)\mathbf{V} = \frac{1}{2}\nabla q^2 - \mathbf{V} \times (\nabla \times \mathbf{V}), \tag{8.14}$$

where $q^2 = \mathbf{V} \cdot \mathbf{V} = u^2 + v^2 + w^2$. Hence, by using this relationship in (8.13) we may write Euler's equation of motion in the following form:

$$\frac{\partial \mathbf{V}}{\partial t} + \frac{1}{2}\nabla q^2 - \mathbf{V} \times (\nabla \times \mathbf{V}) = \mathbf{F} - \frac{1}{\rho}\nabla P. \tag{8.15}$$

If the external $\mathbf{F}$ is conservative, it has a potential V and we have

$$\mathbf{F} = -\nabla V.$$

If the motion of the fluid is irrotational, $\nabla \times \mathbf{V} = \mathbf{0}$ and $\mathbf{V} = \nabla \phi$, where ϕ is a scalar potential. In this case, equation (8.15) can be written as

$$\nabla\left(\frac{\partial \phi}{\partial t}\right) + \frac{1}{2}\nabla q^2 + \nabla V + \frac{1}{\rho}\nabla P = 0 \tag{8.16}$$

or

$$\nabla W = 0, \tag{8.17}$$

where

$$W = \frac{\partial \phi}{\partial t} + \frac{1}{2}q^2 + V + \frac{P}{\rho}.$$

Now if $d\mathbf{R} = \mathbf{i}dx + \mathbf{j}dy + \mathbf{k}dz$ denotes an arbitrary point in the fluid at any instant, we have by the scalar product of ∇W and $d\mathbf{R}$

$$\nabla W \cdot d\mathbf{R} = \frac{\partial W}{\partial x}dx + \frac{\partial W}{\partial y}dy + \frac{\partial W}{\partial z}dz = dW,$$

such that from (8.17)

$$dW = 0$$

and consequently after integration,

$$W = c(t), \tag{8.18}$$

where $c(t)$ is an arbitrary function of time, since we have integrated along an arbitrary path in the fluid at any instant. Thus we have

$$W = \frac{\partial \phi}{\partial t} + \frac{1}{2}q^2 + V + \int \frac{dP}{\rho} = c(t). \tag{8.19}$$

Equation (8.19) is known as Bernoulli's equation. In the special case where ρ is a constant and the motion is steady so that $\partial \phi / \partial t = 0$, this equation becomes

$$\frac{q^2}{2} + V + \frac{P}{\rho} = c, \tag{8.20}$$

where c is now a constant. This equation states that per unit mass of fluid, the sum of the kinetic energy, the potential energy V, and the pressure energy P/ρ has a constant volume c for all points of the fluid.

Remark 8.2   In aerodynamics usually the variations of V are so small that V can be neglected. Therefore, we can write

$$\rho \frac{q^2}{2} + P = P_0, \tag{8.21}$$

where P_0 is the pressure when the fluid is at rest. It can be easily seen that the pressure diminishes as the velocity increases in the form

$$P = P_0 - \rho \frac{q^2}{2}.$$

From equation (8.21), the speed of the object can be determined from the following equation

$$q = \sqrt{\frac{2(P_0 - P)}{\rho}}. \tag{8.22}$$

The physical interpretation is that airplanes are equipped with instruments for measuring P and P_0 so that the speed of the machine relative to the air can be determined by equation (8.22). ∎

8.6 Navier–Stokes equation: equation of vorticity

We know that adding the viscous force term in Euler's equation of motion yields a very ideal equation for real fluids. This equation is known as Navier–Stokes equation. The very simplified form of Navier–Stokes equation is given by

$$\frac{\partial \mathbf{V}}{\partial t} + (\mathbf{V} \cdot \nabla)\mathbf{V} = -\frac{1}{\rho}\nabla P - \nabla \Omega + \nu \nabla^2 \mathbf{V}.$$

Using the vector identity, we obtain $(\mathbf{V} \cdot \nabla)\mathbf{V} = \frac{1}{2}\nabla|\mathbf{V}|^2 - \mathbf{V} \times (\nabla \times \mathbf{V})$, the Navier–Stokes equation reduces to

$$\frac{\partial \mathbf{V}}{\partial t} - \mathbf{V} \times (\nabla \times \mathbf{V}) = -\nabla\left\{\frac{1}{2}|\mathbf{V}|^2 + \frac{P}{\rho} + \Omega\right\} + \nu\nabla^2\mathbf{V}.$$

Let us define the vorticity vector as $\mathbf{W} = \nabla \times \mathbf{V}$, and for an incompressible fluid, ρ is constant and, hence, the continuity equation is given by $\nabla \cdot \mathbf{V} = 0$. With this information, the above equation yields,

$$\frac{\partial \mathbf{V}}{\partial t} - \mathbf{V} \times \mathbf{W} = -\nabla\left\{\frac{1}{2}|\mathbf{V}|^2 + \frac{P}{\rho} + \Omega\right\} - \nu\nabla\mathbf{W}.$$

The last term in the above equation is obtained by the vector identity $\nabla \times (\nabla \times \mathbf{V}) = \nabla(\nabla \cdot \mathbf{V}) - \nabla^2\mathbf{V}$. This simply gives $\nabla \times \mathbf{W} = -\nabla^2\mathbf{V}$. Now take the curl of the above equation and we have

$$\frac{\partial(\nabla \times \mathbf{V})}{\partial t} - \nabla \times (\mathbf{V} \times \mathbf{W}) = -\nu\nabla \times (\nabla \times \mathbf{W}).$$

We know from vector identity that

$$\nabla \times (\mathbf{V} \times \mathbf{W}) = (\mathbf{W} \cdot \nabla)\mathbf{V} - (\nabla \cdot \mathbf{V}) + (\nabla \cdot \mathbf{W})\mathbf{V} - (\mathbf{V} \cdot \nabla)\mathbf{W}$$
$$= (\mathbf{W} \cdot \nabla)\mathbf{V} - (\mathbf{V} \cdot \nabla)\mathbf{W}.$$

Thus using this information, we obtain

$$\frac{\partial \mathbf{W}}{\partial t} - (\mathbf{W} \cdot \nabla)\mathbf{V} + (\mathbf{V} \cdot \nabla)\mathbf{W} = -\nu[\nabla(\nabla \cdot \mathbf{W}) - \nabla^2\mathbf{W}].$$

Hence, after a little reduction, we obtain the vorticity equation as follows:

$$\frac{D\mathbf{W}}{Dt} = \nu\nabla^2\mathbf{W} + (\mathbf{W} \cdot \nabla)\mathbf{V}.$$

This equation is very important in fluid mechanics in a rotating fluid motion.

8.7 Continuity equation and heat conduction

These ideas are readily extended to the analysis of heat conduction. The rate at which heat crosses a unit area normal to the direction of flow depends upon the thermal conductivity κ of the material present and the temperature gradient ∇u. Thus, we have

$$\iint_S (-\kappa \nabla u) \cdot \mathbf{n} \, dS$$

which gives the rate of heat flow across a closed surface S. But if ρ and c denote the density and specific heat capacity, respectively, then

$$\frac{\partial}{\partial t} \iiint_V c \rho u \, dV$$

must be the rate of increase of the heat contained in the volume V enclosed by S. Then by continuity, we must have

$$\iint_S (\kappa \nabla u) \cdot \mathbf{n} \, dS = \frac{\partial}{\partial t} \iiint_V c \rho u \, dV$$

$$= \iiint_V c \rho \frac{\partial u}{\partial t} \, dV, \tag{8.23}$$

assuming that c and ρ are constants. Transforming the left-hand side by the divergence theorem, we obtain

$$\iiint_V \nabla \cdot (\kappa \nabla u) \, dV = \iiint_V c \rho \frac{\partial u}{\partial t} \, dV$$

$$\iiint_V \left\{ \nabla \cdot (\kappa \nabla u) - c\rho \frac{\partial u}{\partial t} \right\} dV = 0. \tag{8.24}$$

Now making the further assumption that κ is constant and since V is arbitrary, we finally obtain

$$\nabla \cdot \nabla u = \nabla^2 u = \frac{c\rho}{\kappa} \frac{\partial u}{\partial t}, \tag{8.25}$$

which is the basic differential equation of heat conduction.

8.8 Poisson's equation

In this section, we consider a problem dealing with a gravitational field. Let $\mathbf{F}$ be the field intensity at a general point $P(x, y, z)$, and let $\Delta \mathbf{F}$ be the contribution to $\mathbf{F}$

due to the infinitesimal mass Δm_i in an infinitesimal volume $\Delta V_1 = \Delta x_1 \Delta y_1 \Delta z_1$ enclosing the point $P_1(x_1, y_1, z_1)$. According to Newton's law of universal gravitation, $\Delta \mathbf{F}$ is a vector whose magnitude is

$$\Delta \mathbf{F} = k \frac{1 \cdot \Delta m_1}{r^2}, \tag{8.26}$$

where $r^2 = (x - x_1)^2 + (y - y_1)^2 + (z - z_1)^2$ and whose direction is opposite to that of the vector

$$\mathbf{R} = (x - x_1)\mathbf{i} + (y - y_1)\mathbf{j} + (z - z_1)\mathbf{k},$$

extending from P_1 to P (see Fig. 8.9).

If we choose $k = -1$ in equation (8.26), then the field intensity at P due to the infinitesimal mass Δm_1 at P_1 can be written in vector form as

$$\Delta \mathbf{F} = -\frac{\Delta m_1}{r^2} \frac{\mathbf{R}}{r} = -\rho(x, y, z)\Delta V_1 \frac{\mathbf{R}}{r^3}, \tag{8.27}$$

where $\rho(x, y, z)$ is the density of the material at point P_1.

Now let S be an arbitrary closed regular surface bounding a volume V, and let I denote the integral over S of the normal component of the force due to all attracting material in the field. We know that $\mathbf{F} = \nabla \phi$ where $\phi(x, y, z)$ is the potential function of the field. Therefore, we have

$$I = \iint_S \mathbf{n} \cdot \mathbf{F} \, dS = \iint_S \mathbf{n} \cdot \nabla \phi \, dS. \tag{8.28}$$

However, I can also be computed by first determining the part of it, ΔI, due to the material in an arbitrary volume element ΔV_1 and then taking into account all such

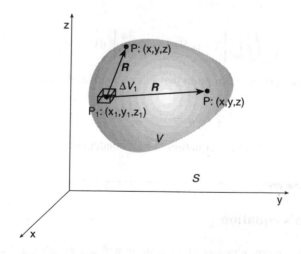

Figure 8.9   Potential at point P due to material in element ΔV_1.

infinitesimal contributions by integrating over the entire field. Thus using equations (8.27) and (8.28), we obtain

$$\Delta I = \iint_S \mathbf{n} \cdot \Delta \mathbf{F} \, dS$$

$$= -\iint_S [\rho(x_1, y_1, z_1)\Delta V_1] \mathbf{n} \cdot \frac{\mathbf{R}}{r^3} dS$$

$$= -\rho(x_1, y_1, z_1)\Delta V_1 \iint_S \mathbf{n} \cdot \frac{\mathbf{R}}{r^3} dS, \tag{8.29}$$

since x_1, y_1 and z_1 are constants with respect to the x, y, z integration over S.

The last integral can, of course, be evaluated by Gauss' theorem (Stokes' theorem). Specifically, if the origin of $\mathbf{R}$, namely the point $P_1(x_1, y_1, z_1)$ is within S, the value of the integral is 4π, otherwise the value of the integral is zero. Hence,

$$\Delta I = \begin{cases} -4\pi\rho(x_1, y_1, z_1)\Delta V_1 & \Delta V_1 \text{ within } S, \\ 0 & \Delta V_1 \text{ outside } S, \end{cases}$$

and integrating only over the volume V bounded by S, we obtain

$$I = \int dI = -4\pi \iiint_V \rho(x_1, y_1, z_1) \, dV_1.$$

Since x_1, y_1, and z_1 are dummy variables, we can write this as

$$I = -4\pi \iint_V \rho(x, y, z) \, dV. \tag{8.30}$$

Equating the two expressions in (8.28) and (8.30), we obtain the following

$$\iint_S \mathbf{n} \cdot \nabla\phi \, dS = -4\pi \iint_V \int \rho(x, y, z) \, dV.$$

We can now apply the Divergence theorem to the integral on the left to get

$$\iint_V \int \nabla \cdot (\nabla\phi) \, dV = -4\pi \iint_V \int \rho(x, y, z) \, dV$$

or

$$\iint_V \int [\nabla \cdot (\nabla\phi) + 4\pi\rho(x, y, z)] \, dV = 0.$$

Since this holds for an arbitrary volume V, it follows that the integrand must vanish identically. This leads to the following equation

$$\nabla^2\phi = -4\pi\rho(x, y, z). \tag{8.31}$$

This equation is known as Poisson's equation.

Remark 8.3   We have thus shown that in regions occupied by matter, the gravitational potential satisfies Poisson's equation. In empty space $\rho(x, y, z) = 0$, and thus in empty space the gravitational potential satisfies Laplace's equation

$$\nabla^2 \phi = 0.$$

∎

8.9 Vectors in electromagnetic theory

The theory of electromagnetic fields and waves is a subject where the basic experimental laws concerning electric and magnetic fields can be fully described by five equations. These five equations, in concise form, utilize the divergence and curl concepts of vector analysis. They have equivalent integral forms involving line, surface and volume integrals as discussed in the previous chapters of this text.

It is also possible, using vector identities, to manipulate the basic equations so as to mathematically predict phenomena such as electromagnetic waves, which are of major importance in, for example, communications engineering. Since electromagnetic field theory is also useful at the 'heavy' end of electrical engineering in connection with electrical machines, it is of interest to the physicist dealing with applications of high-intensity magnetic fields and to the biologist concerned with modelling electrical behaviour of the heart and brain; it is clear that it is a very wide-ranging area of application to discuss.

It is possible to study electromagnetic phenomena without using vector analysis but the mathematical equations quickly become long and involved and the underlying physics is easily lost. With the aid of vector analysis, we can readily represent the basic equations in an elegant and concise manner and, hopefully, understand their physical significance.

8.10 The continuity equation again

Electric current is the flow of electric charge. The electric current across a surface S is defined as the rate at which electric charge flows across S.

If Q denotes the electric charge density, i.e. the amount of electric charge per unit volume, and $\mathbf{V}$ is the velocity of the electric charge, then the vector field

$$\mathbf{J} = Q\mathbf{V}$$

is called the *current density*. The current I crossing a surface element ΔS in an electric conductor is given by

$$I = Q|\mathbf{V}| \cos \theta \Delta S = Q\mathbf{V} \cdot \mathbf{n} \Delta S = \mathbf{J} \cdot \mathbf{n} \Delta S,$$

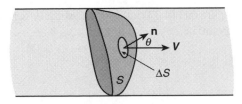

Figure 8.10   Charge through a volume element ΔS.

because this is the amount of charge that will cross ΔS every second (see Fig. 8.10). For a non-elementary surface S, the current will then be given by

$$I = \iint_S \mathbf{J} \cdot \mathbf{n}\,dS = \iint_S \mathbf{J} \cdot d\mathbf{S}, \qquad (8.32)$$

which is recognized as a flux integral. Clearly, the sign of I depends on the sense of the unit normal $\mathbf{n}$. It is usual to denote the normal component of the flux integral by the transformation $d\mathbf{S} = \mathbf{n}dS$. If S is a closed surface, $\mathbf{n}$ is chosen to be the outwardly-directed unit normal. As in the corresponding calculation for fluid flow, the continuity equation for current flow is derived using a simple physical principle, in this case the principle of conservation of electric charge. The total charge q in a volume V is given by the volume integral of the charge density

$$q = \iiint_V \varrho\,dV,$$

so that the rate at which charge decreases in this volume is

$$\frac{\partial q}{\partial t} = -\frac{\partial}{\partial t} \iiint_V \varrho\,dV = -\iiint_V \frac{\partial \varrho}{\partial t}\,dV.$$

But the total electric charge in V can only change if charge is flowing across the bounding surface S. Hence, we must equate the rate of flow of charge across S as given by (8.32) with the rate of decrease of charge in V. In other words,

$$\iint_S \mathbf{J} \cdot \mathbf{n}\,dS = -\iiint_V \frac{\partial \varrho}{\partial t}\,dV.$$

Transforming the left-hand side with the aid of the divergence theorem, we obtain

$$\iiint_V \nabla \cdot \mathbf{J}\,dV = -\iiint_V \frac{\partial \varrho}{\partial t}\,dV$$

or

$$\iiint_V \left(\nabla \cdot \mathbf{J} + \frac{\partial \varrho}{\partial t} \right) dV = 0. \qquad (8.33)$$

However, since we require that (8.33) hold for an arbitrary region V, the integrand must be zero at each point, i.e.

$$\nabla \cdot \mathbf{J} = -\frac{\partial Q}{\partial t},$$

which can be written as

$$\frac{\partial Q}{\partial t} + \nabla \cdot Q\mathbf{V} = 0. \tag{8.34}$$

This is the continuity equation for current flow. As one would expect, it is identical in form with the continuity equation for fluid flow. This is the first of the five equations.

8.11 Maxwell's equations for electromagnetic fields

The remaining four equations of electromagnetism will be discussed in this section. These equations involve the following basic field vectors:

- the electric field $\mathbf{E}$,
- the electric flux density $\mathbf{D}$,
- the magnetic field $\mathbf{H}$,
- the magnetic flux density $\mathbf{B}$,
- the current density $\mathbf{J}$.

and the scalars

- $\epsilon = $ Permittivity,
- $\mu = $ Permeability,
- $\sigma = $ Conductivity,
- $Q = $ Charge density,
- $q = \iiint_V Q \, dV = $ Total charge within volume V,
- $\phi = \iint_S \mathbf{B} \cdot \mathbf{n} \, dS = $ Total magnetic flux passing through surface S,
- $i = \iint_S \mathbf{J} \cdot \mathbf{n} \, dS = $ Total current flowing through surface S.

The field vectors $\mathbf{D}$ and $\mathbf{H}$ are related to $\mathbf{E}$ and $\mathbf{B}$, respectively, by functional relationships which are characteristic of the medium. The simplest situation is where the relationships are

$$\mathbf{D} = \epsilon \mathbf{E} \tag{8.35}$$

and

$$\mathbf{H} = \frac{1}{\mu}\mathbf{B}, \tag{8.36}$$

where ϵ and μ are the scalars previously defined. The quantities just defined (field vectors and scalars) are connected by a number of equations expressing relations discovered experimentally in the early nineteenth century chiefly by Michael Faraday (1791–1867). In particular, we have

1. *Faraday's law*

$$\int_C \mathbf{E} \cdot d\mathbf{R} = -\frac{\partial \phi}{\partial t},\tag{8.37}$$

which asserts that the integral of the tangential component of the electric intensity vector around any closed curve C is equal but opposite in sign to the rate of change of the magnetic flux passing through any surface S surrounded by C.

2. *Ampere's law*

$$\int_C \mathbf{H} \cdot d\mathbf{R} = i,\tag{8.38}$$

which asserts that the integral of the tangential component of the magnetic intensity vector around any closed curve C is equal to the current flowing through any surface spanning C.

3. *Gauss' law for electric fields*

$$\iint_S \mathbf{D} \cdot \mathbf{n}\, dS = q,\tag{8.39}$$

which asserts that the integral of the normal component of the electric flux density over any closed surface S is equal to the total electric charge enclosed by S.

4. *Gauss' law for magnetic fields*

$$\iint_S \mathbf{B} \cdot \mathbf{n}\, dS = 0,\tag{8.40}$$

which asserts that the total magnetic flux passing through any closed surface S is zero. We can now state, in general form, Maxwell's equations:

(a) *Divergence equations*

$$\nabla \cdot \mathbf{D} = q,\tag{8.41}$$

$$\nabla \cdot \mathbf{B} = 0.\tag{8.42}$$

(b) *Curl equations*

$$\nabla \times \mathbf{E} = -\frac{\partial \mathbf{B}}{\partial t},\tag{8.43}$$

$$\nabla \times \mathbf{H} = \mathbf{J}_c + \frac{\partial \mathbf{D}}{\partial t}.\tag{8.44}$$

To show how these are obtained, we begin by applying the divergence theorem to Gauss' first law (8.39) to get:

$$\iiint_V \nabla \cdot \mathbf{D}\, dV = q = \iiint_V Q\, dV$$

from which, since V is arbitrary, it follows that

$$\nabla \cdot \mathbf{D} = Q. \tag{8.45}$$

In the same way, by applying the divergence theorem to Gauss' second law (8.40), we find that

$$\iiint_V \nabla \cdot \mathbf{B}\, dV = 0$$

and since V is arbitrary,

$$\nabla \cdot \mathbf{B} = 0.$$

Next, if we apply Stokes' theorem to Faraday's law (8.37), we have

$$\iint_S \nabla \times \mathbf{E} \cdot \mathbf{n}\, dS = -\frac{\partial \phi}{\partial t}$$

and substituting for ϕ from its definition in terms of $\mathbf{B}$,

$$\iint_S \mathbf{n} \cdot \nabla \times \mathbf{E}\, dS = -\frac{\partial}{\partial t}\left(\iint_S \mathbf{n} \cdot \mathbf{B}\, dS\right) = -\iint_S \mathbf{n} \cdot \frac{\partial \mathbf{B}}{\partial t}\, dS.$$

Since S is an arbitrary surface spanning the arbitrary closed curve C, the last equation can hold if

$$\nabla \times \mathbf{E} = -\frac{\partial \mathbf{B}}{\partial t}.$$

Similarly, if we apply Stokes' theorem to Ampere's law (8.38), we obtain

$$\iint_S \mathbf{n} \cdot \nabla \times \mathbf{H}\, dS = i = \iint_S \mathbf{n} \cdot \mathbf{J}\, dS$$

and again since S is an arbitrary open surface, we conclude that the vectors being integrated over S must be identical, i.e.

$$\nabla \times \mathbf{H} = \mathbf{J}. \tag{8.46}$$

As Maxwell was the first to realize, the current density $\mathbf{J}$ consists of two parts, namely a conduction current density

$$\mathbf{J}_c = \sigma \mathbf{E},$$

due to the flow of electric charges and a displacement (electric flux) current density

$$\mathbf{J}_d = \frac{\partial \mathbf{D}}{\partial t} = \epsilon \frac{\partial \mathbf{E}}{\partial t},$$

due to the time variation of electric field. Thus,

$$\mathbf{J} = \sigma \mathbf{E} + \epsilon \frac{\partial \mathbf{E}}{\partial t}$$

and (8.46) becomes

$$\nabla \times \mathbf{H} = \sigma \mathbf{E} + \frac{\partial \mathbf{E}}{\partial t} \qquad (8.47)$$

or

$$\nabla \times \mathbf{H} = \mathbf{J} + \frac{\partial \mathbf{D}}{\partial t}, \qquad (8.48)$$

where $\mathbf{J}$ is the electric current density or $\mathbf{J_c}$. If we take the curl of (8.43) we obtain

$$\nabla \times (\nabla \times \mathbf{E}) = -\nabla \times \frac{\partial \mathbf{B}}{\partial t} = -\frac{\partial}{\partial t} (\nabla \times \mathbf{B})$$

$$= -\mu \frac{\partial}{\partial t} (\nabla \times \mathbf{H}).$$

If the term $\nabla \times (\nabla \times \mathbf{E})$ is expanded, the last equation becomes

$$\nabla(\nabla \cdot \mathbf{E}) - \nabla^2 \mathbf{E} = -\mu \frac{\partial}{\partial t} (\nabla \times \mathbf{H})$$

and substituting for $\nabla \times \mathbf{H}$ from (8.47)

$$\nabla(\nabla \cdot \mathbf{E}) - \nabla^2 \mathbf{E} = -\mu \frac{\partial}{\partial t} \left(\sigma \mathbf{E} + \epsilon \frac{\partial \mathbf{E}}{\partial t} \right). \qquad (8.49)$$

If the space charge density Q is zero, as it is to a high degree of approximation in both dielectrics and good conductors, then from (8.45) and the relation $\mathbf{D} = \epsilon \mathbf{E}$, we see that

$$\nabla \cdot \mathbf{E} = 0.$$

Therefore, equation (8.49) reduces to

$$\nabla^2 \mathbf{E} = \mu \epsilon \frac{\partial^2 \mathbf{E}}{\partial t^2} + \mu \sigma \frac{\partial \mathbf{E}}{\partial t},$$

which is *Maxwell's equation for the electric intensity vector* $\mathbf{E}$.

In a similar manner, if we take the curl of (8.47), we obtain

$$\nabla \times (\nabla \times \mathbf{H}) = \nabla \times \left(\sigma \mathbf{E} + \epsilon \frac{\partial \mathbf{E}}{\partial t} \right)$$

and expanding the left-hand side,

$$\nabla(\nabla \cdot \mathbf{H}) - \nabla^2 \mathbf{H} = \sigma \nabla \times \mathbf{E} + \epsilon \nabla \times \frac{\partial \mathbf{E}}{\partial t}$$

$$= \sigma \nabla \mathbf{E} + \epsilon \frac{\partial}{\partial t} (\nabla \times \mathbf{E}).$$

Now substituting for $\nabla \times \mathbf{E}$ from (8.43)

$$\nabla(\nabla \cdot \mathbf{H}) - \nabla^2 \mathbf{H} = \sigma \left(-\frac{\partial \mathbf{B}}{\partial t}\right) + \epsilon \left(-\frac{\partial^2 \mathbf{B}}{\partial t^2}\right).$$

But $\mathbf{B} = \mu \mathbf{H}$, by definition. Hence, (8.42) implies that $\nabla \cdot \mathbf{H} = 0$ and, therefore, the last equation reduces to

$$\nabla^2 \mathbf{H} = \mu \epsilon \frac{\partial^2 \mathbf{H}}{\partial t^2} + \mu \sigma \frac{\partial \mathbf{H}}{\partial t},$$

which is *Maxwell's equation for the magnetic intensity vector* $\mathbf{H}$.

For a perfect dielectric, $\sigma = 0$. Hence, in this case Maxwell's equations reduce to the one-dimensional wave equations

$$\nabla^2 \mathbf{E} = \mu \epsilon \frac{\partial^2 \mathbf{E}}{\partial t^2} \quad \text{and} \quad \nabla^2 \mathbf{H} = \mu \epsilon \frac{\partial^2 \mathbf{H}}{\partial t^2}.$$

On the other hand, in a good conductor, the terms arising from the displacement current, i.e. the terms containing the second time derivatives, are negligible, and Maxwell's equations reduce to

$$\nabla^2 \mathbf{E} = \mu \sigma \frac{\partial \mathbf{E}}{\partial t} \quad \text{and} \quad \nabla^2 \mathbf{H} = \mu \sigma \frac{\partial \mathbf{H}}{\partial t},$$

which are examples of the three-dimensional heat equation.

Both the curl equations (8.43), (8.44) relate an electric vector ($\mathbf{E}$ or $\mathbf{D}$) with a magnetic vector ($\mathbf{B}$ or $\mathbf{H}$), justifying the use of the term 'electromagnetic'. However, for fields which are classified as static or time-dependent, the time-derivative terms are zero and the curl equations become

$$\nabla \times \mathbf{E} = 0, \tag{8.50}$$

$$\nabla \times \mathbf{H} = \mathbf{J}. \tag{8.51}$$

That is, the electric and magnetic vectors are separate or *uncoupled* in the static situation. The theory of static fields involves the study of equations (8.50), (8.51) and (8.41), (8.42), which are the same for static and time varying fields.

The interpretation of (8.41) for electric fields is that a positive charge is a *source* of the vector $\mathbf{D}$ and a negative charge is a *sink* ($\nabla \cdot \mathbf{D} > 0$ when $q > 0$ and $\nabla \cdot \mathbf{D} < 0$ when $q < 0$). At points not occupied by any electric charge, $\nabla \cdot \mathbf{D}$ is zero. Equation (8.50), for static fields, tells us that an electrostatic field $\mathbf{E}$ is an example of a conservative vector field so that

$$\int_C \mathbf{E} \cdot d\mathbf{R} = 0,$$

for any closed path C. Physically, since $\mathbf{E}$ is the force on unit charge, the zero value for this line integral tells us that the work done in moving a unit charge around a closed path is always zero. It follows that we can neither gain nor lose energy by such a process. Also, since $\mathbf{E}$ is conservative, we can represent $\mathbf{E}$ as the gradient of a scalar potential ϕ. Thus, we can write

$$\mathbf{E} = -\nabla\phi \text{ (static fields).} \tag{8.52}$$

The advantage of using ϕ in electrostatic field problems is that it is usually easier to calculate a scalar quantity than a vector. To actually find ϕ [and then $\mathbf{E}$, using (8.52)], we must determine the equation satisfied by ϕ. This is readily done using (8.41) and assuming that the medium is such that $\mathbf{D} = \epsilon\mathbf{E}$ holds:

$$\nabla \cdot \mathbf{D} = \nabla \cdot (\epsilon\mathbf{E}) = \epsilon\nabla \cdot \mathbf{E} = -\epsilon\nabla \cdot \nabla\phi = Q$$

or

$$\nabla^2\phi = -\frac{Q}{\epsilon}. \tag{8.53}$$

Equation (8.53), which is called *Poisson's equation*, is the relation between the potential at a point and the electric charge density at that point. In particular, at points where no charge is present ($q = 0$), the potential satisfies

$$\nabla^2\phi = 0, \tag{8.54}$$

which is *Laplace's equation*. Practical electrostatic problems often involve solving Laplace's equation in a region with known boundary conditions, for example calculating the potential and field between the conductors in a capacitor. The equation satisfied by the electric field $\mathbf{E}$ at points where $Q = 0$ is, of course

$$\nabla \cdot \mathbf{E} = 0. \tag{8.55}$$

Example 8.1   Given the vector field

$$\mathbf{F} = \frac{K}{r^n}\mathbf{r} \ (r \neq 0)$$

(in spherical coordinates), deduce the value of n that makes the field solenoidal. Show that $\mathbf{F}$ is conservative for any value of n. Give an interpretation for $\mathbf{F}$ in an electrostatic context.   □

Solution   Using spherical coordinates, the divergence is

$$\nabla \cdot \mathbf{F} = \frac{1}{r^2}\frac{\partial}{\partial r}\left(\frac{Kr^2}{r^n}\right) = \frac{K}{r^2}(2 - n)r^{1-n} = K\left(\frac{2 - n}{r^{n+1}}\right),$$

which is solenoidal ($\nabla \cdot \mathbf{F} = 0$) if $n = 2$. The curl in spherical coordinates is given by

$$\nabla \times \mathbf{F} = \frac{1}{r^2 \sin \theta} \begin{vmatrix} \mathbf{r} & r\boldsymbol{\theta} & r \sin \theta \boldsymbol{\phi} \\ \dfrac{\partial}{\partial r} & \dfrac{\partial}{\partial \theta} & \dfrac{\partial}{\partial \phi} \\ \dfrac{K}{r^n} & 0 & 0 \end{vmatrix}$$

$$= \mathbf{0} \quad \text{(all components zero)}.$$

Since $\mathbf{F}$ has zero divergence when $n = 2$ and also zero curl, it follows from (8.50) and (8.55) that

$$\mathbf{F} = \frac{K}{r^2} \mathbf{r}$$

is a possible electrostatic field at any point $r \neq 0$. In fact, $\mathbf{F}$ represents the field $\mathbf{E}$ due to a point charge at $r = 0$, the value of the constant K depending on the system of units being employed. ■

Turning now to equations (8.42) and (8.51) for the magnetostatic field, the first of these,

$$\nabla \cdot \mathbf{B} = 0$$

tells us that a magnetic field is always solenoidal, so that magnetic field lines neither start nor end. This implies that isolated magnetic poles do not exist (i.e. there is no magnetic equivalent of an electric charge). The equation

$$\nabla \times \mathbf{H} = \mathbf{J}$$

links the magnetic field $\mathbf{H}$ and the electric current density $\mathbf{J}$ at a point. Since $\nabla \times \mathbf{H}$ is only zero at points where $\mathbf{J} = 0$, that is, outside current-carrying conductors, it is clear that $\mathbf{H}$, unlike $\mathbf{E}$, is in general a non-conservative vector field. Therefore, we cannot normally represent $\mathbf{H}$ as the gradient of a scalar potential, although a magnetic scalar potential function is occasionally used to solve magnetic field problems in regions where $\mathbf{J}$ is zero everywhere.

Example 8.2   Deduce the corresponding integral forms of the Maxwell equations (8.41), (8.42) and (8.50), (8.51) for static fields, and interpret the results.   □

Solution   For the divergence equations (8.41) and (8.42), we integrate both sides of each equation over an arbitrary volume V bounded by a surface S and use the divergence theorem. In the electrostatic case we obtain

$$\iiint_V \nabla \cdot \mathbf{D} \, dV = \iiint_V q \, dV$$

or

$$\iint_S \mathbf{D} \cdot \mathbf{n} \, dS = Q, \tag{8.56}$$

where $Q = \iiint_V d \, dV$ is the total charge in the volume V. Equation (8.56) tells us that the electric flux over a closed surface equals the charge inside that surface. This is known as *Gauss' law*, and it can be used for finding $\mathbf{D}$ (or $\mathbf{E}$) in the field problems where a high degree of symmetry is present. Similarly, equation (8.42) gives

$$\iiint_V \nabla \cdot \mathbf{B} \, dV = 0 \quad \text{for any volume } V,$$

which transforms to

$$\iint_S \mathbf{B} \cdot \mathbf{n} \, dS = 0. \tag{8.57}$$

This is Gauss' law in magnetostatics. It tells us that the magnetic flux over any closed surface is zero, confirming the non-existence of sources or sinks for magnetic fields. In fact, (8.57) holds for any magnetic field, static or time-varying, since the result $\nabla \cdot \mathbf{B} = 0$ is valid in both cases.

For the curl equations (8.50) and (8.51), we find the flux of both sides over an arbitrary surface S bounded by a closed curve C. The magnetic field equation (8.51) gives

$$\iint_S \nabla \times \mathbf{H} \cdot \mathbf{n} \, dS = \iint_S \mathbf{J} \cdot \mathbf{n} \, dS$$

or using Stokes' theorem to transform the left-hand side

$$\oint_C \mathbf{H} \cdot d\mathbf{R} = I, \tag{8.58}$$

where we have also used (8.32) to identify the right-hand side as the electric current enclosed by the path C. Equation (8.58) is known as *Ampere's law in magnetostatics*; it can be used to find the magnitude of a magnetic field in a problem where a high degree of symmetry is present. Similarly, for the electrostatic field, equation (8.50) transforms (by Stokes' theorem) into

$$\oint_C \mathbf{E} \cdot d\mathbf{R} = 0,$$

which merely confirms that the electrostatic field is conservative.   ∎

8.12 The Poynting vector

The Poynting vector $\mathbf{P}$ at any point in an electromagnetic field is defined as $\mathbf{P} = \mathbf{E} \times \mathbf{H}$. The divergence of this vector turns out to give us results of considerable significance in electrical engineering. Thus we have

$$\nabla \cdot \mathbf{P} = \nabla \cdot (\mathbf{E} \times \mathbf{H})$$
$$= \mathbf{H} \cdot (\nabla \times \mathbf{E}) - \mathbf{E} \cdot (\nabla \times \mathbf{H}),$$

by a standard vector identity (see 4.24). Using Maxwell's equations (8.43) and (8.44), we obtain

$$\nabla \cdot \mathbf{P} = \mathbf{H} \cdot \left(-\frac{\partial \mathbf{B}}{\partial t}\right) - \mathbf{E} \cdot \left(\mathbf{J}_c + \frac{\partial \mathbf{D}}{\partial t}\right).$$

Therefore,

$$\nabla \cdot \mathbf{P} + \mathbf{H} \cdot \frac{\partial \mathbf{B}}{\partial t} + \mathbf{E} \cdot \frac{\partial \mathbf{D}}{\partial t} = -\mathbf{E} \cdot \mathbf{J}_c.$$

We know that $\mathbf{B} = \mu\mathbf{H}$ and $\mathbf{D} = \epsilon\mathbf{E}$, and using these relations yields

$$\nabla \cdot \mathbf{P} + \mu\mathbf{H} \cdot \frac{\partial \mathbf{H}}{\partial t} + \epsilon\mathbf{E} \cdot \frac{\partial \mathbf{E}}{\partial t} = -\mathbf{E} \cdot \mathbf{J}_c.$$

We know that $\mathbf{H} \cdot \partial\mathbf{H}/\partial t = \frac{1}{2}\partial|\mathbf{H}|^2/\partial t$, and using this vector identity in the above equation we find

$$\nabla \cdot \mathbf{P} + \frac{\partial \eta}{\partial t} = -\mathbf{E} \cdot \mathbf{J}_c,$$

where $\eta = \frac{1}{2}\mu\partial|\mathbf{H}|^2/\partial t + \frac{1}{2}\epsilon\partial|\mathbf{E}|^2/\partial t$, and it is known as the energy density (energy per unit volume) in the electromagnetic field.

In the special case of perfect insulation, the conduction current density $\mathbf{J}_c$ would be zero, and in which case, we have

$$\nabla \cdot \mathbf{P} + \frac{\partial \eta}{\partial t} = 0,$$

which has exactly the same form as the continuity equation

$$\nabla \cdot \mathbf{J} + \frac{\partial Q}{\partial t} = 0.$$

This similarity suggests that just as the current density vector $\mathbf{J}$ is associated with flow of electric charge of density Q, so the Poynting vector $\mathbf{P}$ can be associated with flow of energy of density η. Indeed, we can obtain an energy balance equation from the above equation by integrating both sides over a volume V.

$$\iiint_V \nabla \cdot \mathbf{P} dV = - \iiint_V \frac{\partial \eta}{\partial t} dV,$$

which can be rewritten by using the divergence theorem on the left-hand side as follows:

$$\iint_S \mathbf{P} \cdot \mathbf{n} dS = - \iiint_V \frac{\partial \eta}{\partial t} dV.$$

Using the interpretation of η as an energy density, the right-hand side of the above equation is the rate of change of electromagnetic energy within the volume V. Hence, the left-hand side must be the rate at which energy crosses the surface S. Thus the Poynting vector $\mathbf{P}$ has the significance of being a power density, provided the interpretation implied in the above equation is used, namely that the flux integral of $\mathbf{P}$ gives us the power crossing S.

If we want to integrate the full energy density equation in the electromagnetic field over a volume V by a similar process, we obtain by using divergence theorem immediately,

$$\iint_S \mathbf{P} \cdot \mathbf{n} dS = - \iiint_V \frac{\partial \eta}{\partial t} dV - \iiint_V \mathbf{E} \cdot \mathbf{J}_c dV$$

or simply we can rewrite the equation as

$$- \iiint_V \frac{\partial \eta}{\partial t} dV = \iint_S \mathbf{P} \cdot \mathbf{n} dS + \iiint_V \mathbf{E} \cdot \mathbf{J}_c dV.$$

Since the left-hand side here is the loss of energy within V and the first term on the right hand side is the power (rate of energy flow) outwards across S, it follows that the second term must also contribute to a loss of electromagnetic energy. It is, in fact, the rate at which energy is distributed by the passage of current — the Joules heating law. For many materials, the conduction current density vector $\mathbf{J}_c$ is related to the electric field $\mathbf{E}$ by

$$\mathbf{J}_c = \sigma \mathbf{E},$$

where σ is called the electrical conductivity of the medium. This equation is, in fact, a field form of the well-known Ohm's law. It follows from the last two equations that the Joule heating loss occurs at a rate $\sigma |\mathbf{E}|^2$ per unit volume.

8.13 Waves solutions of Maxwell's equations

One consequence of the Maxwell equations for time-varying electromagnetic fields is the prediction that variations in the electric and magnetic fields $\mathbf{E}$ and $\mathbf{H}$ propagate as waves. The formal analysis leading to this prediction is readily carried out using the tools of vector analysis. The experimental verification of the predictions by Hertz was powerful evidence for the validity of the Maxwell equations. The simplest model to consider is where the Maxwell equations (8.41)–(8.44) take on the forms

$$\nabla \cdot \mathbf{D} = 0,$$

$$\nabla \cdot \mathbf{B} = 0,$$

$$\nabla \times \mathbf{E} = -\frac{\partial \mathbf{B}}{\partial t},$$

$$\nabla \times \mathbf{H} = \frac{\partial \mathbf{D}}{\partial t},$$

or, eliminating $\mathbf{D}$ and $\mathbf{B}$,

$$\nabla \cdot \mathbf{E} = 0, \tag{8.59}$$

$$\nabla \cdot \mathbf{H} = 0, \tag{8.60}$$

$$\nabla \times \mathbf{E} = -\mu \frac{\partial \mathbf{H}}{\partial t}, \tag{8.61}$$

$$\nabla \times \mathbf{H} = \epsilon \frac{\partial \mathbf{E}}{\partial t}. \tag{8.62}$$

From these equations it is evident that we are considering a linear homogeneous medium with no electric charges or electric currents at any point. The symmetry of the two divergence equations (8.59), (8.60) and the two curl equations (8.61), (8.62) is readily seen.

To obtain predictions of wave behaviour, we take the curl of both sides of equation (8.61) and use the identity for $\nabla \times \nabla \times \mathbf{A}$. We obtain

$$\nabla(\nabla \cdot \mathbf{E}) - \nabla^2 \mathbf{E} = -\mu \frac{\partial}{\partial t} (\nabla \times \mathbf{H})$$

or

$$\nabla^2 \mathbf{E} = \mu\epsilon \frac{\partial^2 \mathbf{E}}{\partial t^2}, \tag{8.63}$$

using (8.59) and (8.62). A similar procedure using the curl of equation (8.62) yields

$$\nabla^2 \mathbf{H} = \mu\epsilon \frac{\partial^2 \mathbf{H}}{\partial t^2}. \tag{8.64}$$

An alternate and more compact form of the wave equation is

$$\frac{\partial^2 \mathbf{E}}{\partial t^2} = c^2 \nabla^2 \mathbf{E}, \tag{8.65}$$

where $c^2 = 1/\mu_0\varepsilon_0$. Equations (8.63) and (8.64) are both vector equations and are thus each equivalent to three scalar equations. In Cartesian form, these equations are

$$\nabla^2 E_x = \mu\epsilon \frac{\partial E_x}{\partial t^2} \quad \text{and} \quad \nabla^2 H_x = \mu\epsilon \frac{\partial H_x}{\partial t^2}, \tag{8.66}$$

with similar relationships for the y- and z-components. Equations (8.66) and (8.65) are called *wave equations* since their solutions are waves travelling at speed c. The constant c is the speed of light and the waves are known as electromagnetic waves.

Many different solutions, known as *plane waves*, are considered here. In this case, the magnitudes of the vectors $\mathbf{E}$ and $\mathbf{H}$ are assumed to be constant at all points on a plane perpendicular to a given direction, which we will take to be the z-axis of a Cartesian coordinate system. Mathematically, this implies that all partial derivatives with respect to x and y may be set to zero.

Example 8.3   Suppose that $\mathbf{E} = (f(x, t), 0, 0)$ in a Cartesian coordinate system so that the wave equation becomes

$$\frac{\partial^2 f}{\partial t^2} = c^2 \frac{\partial^2 f}{\partial x^2}. \tag{8.67}$$

Show that $f(x, t) = \sin k(x - ct)$ is a solution for any value of the constant k and interpret this solution physically.   □

Solution   For the function $f(x, t) = \sin k(x - ct)$,

$$\frac{\partial^2 f}{\partial t^2} = -(kc)^2 \sin k(x - ct) \quad \text{and} \quad \frac{\partial^2 f}{\partial x^2} = -k^2 \sin k(x - ct),$$

so $f(x, t) = \sin k(x - ct)$ obeys (8.67). Physically, this solution corresponds to a sine wave travelling to the right at speed c. At $t = 0$, $f(x, 0) = \sin kx$ which is a sine wave which has $f = 0$ at $x = 0$. At a later time t, the point at which $f = 0$ has moved to the position $x = ct$, so the wave has moved to the right a distance ct (Fig. 8.11). The speed of the wave is, therefore, c. Similarly, it can be shown that the function $g(x, t) = \sin k(x + ct)$ is a solution of equation (8.67), representing a sine wave travelling to the left at speed c. In Fig. 8.11, the solid line represents the solution at $t = 0$, while the dashed line represents the solution at a later time.   ■

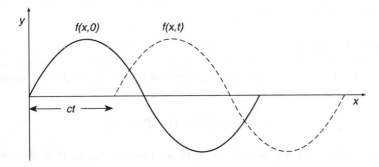

Figure 8.11   The solution $f(x, t) = \sin k(x - ct)$ of the wave equation.

Example 8.4   Show that $\mathbf{E} = \mathbf{E}_0 f(\mathbf{k} \cdot \mathbf{x} - \omega t)$ is a solution to the wave equation (8.65), where $\mathbf{E}_0$ and $\mathbf{k}$ are constant vectors, f is any function and $\omega = \pm c|\mathbf{k}|$. $\square$

Solution   Let $u = \mathbf{k} \cdot \mathbf{x} - \omega t$ so that the solution under consideration is $\mathbf{E} = \mathbf{E}_0 f(u)$. Then,

$$\frac{\partial \mathbf{E}}{\partial t} = \mathbf{E}_0 \frac{\mathrm{d} f}{\mathrm{d} u} \frac{\partial u}{\partial t} = -\omega \mathbf{E}_0 \frac{\mathrm{d} f}{\mathrm{d} u}$$

and, similarly,

$$\frac{\partial^2 \mathbf{E}}{\partial t^2} = -\omega \mathbf{E}_0 \frac{\mathrm{d}^2 f}{\mathrm{d} u^2} \frac{\partial u}{\partial t} = \omega^2 \mathbf{E}_0 \frac{\mathrm{d}^2 f}{\mathrm{d} u^2}.$$

To find $\nabla^2 \mathbf{E}$, note that since $\mathbf{E}_0$ is constant,

$$\nabla^2 (\mathbf{E}_0 f(u)) = \mathbf{E}_0 \nabla^2 f(u) = \mathbf{E}_0 \nabla \cdot (\nabla f(u)).$$

The gradient of $f(u)$ is

$$\nabla f(u) = \frac{\mathrm{d} f}{\mathrm{d} u} \nabla (\mathbf{k} \cdot \mathbf{x} - \omega t) = \frac{\mathrm{d} f}{\mathrm{d} u} \mathbf{k}.$$

and taking the divergence of this gives

$$\nabla^2 f(u) = \nabla \left(\frac{\mathrm{d} f}{\mathrm{d} u} \right) \cdot \mathbf{k} = \frac{\mathrm{d}^2 f}{\mathrm{d} u^2} \nabla u \cdot \mathbf{k} = \frac{\mathrm{d}^2 f}{\mathrm{d} u^2} |\mathbf{k}|^2.$$

Therefore, the wave equation is obtained provided that

$$\omega^2 \mathbf{E}_0 \frac{\mathrm{d}^2 f}{\mathrm{d} u^2} = c^2 \mathbf{E}_0 \frac{\mathrm{d}^2 f}{\mathrm{d} u^2} |\mathbf{k}|^2 \quad \text{which gives} \quad \omega^2 = c^2 |\mathbf{k}|^2.$$

So the function f is arbitrary and the only condition is that the frequency of the wave ω must be related to the speed c and the constant vector $\mathbf{k}$ (which is known as the wave vector) by $\omega = \pm c|\mathbf{k}|$.   ■

The Maxwell equation $\nabla \cdot \mathbf{E} = 0$ reduces, for this case, to $\partial E_z / \partial z = 0$, which tells us that E_z does not vary with z. Hence E_z is constant in space and we can, without loss of generality, take it to be zero. Therefore, we need only consider an electric field vector of the form

$$\mathbf{E}(z, t) = E_x(z, t)\mathbf{i} + E_y(z, t)\mathbf{j}$$

and, by similar reasoning since $\nabla \cdot \mathbf{H} = 0$, we assume a magnetic field vector

$$\mathbf{H}(z, t) = H_x(z, t)\mathbf{i} + H_y(z, t)\mathbf{j}.$$

The equations satisfied by the components of **E** and **H** are, using (8.66)

$$\frac{\partial^2 E_x}{\partial z^2} = \mu\epsilon \frac{\partial^2 E_x}{\partial t^2},$$
(8.68)

$$\frac{\partial^2 H_x}{\partial z^2} = \mu\epsilon \frac{\partial^2 H_x}{\partial t^2},$$
(8.69)

and similarly for the y-components. These equations are all examples of the one-dimensional wave equation. It is easy to show, by substitution, that

$$E_x = E_{x0} f(z - vt)$$
(8.70)

(where $v = 1/\sqrt{(\mu\epsilon)}$) is one possible solution of equation (8.68), where f is any well-behaved (differentiable) function of the variable $(z - vt)$ and E_{x0} is a constant. Physically, such a solution represents a wave travelling without change of shape in the positive z-direction with a speed of v (see Fig. 8.12).

Similarly, one solution for the y-component of **E** is

$$E_y = E_{y0} f(z - vt).$$

We see that the **E** vector in a plane electromagnetic wave is normal or transverse to the direction in which the wave is propagating. We now calculate the corresponding components of **H** using the Maxwell equation (8.61) and the Cartesian form of the curl:

$$\nabla \times \mathbf{E} = -\frac{\partial E_y}{\partial z}\mathbf{i} + \frac{\partial E_x}{\partial z}\mathbf{j}.$$

Hence, since

$$\frac{\partial \mathbf{H}}{\partial t} = \frac{\partial H_x}{\partial t}\mathbf{i} + \frac{\partial H_y}{\partial t}\mathbf{j},$$

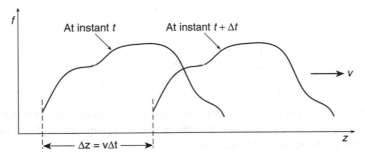

Figure 8.12   Wave travelling with velocity v.

we obtain, on comparing components

$$\frac{\partial E_y}{\partial z} = \mu \frac{\partial H_x}{\partial t} \tag{8.71}$$

$$\frac{\partial E_x}{\partial z} = -\mu \frac{\partial H_y}{\partial t}. \tag{8.72}$$

But from (8.70),

$$\frac{\partial E_x}{\partial z} = \frac{\partial E_x}{\partial (z - vt)} \frac{\partial (z - vt)}{\partial z} = \frac{\partial E_x}{\partial (z - vt)}$$

and

$$\frac{\partial E_x}{\partial t} = \frac{\partial E_x}{\partial (z - vt)} \frac{\partial (z - vt)}{\partial t} = -v \frac{\partial E_x}{\partial (z - vt)}$$

from which

$$\frac{\partial E_x}{\partial z} = -\frac{1}{v} \frac{\partial E_x}{\partial t}.$$

Hence (8.72) becomes

$$\frac{\partial H_y}{\partial t} = \frac{1}{v\mu} \frac{\partial E_x}{\partial t} = \sqrt{\left(\frac{\epsilon}{\mu}\right)} \frac{\partial E_x}{\partial t}$$

and integrating

$$H_y = \sqrt{\left(\frac{\epsilon}{\mu}\right)} E_x = \sqrt{\left(\frac{\epsilon}{\mu}\right)} E_{x0} f(z - vt).$$

Similar reasoning applied to (8.71) shows that

$$H_x = -\sqrt{\left(\frac{\epsilon}{\mu}\right)} E_y = -\sqrt{\left(\frac{\epsilon}{\mu}\right)} E_{y0} f(z - vt).$$

We see that variations of $\mathbf{H}$, like those of $\mathbf{E}$, propagate as transverse travelling waves with speed $v = 1/\sqrt{(\epsilon\mu)}$. The ratio

$$\frac{|\mathbf{E}|}{|\mathbf{H}|} = \frac{\sqrt{(E_x^2 + E_y^2)}}{\sqrt{(H_x^2 + H_y^2)}} = \frac{\sqrt{(E_x^2 + E_y^2)}}{\sqrt{\left(\frac{\epsilon}{\mu} E_x^2 + \frac{\epsilon}{\mu} E_y^2\right)}} = \sqrt{\left(\frac{\mu}{\epsilon}\right)}$$

depends, like the speed, on the electrical constants μ and ϵ. This ratio is referred to as the *characteristic impedance Z* of the medium of propagation. For a vacuum, or free space, the numerical values are

$$\epsilon = \epsilon_0 = 8.85 \times 10^{-12} \, \text{Fm}^{-1} \quad \text{and} \quad \mu = \mu_0 = 4\pi \times 10^{-7} \, \text{Hm}^{-1},$$

from which

$$Z = z_0 = \sqrt{\left(\frac{\mu_0}{\epsilon_0}\right)} \approx 120\pi \text{ ohms} \quad \text{and} \quad v = \frac{1}{\sqrt{(\epsilon\mu)}} \approx 3 \times 10^8 \text{ ms}^{-1}.$$

We see that the predicted speed of propagation v the electromagnetic waves in free space is equal to the experimental speed of light. This result was early evidence that light waves are electromagnetic in nature.

Finally, we can easily show that $\mathbf{E}$ and $\mathbf{H}$ are mutually perpendicular vectors and that $\mathbf{E} \times \mathbf{H}$ is a vector in the direction of propagation of the electromagnetic wave. The former is shown by noting that

$$\mathbf{E} \cdot \mathbf{H} = E_x H_x + E_y H_y = -E_x \sqrt{\left(\frac{\epsilon}{\mu}\right)} E_y + E_y \sqrt{\left(\frac{\epsilon}{\mu}\right)} E_x = 0,$$

while

$$\mathbf{E} \times \mathbf{H} = \begin{vmatrix} \mathbf{i} & \mathbf{j} & \mathbf{k} \\ E_x & E_y & 0 \\ H_x & H_y & 0 \end{vmatrix}$$

$$= (E_x H_y - E_y H_x)\mathbf{k}$$

$$= (E_x^2 + E_y^2)\sqrt{(\epsilon\mu)}\mathbf{k}$$

$$= \frac{|\mathbf{E}|^2}{Z}\mathbf{k},$$

so that $\mathbf{E} \times \mathbf{H}$ is indeed directed along the propagation direction (positive z).

Example 8.5
(a) A plane sinusoidal electromagnetic wave in free space has an electric vector

$$\mathbf{E} = E_0 \cos(\beta z - \omega t)\mathbf{i}.$$

Deduce the corresponding magnetic field $\mathbf{H}$ and show that the phase velocity of the wave is given by $v = 1/\sqrt{(\mu_0\epsilon_0)}$.

(b) If $E_0 = 200\pi$ Vm^{-1}, find the amplitude of the magnetic field and the value of the angular frequency ω of the wavelength is 1.89 m. □

Solution In this problem we are considering perhaps the most important type of plane wave, namely one where the general function f of the earlier theory is sinusoidal, so that instead of $f(z - vt)$ we have $\cos 2\pi(z - vt)/\lambda$, where λ is called the *wavelength*. The name arises because, for a given t,

$$\cos\frac{2\pi}{\lambda}[(z + \lambda) - vt] = \cos\frac{2\pi}{\lambda}(z - vt),$$

i.e. the wave repeats after a distance λ. Also if, at a given value of z, t increases by λ/v,

$$\cos \frac{2\pi}{\lambda}\left\{z - v\left(t + \frac{\lambda}{v}\right)\right\} = \cos\left\{\frac{2\pi}{\lambda}(z - vt) - 2\pi\right\} = \cos \frac{2\pi}{\lambda}(z - vt).$$

That is, the period T of the oscillation at a point is λ/v and the frequency f is $1/T = v/\lambda$. The *angular frequency* ω is defined as $2\pi f$. If we also put $2\pi/\lambda = \beta$, we obtain

$$\cos \frac{2\pi}{\lambda}(z - vt) = \cos(\beta z - \omega t) \text{ [or, equivalently, } \cos(\omega t - \beta z)]$$

which is the given form.

(a)   We see that, for the given field $\mathbf{E}$,

$$\nabla \cdot \mathbf{E} = \frac{\partial E_x}{\partial x} + \frac{\partial E_y}{\partial y} + \frac{\partial E_z}{\partial z} = \frac{\partial E_x}{\partial x} = 0,$$

so that $\mathbf{E}$ does satisfy the Maxwell equation (8.59). To find $\mathbf{H}$ we use (8.61), that is

$$\frac{\partial \mathbf{H}}{\partial t} = -\frac{1}{\mu}\nabla \times \mathbf{E} = -\frac{1}{\mu}\begin{vmatrix} \mathbf{i} & \mathbf{j} & \mathbf{k} \\ \dfrac{\partial}{\partial x} & \dfrac{\partial}{\partial y} & \dfrac{\partial}{\partial z} \\ E_0 \cos(\omega t - \beta z) & 0 & 0 \end{vmatrix}$$

$$= -\frac{E_0}{\mu}\beta \sin(\omega t - \beta z)\mathbf{j}.$$

Integrating with respect to t,

$$\mathbf{H} = \frac{\beta E_0}{\omega \mu} \cos(\omega t - \beta z)\mathbf{j}$$

$$= H_0 \cos(\omega t - \beta z)\mathbf{j}.$$

Thus, as in the general theory, we have an $\mathbf{H}$ vector with a y-component associated with an $\mathbf{E}$ vector with an x-component, both transverse to the direction of propagation. Clearly,

$$\nabla \cdot \mathbf{H} = \frac{\partial H_y}{\partial y} = 0,$$

i.e. $\mathbf{H}$ satisfies (8.60). Also,

$$\nabla \times \mathbf{H} = \begin{vmatrix} \mathbf{i} & \mathbf{j} & \mathbf{k} \\ \dfrac{\partial}{\partial x} & \dfrac{\partial}{\partial y} & \dfrac{\partial}{\partial z} \\ 0 & H_0 \cos(\omega t - \beta z) & 0 \end{vmatrix}$$

$$= \beta H_0 \sin(\omega t - \beta z)\mathbf{i}.$$

But $\partial \mathbf{E}/\partial t = \omega E_0 \sin(\omega t - \beta z)\mathbf{i}$, so that for the final Maxwell equation (8.62) to be satisfied we must have

$$\beta H_0 = \frac{\beta^2 E_0}{\omega \mu} = \epsilon_0 \omega E_0$$

or

$$\frac{\omega^2}{\beta^2} = \frac{1}{\epsilon_0 \mu_0}. \tag{8.73}$$

But for a sinusoidal wave of the form

$$\mathbf{E} = E_0 \cos(\omega t - \beta z)\mathbf{i},$$

the *phase velocity* is determined by putting

$$\omega t - \beta z = \text{constant} \quad \text{or} \quad \frac{dz}{dt} = \frac{\omega}{\beta}.$$

But we have shown in the general theory that $1/\sqrt{(\mu_0 \epsilon_0)}$ is the speed of a plane electromagnetic wave in free space, so that (8.73) is valid and (8.62) is indeed satisfied by our calculated field $\mathbf{H}$.

(b)   We know from the general theory that the amplitudes of the $\mathbf{E}$ and $\mathbf{H}$ vectors are linked in free space by

$$\frac{E_0}{H_0} = Z_0 = 120\pi \text{ ohms.}$$

Here, $H_0 = 200\pi/120\pi = 1.667 \,\text{Am}^{-1}$. The frequency of the sinusoidal oscillations is

$$f = \frac{v}{\lambda} = \frac{3 \times 10^8}{1.89} = 1.587 \times 10^8 \text{ Hz,}$$

so the required angular frequency is

$$\omega = 2\pi f = 9.97 \times 10^8 \text{ rad s}^{-1}.$$

■

8.14 Ocean wave interactions

There are several application areas of vector analysis as we have seen thus far. In this section, we briefly present the application of vector analysis to the theory of ocean waves. Much of the research in this area relies heavily on vector analysis to establish results.

First order ocean waves can be regarded as the superposition of free and independent spectral components. However, at a higher order, there exist interactions between spectral components which result in transfer of energy between the components (Phillips 1960). The lowest order at which such interactions may occur involves the interaction of three waves, i.e. triad interactions. In such cases, two waves interact nonlinearly and transfer energy to a third component. Such interactions are important in finite depth conditions but are insignificant in deep water (Hasselmann 1962).

A general perturbation theory for the nonlinear resonant interaction of waves in a random sea was developed by Hasselmann (1962, 1963). He found that a set of four waves, called a quadruplet, could exchange energy when the following resonance conditions were satisfied:

$$\mathbf{k}_1 + \mathbf{k}_2 = \mathbf{k}_3 + \mathbf{k}_4, \tag{8.74}$$

$$\omega_1 + \omega_2 = \omega_3 + \omega_4, \tag{8.75}$$

where ω_j is the radian frequency and $\mathbf{k}_j$ the wavenumber vector ($j = 1, 2, 3, 4$). The wavenumber and frequency are related through the dispersion relationship $\omega^2 = gk \tanh kh$ for finite depth ocean and $\omega^2 = gk$ for infinite depth ocean. The four interacting wave components described by (8.74) and (8.75) form a quadruplet.

Hasselmann (1963) described the nonlinear interactions between wave quadruplets in terms of their action density, n, where $n(\mathbf{k}) = Q(\mathbf{k})/\omega$ and $Q(\mathbf{k})$ is the wavenumber spectrum. Hasselmann assumed that the waves were only weakly nonlinear and, hence, applied a perturbation technique to determine the rate of change in action density at $\mathbf{k}_1$ due to all quadruplet interactions involving $\mathbf{k}_1$. The rate of change $\partial n_1/\partial t$ which is directly related to $S_{nl}(f, \theta)$ is given by

$$\frac{\partial n_1}{\partial t} = S_{nl}$$

$$= \iiint G(\mathbf{k}_1, \mathbf{k}_2, \mathbf{k}_3, \mathbf{k}_4) \times \delta(\mathbf{k}_1 + \mathbf{k}_2 - \mathbf{k}_3 - \mathbf{k}_4)$$

$$\times \delta(\omega_1 + \omega_2 - \omega_3 - \omega_4)$$

$$\times [n_1 n_2(n_4 - n_2) + n_2 n_4(n_3 - n_1)] d\mathbf{k}_2 \, d\mathbf{k}_3 \, d\mathbf{k}_4, \tag{8.76}$$

where $n_j = n(\mathbf{k}_j)$ is the action density at wavenumber $\mathbf{k}_j$ and G is a coupling coefficient.

The delta function in (8.76) ensures that the contribution to the integral only occurs for quadruplets that satisfy the resonance conditions (8.74) and (8.75). The integral equation (8.76) is primarily due to Boltzmann and attributed to his name as the Boltzmann integral for wind waves, in analogy to similar expressions used in theoretical physics, to describe the rate of change of particle density distributions in a system of interacting particles.

To extract maximum energy from the interactions of nonlinear ocean waves in deep water, the Boltzmann equation (8.76) plays a very important role. The resonant conditions in equations (8.74) and (8.75) for frequencies and wavenumbers are extremely important in wave energy studies. Knowledge of vector analysis and Dirac's delta function are absolutely necessary. Waves interacting with each other with different frequencies and subsequently with wavenumbers give rise to an integral formulation developed by Boltzmann. Boltzmann introduced the Dirac delta as a function of frequencies and wavenumbers in equation (8.76). In one-dimensional form the Dirac delta is defined as (see Lighthill 1958)

$$\delta(x - x_0) = \begin{cases} \infty & \text{if } x = x_0 \\ 0 & \text{otherwise} \end{cases}$$

and this Dirac delta function has the property

$$\int_{-\infty}^{\infty} \delta(x - x_0) f(x) dx = f(x_0),$$

for any suitable continuous function $f(x)$. It is worth noting that no function in the ordinary sense has this property but one can imagine a sequence of functions which have progressively taller and thinner peaks at $x = x_0$, with the area under the curve remaining equal to 1, while the value of $\delta(x - x_0)$ function tend to zero at every point except at $x = x_0$ where it tends to infinity. In the limit, the sequence would have the property given above. This is evident if we consider that $f(x) = 1$ such that $\int_{-\infty}^{\infty} \delta(x - x_0) dx = 1$. This definition of delta function can be readily extended to any dimensions, in another words, it can be a function of a vector quantity. To be specific, we can write the property of the delta function for the three-dimensional case

$$\int_{-\infty}^{\infty} \int_{-\infty}^{\infty} \int_{-\infty}^{\infty} \delta(x - x_0, y - y_0, z - z_0) f(x, y, z) dx dy dz = f(x_0, y_0, z_0).$$

Again, one might 'differentiate' $\delta(x)$ to obtain a function $\delta'(x)$ with the property

$$\int_{-\infty}^{\infty} \delta'(x) f(x) dx = - \int_{-\infty}^{\infty} \delta(x) f'(x) dx = -f'(0)$$

for any continuously differentiable function $f(x)$. Physically, $\delta(x)$ can be regarded as that distribution of charge along the x-axis which one speaks of as a unit point charge at the origin. Similarly, $\delta'(x)$ corresponds to a dipole of unit electric moment, since as a special case of the above property we have

$$\int_{-\infty}^{\infty} x\delta'(x)dx = -1.$$

Thus, these generalized functions correspond to familiar physical idealizations. In our wave energy problem, $\delta(\omega_1 + \omega_2 - \omega_3 - \omega_4)$ is a function of frequencies, and $\delta(\mathbf{k}_1 + \mathbf{k}_2 - \mathbf{k}_3 - \mathbf{k}_4)$ is a function of four vectors consisting of wavenumbers. With the property of delta function, the Boltzmann integral can be reduced to a simple integral formula.

The resonant conditions define not only the frequencies of the spectral components that can interact nonlinearly but also their propagation directions, since (8.76) is a vector expression. This relationship can be represented in the form of a vector diagram as shown in Fig. 8.13. To contribute to the integral expression (8.76), the frequencies associated with each of these wavenumbers must also be satisfied.

Hence, all components of the spectrum are potentially coupled and energy can be exchanged, not only between components of different frequencies, but also amongst components propagating in different directions. The nonlinear energy transfer represented by (8.76) conserves both the total energy and momentum of the wave field, merely redistributing it among the spectrum. As a consequence of the symmetry of the resonance conditions with respect to the pairs of wavenumbers $(\mathbf{k}_1, \mathbf{k}_2)$ and $(\mathbf{k}_3, \mathbf{k}_4)$, the quadruplet interactions also conserve the wave action. The absolute value is equal for all wavenumbers within the quadruplet.

$$\frac{d\eta_1}{dt} = \frac{d\eta_2}{dt} = -\frac{d\eta_3}{dt} = -\frac{d\eta_4}{dt}. \tag{8.77}$$

This result is known as the property of detailed balance. In a more general form, detailed balance states that the absolute value of the change in action density, $|\Delta\eta_j|$, is the same for all components in a resonance set of wavenumbers.

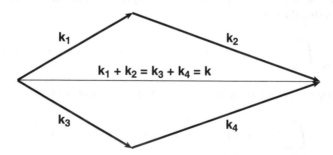

Figure 8.13   Wavenumber vectors satisfying resonance condition.

8.15 Solution techniques for S_{nl}

Numerical techniques have been proposed for solution of (8.76). The major goals of all these techniques have been to:

(a)   improve computational speed;
(b)   to provide a better physical understanding of the source term.

The significant computer time required for the solution arises for two reasons:

1.   the fact that (8.76) represents a six dimensional integral,
2.   the complexity of the complicated coupling coefficient, G.

It was suggested by many previous workers that it is important to first remove the delta functions by projecting the six-dimensional integral onto the three-dimensional resonance space defined by (8.74) and (8.75). Although the details of the various projections differ, the goal was the same. The integral is then solved within the transformed space.

A conceptual difficulty arises in attempting to visualize the multi-dimensional integration space. It would add significantly to the understanding of the problem, if the wavenumber quadruplets which contribute significantly to the integral could be visualized in some simple form. Based on a suggestion by M.S. Longuet-Higgins, Hasselmann (1963) proposed the so called *figures of eight* graphical representation of the interaction space. Figure 8.14 shows the interaction space for a given value $\mathbf{k} = \mathbf{k}_1 + \mathbf{k}_2 = \mathbf{k}_3 + \mathbf{k}_4$. Each of the curves is for a specific value of γ. For deep water conditions, where the dispersion relation $\omega^2 = gk\tanh kh$ reduces to $\omega^2 = gk$, resonance conditions (8.74) and (8.75) become

$$\mathbf{k}_1 + \mathbf{k}_2 = \mathbf{k}_3 \mid \mathbf{k}_4 = \mathbf{k}, \tag{8.78}$$

$$\omega_1 + \omega_2 = \omega_3 + \omega_4 = \gamma\omega, \tag{8.79}$$

where γ is a numerical constant. Equation (8.79) can be rewritten as

$$\sqrt{\mathbf{k}_1} + \sqrt{\mathbf{k}_2} = \sqrt{\mathbf{k}_3} + \sqrt{\mathbf{k}_4} = \gamma\sqrt{\mathbf{k}}. \tag{8.80}$$

Normalizing, such that $\mathbf{k} = 1$ and representing the angle between $\mathbf{k}_1$ and $\mathbf{k}_2$ as θ yields,

$$\cos\theta = \frac{1 + \mathbf{k}_1^2 - (\gamma - \sqrt{\mathbf{k}_1})^4}{2\mathbf{k}_1}. \tag{8.81}$$

For a fixed value of γ, the locus of all values of $\mathbf{k}_1$ which satisfy the resonance conditions is given by (8.80). The corresponding locus of $\mathbf{k}_2$ is given by (8.81). Typical integration curves for a given value of γ are illustrated in Fig. 8.14.

A simple and perhaps the most straightforward representation of the integration space was presented by Webb (1978). For fixed $\mathbf{k}_1$ and $\mathbf{k}_3$, the locus of the possible

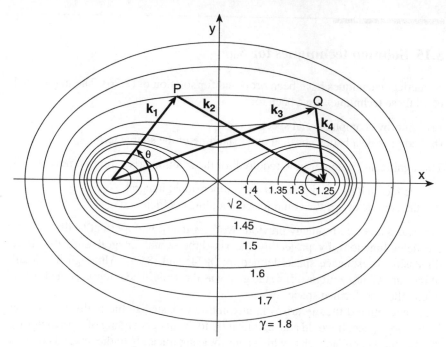

Figure 8.14   The so called *figures of eight* diagram.

values of $\mathbf{k}_2$ and $\mathbf{k}_4$ which close the vector polygon represented by (8.74) and will also satisfy (8.75), trace out two egg-shaped figures. Figure 8.14 shows the loci of $\mathbf{k}_2$ and $\mathbf{k}_4$ for one combination of $\mathbf{k}_1$ and $\mathbf{k}_3$. For this combination of $\mathbf{k}_1$ and $\mathbf{k}_3$, only wavenumber components which lie on these loci will satisfy the resonance conditions and contribute to the integral.

The direct integration techniques mentioned above provide valuable insight into the basic properties of the nonlinear source term S_{nl}. However, they were computationally too expensive for use in wave prediction models. As a result, approximation for narrow-peak spectra were proposed by Longuet-Higgins (1976) and many other researchers.

The solution technique proposed by Webb (1978) reduces the multi-dimensional integral represented by (8.76) to a series of line integrals around the *egg-shaped* loci resulting from the projection used to remove the delta functions. Although this operation simplifies the mathematical formulation, the number of possible loci along which line integrals need to be preformed is enormous. Figure 8.15 shows these loci for the values $\mathbf{k}_1 = 0.14$ rad/m, $\mathbf{k}_3 = 0.20$ rad/m. The $\mathbf{k}_1$ value is in the x-direction and $\mathbf{k}_4$ is 45° clockwise. The solid line represents the loci of $\mathbf{k}_2$ while the dashed one represents that for $\mathbf{k}_4$.

Webb (1978) introduced a transformation $T(\mathbf{k}_1, \mathbf{k}_3)$

$$\frac{\mathrm{d}n_1}{\mathrm{d}t} = \int T(\mathbf{k}_1, \mathbf{k}_3)\mathrm{d}\mathbf{k}_3, \qquad (8.82)$$

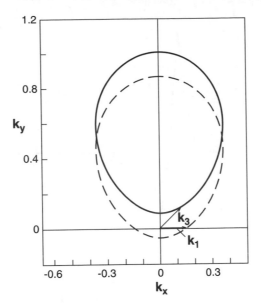

Figure 8.15   Loci of vectors.

where

$$T(\mathbf{k}_1, \mathbf{k}_3) = 2 \iint G(\mathbf{k}_1, \mathbf{k}_2, \mathbf{k}_3, \mathbf{k}_4) \times \delta(\mathbf{k}_1 + \mathbf{k}_2 - \mathbf{k}_3 - \mathbf{k}_4)$$

$$\times \, \delta(\omega_1 + \omega_2 - \omega_3 - \omega_4)$$

$$\times \, [n_1 n_2 (n_4 - n_2) + n_2 n_4 (n_3 - n_1)]$$

$$\times \, \Theta(|\mathbf{k}_1 - \mathbf{k}_4| - |\mathbf{k}_1 - \mathbf{k}_3|) \mathrm{d}\mathbf{k}_2 \, \mathrm{d}\mathbf{k}_4, \qquad (8.83)$$

where n_i is the action density $n(\mathbf{k}_i)$ at wavenumber $\mathbf{k}_i$, ω_i is the angular frequency at $\mathbf{k}_i$, the $\delta(\cdots)$ is the Dirac delta function and the term G is the coupling coefficient (Webb 1978; Tracy and Resio 1982) and

$$\Theta(x) = \begin{cases} 1 & \text{if } x > 0 \\ 0 & \text{if } x \le 0, \end{cases} \qquad (8.84)$$

where $x = |\mathbf{k}_1 - \mathbf{k}_4| - |\mathbf{k}_1 - \mathbf{k}_3|$.

Applying resonance conditions (8.74) and (8.75), equation (8.83) can be solved. Finally, in polar coordinates the nonlinear energy transfer (8.82) can be computed from the following equation

$$\frac{\mathrm{d}n_1}{\mathrm{d}t} = \int_0^\infty \int_0^{2\pi} T(\mathbf{k}_1, \mathbf{k}_3) k_3 \mathrm{d}\theta_3 \mathrm{d}k_3. \qquad (8.85)$$

This implies that the nonlinear energy transfer must be computed over the entire two-dimensional spectrum. Using the method developed by Susilo *et al.* (2007) the

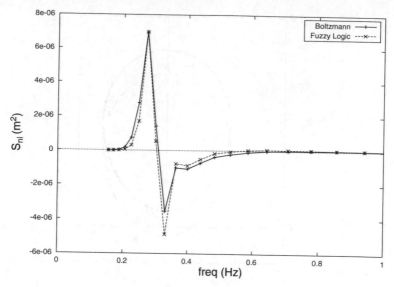

Figure 8.16   Numerical solution of S_{nl}.

computed S_{nl} are shown in Fig. 8.16. This figure shows the comparison between the full Boltzmann method and the present one in one-dimension.

8.16 Graphical simulation of a vector field

This section is concerned with the graphical simulation of a vector field. We will demonstrate here three plots of a vector field as illustrated in *The Mathematica* by Stephen Wolfram. In the first simulation, we plot the vector field in two dimensions given by $\mathbf{A} = \mathbf{i} \sin x + \mathbf{j} \cos x + 0\mathbf{k}$ (Fig. 8.17). In the second simulation, we plot the gradient field of a scalar *potential* function such as the velocity potential given by $\mathbf{V} = \mathbf{i} \partial\phi/\partial x + \mathbf{j} \partial\phi/\partial y + 0\mathbf{k}$ in two dimensions where $\phi(x, y) = x^2 + y^2$ (Fig. 8.18).

The third example is the plot of the Hamiltonian vector field which is defined as $\partial\phi/\partial y$ and $-\partial\phi/\partial x$ (Fig. 8.19). Given a scalar-valued function ϕ on the $x - y$ plane, there are two common ways of constructing a vector field. The function's gradient vector field has first and second components given by $\partial\phi/\partial x$ and $\partial\phi/\partial y$, respectively. The two components of the function's Hamiltonian vector field are $\partial\phi/\partial y$ and $-\partial\phi/\partial x$.

The gradient field gives the direction in which the function is increasing most rapidly. If one treats the function as the Hamiltonian of a mechanical system, the Hamiltonian vector field gives the equation of motion in phase space. The package for plotting the vector field can be found in Mathematica as discussed in the text

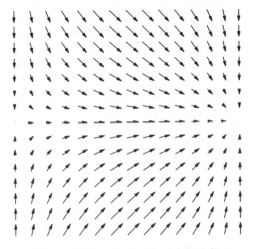

Figure 8.17   Plot of a vector field.

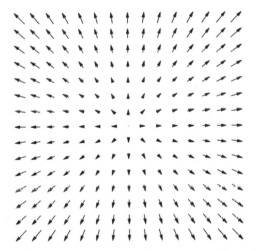

Figure 8.18   Gradient field plot of $\phi = x^2 + y^2$.

The Mathematica. We simply use the Mathematica commands as illustrated by Wolfram (1996). Interested readers are referred to Wolfram's work.

The following package plots two-dimensional vector fields. For doing vector field plots in three-dimensions, one uses the command

```
Graphics'PlotField3D'
```

before gaining access to the various plotting commands. The following is a summary of the commands which can now be used.

1. `PlotVectorField[{phi_x, phi_y$}, {x, xmin, xmax}, {y, ymin,ymax}]`

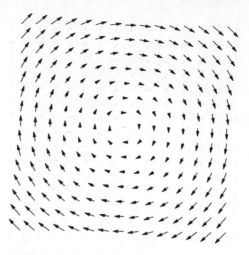

Figure 8.19   Hamiltonian vector field.

This plots the vector field given by the vector-valued function in the range specified.

2. `PlotGradientField[phi, {x, xmin, xmax},`
   `{y, ymin, ymax}]`
   This command plots the gradient vector field of the scalar-valued function ϕ.

3. `PlotHamiltonianField[phi, {x, xmin, xmax},`
   `{y, ymin, ymax}]`
   This command plots the Hamiltonian vector field of the scalar-valued function ϕ.

1. *Plotting vector fields in two-dimensions*
   (a) We can load the package with the following command at the Mathematica prompt:

   $$In[1]:= < Graphics'PlotField'$$

   Once the package is loaded, then the vector field can be plotted or manipulated as desired.
   (b) As an example, we can give the two components of the vector field as given by $\sin[x]$ and $\cos[y]$.

   $$In[2]:= PlotVectorField[sin[x], cos[y]\},$$
   $$\{x, 0, pi\},\{y, 0, pi\}]$$

2. *PlotGradientField and PlotHamiltonianField*
   These commands compute the formulae for the partial derivatives of the scalar-valued function given as an argument. This means that the functions one gives

must be such that Mathematica can compute their derivatives, i.e. continuously differentiable functions.

(a)   Here is the gradient field of the potential

$$\phi(x, y) = x^2 + y^2.$$

```
In[3]:= PlotGradientField[x^2 + y^2,
        {x, -3, 3},{y, -3, 3}]
```

(b)   In the plane, the gradient and Hamiltonian vector fields of the same function are always orthogonal.

```
In[4]:= PlotHamiltonianField[x^2 + y^2,
        {x, -3, 3},{y, -3, 3}]
```

8.17 A boundary-initial value problem of the heat equation

In this section, we shall investigate the question whether a solution of the heat equation satisfying prescribed boundary and initial conditions over a given region is necessarily unique.

Let us suppose that we are to solve the three-dimensional heat equation

$$\frac{\partial u}{\partial t} = \alpha \nabla^2 u,$$

through a region V bounded by a closed surface S, subject to the boundary condition $u = f(x, y, z, t)$ on S and the initial condition $u(x, y, z, 0) = g(x, y, z)$ through V.

Let us consider that we have two solutions of the problem, u_1 and u_2, each of which, with its derivatives through the second, is continuous in V.

If we define a new function, $w(x, y, z, t) = u_2(x, y, z, t) - u_1(x, y, z, t)$, it is clear from the linearity of the heat equation that w also satisfies this equation. Moreover, w obviously assumes boundary and initial conditions which are identically zero.

Now consider the volume integral

$$J(t) = \frac{1}{2} \iiint_V w^2(x, y, z, t) dV \qquad t \geq 0,$$

Clearly, $J(t)$ is a continuous function of t which is always equal to or greater than zero, since its integrand is everywhere nonnegative. Also, since $w = 0$ when $t = 0$, it follows that $J(0) = 0$. Now differentiating with respect to t inside the integral sign, we obtain

$$J'(t) = \frac{1}{2} \iiint_V 2w \frac{\partial w}{\partial t} dV$$

and thus, since w satisfies the heat equation, on substituting for $\partial w/\partial t$ we have

$$J'(t) = \alpha \iiint_V w\nabla^2 w \, dV. \tag{8.86}$$

We know from vector identities and the divergence theorem that

$$\iiint_V (w\nabla^2 w + \nabla w \cdot \nabla w) \, dV = \iint_S w\nabla w \cdot \mathbf{n} \, dS.$$

Because of the surface boundary condition $w = 0$, the right-hand side must be zero, and we have

$$\iiint_V w\nabla^2 w \, dV = - \iiint_V \nabla w \cdot \nabla w \, dV.$$

Hence, substituting into equation (8.86)

$$J'(t) = -\alpha \iiint_V \nabla w \cdot \nabla w \, dV$$

$$= -\alpha \iiint_V \left\{ \left(\frac{\partial w}{\partial x}\right)^2 + \left(\frac{\partial w}{\partial y}\right)^2 + \left(\frac{\partial w}{\partial z}\right)^2 \right\} dV.$$

Since $\alpha > 0$, the above result shows that $J'(t) \leq 0$ for $t \geq 0$. Now by the law of the mean,

$$\frac{J(t) - J(0)}{t} = J'(t) \qquad 0 < t_1 < t$$

or

$$J(t) = J(0) + t J'(t) \qquad 0 < t_1 < t.$$

But we have already verified that $J(0) = 0$. Hence, the last equation reduces to $J(t) = t J'(t)$ which implies that $J(t) \leq 0$ for $t \geq 0$ since we have just proved that J' is nonnegative for $t \geq 0$. However, as we observed earlier, the definition of $J(t)$ shows that $J(t) \geq 0$ for $t \geq 0$. The only way in which these inequalities can simultaneously be fulfilled is for $J(t)$ to vanish identically.

Hence, $w(x, y, z, t) \equiv u_2(x, y, z, t) - u_1(x, y, z, t) \equiv 0$, or simply $u_2(x, y, z, t) = u_1(x, y, z, t)$.

Thus, *in boundary region, twice differentiable solutions of the heat equation satisfying prescribed surface and initial temperature conditions are unique.*

8.18 Irrotational flow of an incompressible and inviscid fluid over a stationary ellipsoid

The flows for which the vorticity vector vanishes, that, $\nabla \times \mathbf{V} = 0$ automatically everywhere in the flow field are said to be irrotational. This implies that there exists a scalar potential ϕ such that $\mathbf{V} = \nabla\phi$ in which case the scalar function ϕ is called the velocity potential. By using the incompressibily of fluid condition which is known as the continuity condition, that is, $\nabla \cdot \mathbf{V} = 0$ yields the Laplace equation, $\nabla \cdot \nabla\phi = \nabla^2\phi = 0$.

In this section, we briefly present a classical solution of this Laplace equation satisfied by ϕ for the flow induced by an ellipsoidal structure that translates along the x-axis. This analysis is entirely due to Lamb (1993). The harmonic function presented by Lamb represents the solution to Laplace's equation expressed in terms of a special system of orthogonal curvilinear coordinates known as ellipsoidal coordinates.

In the applications to which we now proceed the triple orthogonal system consists of a confocal quadrics

$$\frac{x^2}{a^2 + \theta} + \frac{y^2}{b^2 + \theta} + \frac{z^2}{c^2 + \theta} - 1 = 0 \quad a > b > c, \tag{8.87}$$

whose properties are explained in the books on solid geometry. Here a, b, c are constant and θ is a variable parameter. Through any given point (x, y, z) there pass three surfaces of the system, corresponding to the three roots of equation (8.87), considered as a cubic in θ. If $a > b > c$, one of these roots (λ, say) will lie between ∞ and $-c^2$, another (μ) between $-c^2$ and $-b^2$ and the third (ν) between $-b^2$ and $-a^2$. The surfaces λ, μ, ν are, therefore, ellipsoids, hyperboloids of one sheet, and hyperboloids of two sheets, respectively.

It follows immediately from this definition of λ, μ, ν, that

$$\frac{x^2}{a^2 + \theta} + \frac{y^2}{b^2 + \theta} + \frac{z^2}{c^2 + \theta} - 1 = \frac{(\lambda - \theta)(\mu - \theta)(\nu - \theta)}{(a^2 + \theta)(b^2 + \theta)(c^2 + \theta)}, \tag{8.88}$$

identically, for all values of θ. Hence, multiplying by $a^2 + \theta$ and afterwards putting $\theta = -a^2$, we obtain the first of the following equations:

$$x^2 = \frac{(a^2 + \lambda)(a^2 + \mu)(a^2 + \nu)}{(a^2 - b^2)(a^2 - c^2)},$$

$$y^2 = \frac{(b^2 + \lambda)(b^2 + \mu)(b^2 + \nu)}{(b^2 - c^2)(b^2 - a^2)}, \tag{8.89}$$

$$z^2 = \frac{(c^2 + \lambda)(c^2 + \mu)(c^2 + \nu)}{(c^2 - a^2)(c^2 - b^2)}.$$

These give

$$\frac{\partial x}{\partial \lambda} = \frac{1}{2}\frac{x}{a^2 + \lambda},$$

$$\frac{\partial y}{\partial \lambda} = \frac{1}{2}\frac{y}{b^2 + \lambda}, \tag{8.90}$$

$$\frac{\partial z}{\partial \lambda} = \frac{1}{2}\frac{z}{c^2 + \lambda}.$$

If x, y, z be functions of the parameters λ, μ, ν such that the surfaces $\lambda = $ constant, $\mu = $ constant, $\nu = $ constant, are mutually orthogonal at their intersections, and if we write

$$
\begin{aligned}
h_1^2 &= \left(\frac{\partial x}{\partial \lambda}\right)^2 + \left(\frac{\partial y}{\partial \lambda}\right)^2 + \left(\frac{\partial z}{\partial \lambda}\right)^2, \\
h_2^2 &= \left(\frac{\partial x}{\partial \mu}\right)^2 + \left(\frac{\partial y}{\partial \mu}\right)^2 + \left(\frac{\partial z}{\partial \mu}\right)^2, \\
h_3^2 &= \left(\frac{\partial x}{\partial \nu}\right)^2 + \left(\frac{\partial y}{\partial \nu}\right)^2 + \left(\frac{\partial z}{\partial \nu}\right)^2,
\end{aligned}
\tag{8.91}
$$

the direction-cosines of the normals to the three surfaces which pass through (x, y, z), we get

$$
\begin{aligned}
&\left\{\frac{1}{h_1}\frac{\partial x}{\partial \lambda}, \frac{1}{h_1}\frac{\partial y}{\partial \lambda}, \frac{1}{h_1}\frac{\partial z}{\partial \lambda}\right\}, \\
&\left\{\frac{1}{h_2}\frac{\partial x}{\partial \mu}, \frac{1}{h_2}\frac{\partial y}{\partial \mu}, \frac{1}{h_2}\frac{\partial z}{\partial \mu}\right\}, \\
&\left\{\frac{1}{h_3}\frac{\partial x}{\partial \nu}, \frac{1}{h_3}\frac{\partial y}{\partial \nu}, \frac{1}{h_3}\frac{\partial z}{\partial \nu}\right\},
\end{aligned}
\tag{8.92}
$$

respectively. It follows that the lengths of the linear elements drawn in the directions of these normals will be $h_1\delta\lambda$, $h_2\delta\mu$, $h_3\delta\nu$. Here h_1, h_2, h_3 are known as the scale factors.

Remark 8.4   We may note that if λ, μ, ν are regarded as functions of x, y, z, the direction-cosines of the three line-elements considered above can be expressed in the form

$$
\begin{aligned}
&\left\{h_1\frac{\partial \lambda}{\partial x}, h_1\frac{\partial \lambda}{\partial y}, h_1\frac{\partial \lambda}{\partial z}\right\}, \\
&\left\{h_2\frac{\partial \mu}{\partial x}, h_2\frac{\partial \mu}{\partial y}, h_2\frac{\partial \mu}{\partial z}\right\}, \\
&\left\{h_3\frac{\partial \nu}{\partial x}, h_3\frac{\partial \nu}{\partial y}, h_3\frac{\partial \nu}{\partial z}\right\},
\end{aligned}
\tag{8.93}
$$

from which, and from (8.92), various interesting relations can be derived.

The square of the scale factor h_1 can be obtained from the first equation of (8.91) as

$$
h_1^2 = \frac{1}{4}\left\{\frac{x^2}{(a^2 + \lambda)^2} + \frac{y^2}{(b^2 + \lambda)^2} + \frac{z^2}{(c^2 + \lambda)^2}\right\}.
\tag{8.94}
$$

Similarly the square of the other two scale factors can be determined. These scale factors can be simplified considerably by using a little trick. If we differentiate (8.88) with respect to θ and afterwards put $\theta = \lambda$, we deduce the first of the following three relations in ellipsoidal coordinates:

$$h_1^2 = \frac{1}{4} \left\{ \frac{(\lambda - \mu)(\lambda - v)}{(a^2 + \lambda)(b^2 + \lambda)(c^2 + \lambda)} \right\},$$

$$h_2^2 = \frac{1}{4} \left\{ \frac{(\mu - v)(\mu - \lambda)}{(a^2 + \mu)(b^2 + \mu)(c^2 + \mu)} \right\}, \qquad (8.95)$$

$$h_3^2 = \frac{1}{4} \left\{ \frac{(v - \lambda)(v - \mu)}{(a^2 + v)(b^2 + v)(c^2 + v)} \right\}.$$

■

Note: For our present analysis the following relations will also be useful:

$$\frac{\partial \lambda}{\partial x} = \frac{1}{h_1^2} \frac{\partial x}{\partial \lambda} = \frac{x}{2h_1^2(a^2 + \lambda)},$$

$$\frac{\partial \lambda}{\partial y} = \frac{1}{h_1^2} \frac{\partial y}{\partial \lambda} = \frac{x}{2h_1^2(b^2 + \lambda)}, \qquad (8.96)$$

$$\frac{\partial \lambda}{\partial z} = \frac{1}{h_1^2} \frac{\partial z}{\partial \lambda} = \frac{x}{2h_1^2(c^2 + \lambda)}.$$

The Laplacian (∇^2) of the scalar function ϕ in ellipsoidal coordinates can be written in the form

$$\nabla^2 \phi = \frac{1}{h_1 h_2 h_3} \left\{ \frac{\partial}{\partial \lambda} \left(\frac{h_2 h_3}{h_1} \frac{\partial \phi}{\partial \lambda} \right) + \frac{\partial}{\partial \mu} \left(\frac{h_3 h_1}{h_2} \frac{\partial \phi}{\partial \mu} \right) + \frac{\partial}{\partial v} \left(\frac{h_1 h_2}{h_3} \frac{\partial \phi}{\partial v} \right) \right\}.$$

$$(8.97)$$

Equating this to zero, we obtain the Laplace equation that is the general expression of continuity equation given in ellipsoidal coordinates. The particular solutions of the transformed equation $\nabla^2 \phi = 0$ which first present themselves are those in which ϕ is a function of one (only) of the variables λ, μ, v. Thus ϕ may be a function of λ alone, provided

$$\frac{h_2 h_3}{h_1} \frac{d\phi}{d\lambda} = \text{constant}. \qquad (8.98)$$

Using the expressions of the scale factors, and after integration with respect to λ, we obtain

$$\phi = C \int_\lambda^\infty \frac{d\lambda}{\Delta}, \qquad (8.99)$$

where C is a constant and $\Delta = \sqrt{(a^2 + \lambda)(b^2 + \lambda)(c^2 + \lambda)}$, the additive constant which attaches to ϕ being chosen so as to make ϕ vanish for $\lambda = \infty$.

In this solution, which corresponds to $\phi = A/r$ in spherical harmonics, the equipotential surfaces are the confocal ellipsoids, and the motion in the space external to any one of these (say that for which $\lambda = 0$) is that due to a certain arrangement of simple sources over it. The velocity at any point is given by the formula $-h_1 d\phi/d\lambda = Ch_1/\Delta$. For more information about this subject the reader is referred to Lamb (1993).

8.19 Summary

In this chapter we have discussed the following application problems:

- Navier–Stokes equations
- Euler's equations
- Bernoulli's equation
- Poisson's equation
- Heat equations
- Maxwell's electromagnetic theory
- Maxwell's wave and heat equations
- Nonlinear wave-wave interactions
- Uniqueness of solutions of the heat equation
- Use of Green's theorem, Stokes theorem and divergence theorem
- Irrotational flow over an ellipsoid: use of orthogonal curvilinear coordinates.

Exercises

1. What is the potential function for a central-force field in which the attraction on a particle varies as the square of the distance from the origin? Inversely as the distance from the origin?

2. What is the potential function for the gravitational field of a uniform circular disk at any point on the axis of the disk?

3. What is the potential function for the gravitational field of a uniform sphere of radius a and mass M? Show that the attraction of the sphere at a point P a distance r from the centre of the sphere is

$$
\mathbf{F} = \begin{cases} -\dfrac{M\mathbf{R}}{a^3} & r \le a, \\[2mm] -\dfrac{M\mathbf{R}}{r^3} & r \ge a. \end{cases}
$$

4. If a conductor is defined to be a body in whose interior the electric field is everywhere zero, show that any charge on a conductor must be located entirely on its surface.

5. In cylindrical coordinates, the electric flux density vector $\mathbf{D}$ in a region is given by

$$\mathbf{D} = 4\rho^2 \boldsymbol{\rho} \, C \, m^{-2}.$$

Verify that the total flux of $\mathbf{D}$ over the surface of the cylinder $\rho = 2, |z| < 5$ is equal to the total charge inside the cylinder.

6. Calculate the total magnetic flux crossing the $z = 0$ plane for $\rho \le 5 \times 10^{-2}$ m when

$$\mathbf{B} = \frac{0.2}{\rho} \sin^2 \phi \mathbf{k} \, Wb \, m^{-2}.$$

7. Show that the fields

$$\mathbf{E} = E_0 \sin x \sin t \mathbf{j} \quad \text{and} \quad \mathbf{H} = \frac{E_0}{\mu_0} \cos x \cos t \mathbf{k},$$

satisfy only three of the four Maxwell equations and, thus, are both valid solutions.

8. In free space, a time-dependent magnetic vector potential $\mathbf{A} = \exp(-2\rho)\mathbf{k}$ is present. Calculate expressions for the vector fields $\mathbf{B}, \mathbf{H}$ and $\mathbf{J}$. Also calculate the total current crossing the $z = 0$ plane between $\rho = 0$ and $\rho = 10$ mm.

9. An electric charge of 3 C is moved from the point $P_1(1, 2, 0)$ to the point P_2 $(2, 8, 0)$ (distances are in meters). Find the work done if the electric field is

$$\mathbf{E} = 4xy\mathbf{i} + 2x^2\mathbf{j} \, V \, m^{-1}$$

(a) by direct integration, and
(b) by finding the associated scalar potential ϕ.

10. Electric charge is distributed within a spherical region of radius a such that the electric field is

$$\mathbf{E} = Ar^3\mathbf{r} \, V \, m^{-1},$$

where A is a constant. Calculate
(a) the total charge within the sphere,
(b) the total electrostatic energy stored within this region. (Note: $\frac{1}{2}\epsilon_0|\mathbf{E}|^2$ is the electrostatic energy density.)

11. The fields in a spherical wave at a large distance r from a dipole antenna in free space, using spherical coordinates,

$$\mathbf{E} = \frac{150}{r} \sin \theta \cos(\omega t - \beta r)\boldsymbol{\theta} \, V \, m^{-1},$$

$$\mathbf{H} = \frac{0.2}{r} \sin \theta \cos(\omega t - \beta r)\boldsymbol{\phi} \, A \, m^{-1}.$$

Calculate the instantaneous and average powers crossing a hemisphere $0 < \theta < \pi/2$ at a distance of 1 km from the antenna.

12. Show by substitution, that the fields

$$\mathbf{E} = E_0 \cos(\omega t + \beta z)\mathbf{i} \quad \text{and} \quad \mathbf{H} = -\frac{E_0}{Z} \cos(\omega t + \beta z)\mathbf{j}$$

(where $Z = \sqrt{(\mu/\epsilon)}$), satisfy Maxwell's four equations for a perfect charge-free dielectric characterized by μ and ϵ.

13. The displacement current density in a perfect dielectric with $\epsilon = 4\epsilon_0$ and $\mu = 5\mu_0$ is given by

$$\frac{\partial \mathbf{D}}{\partial t} = 2\cos(\omega t - 5z)\mathbf{i} \ \mu\text{Am}^{-2}.$$

Find expressions for the fields $\mathbf{D}$, $\mathbf{E}$, $\mathbf{B}$ and $\mathbf{H}$, ignoring all constant terms. Recalculate $\partial \mathbf{D}/\partial t$, and, hence, find the value of ω.

14. In free space, an electromagnetic wave has an $\mathbf{E}$ vector

$$\mathbf{E}(z,t) = 150\sin(\omega t - \beta z)\mathbf{i} \ \text{Vm}^{-1}.$$

Calculate the total average power passing through a rectangular area of sides 30 mm and 15 mm in the $z = 0$ plane.

15. In a cylindrical conductor or radius 2 mm, the current density $\mathbf{J}$ is given by

$$\mathbf{J} = 10^3 e^{-400\rho}\mathbf{k} \ \text{Am}^{-2}.$$

Calculate the total current I in the conductor.

16. In spherical coordinates, a magnetic field $\mathbf{H}$ is given by

$$\mathbf{H} = 10^6 r \sin\theta \boldsymbol{\phi} \ \text{Am}^{-1}.$$

Deduce the r-component of $\nabla \times \mathbf{H}$, and, hence, the current flowing across the spherical 'cap'

$$r = 1 \text{ mm}, \quad 0 < \theta < \pi/6, \quad 0 < \phi < 2\pi,$$

17. Show directly from Maxwell's equations that the charge density ρ and the electric current density $\mathbf{j}$ obey the conservation law

$$\frac{\partial \rho}{\partial t} + \nabla \cdot \mathbf{j} = 0.$$

18. Derive the formula for the electric field $\mathbf{E}$ due to a point charge Q using Gauss's law.

19. Show from Maxwell's equations in a vacuum that the magnetic field $\mathbf{B}$ obeys the wave equation.

20. For the electromagnetic wave in which the electric field is given by $\mathbf{E} = \mathbf{E}_0 f(\mathbf{k} \cdot \mathbf{x} - \omega t)$, calculate the corresponding magnetic field $\mathbf{B}$. What can be deduced about the directions of the vectors $\mathbf{E}$ and $\mathbf{B}$?

21. Using Maxwell's equations in a vacuum, obtain an equation in the form of a conservation law for the rate of change of the energy $w = |\mathbf{B}|^2/2 + |\mathbf{E}|^2/2c^2$ of an electromagnetic wave.

Appendix A

Answers to some selected exercises

Chapter 1

1. (i) $\cos\theta = (\mathbf{A}\cdot\mathbf{B})/(|\mathbf{A}||\mathbf{B}|) = -42/42 = -1$; hence $\theta = \pi$, which implies that $\mathbf{A}$ is parallel to $\mathbf{B}$. (ii) $\sin\theta = (\mathbf{A}\times\mathbf{B})/(|\mathbf{A}||\mathbf{B}|) = 0/42 = 0$; hence $\theta = 0$, which implies that $\mathbf{A}$ is parallel to $\mathbf{B}$ or one lies over the other.

2. $\mathbf{C} = \mathbf{A}\times\mathbf{B}$ is a vector perpendicular to both $\mathbf{A}$ and $\mathbf{B}$. Hence, $\mathbf{C} = 0\mathbf{i} + 2\mathbf{j} + 2\mathbf{k}$. The unit vector $= \mathbf{C}/|\mathbf{C}| = \frac{1}{2\sqrt{2}}(0\mathbf{i} + 2\mathbf{j} + 2\mathbf{k})$.

3. $\cos\theta = (\mathbf{A}\cdot\mathbf{B})/(|\mathbf{A}||\mathbf{B}|) = [(2\mathbf{i} - 3\mathbf{j} + \mathbf{k})\cdot(3\mathbf{i} - \mathbf{j} - 2\mathbf{k})]/[(\sqrt{14})(\sqrt{14})] = \frac{1}{2}$. Therefore, $\theta = \cos^{-1}(1/2) = \pi/3$.

4. The vector which is parallel to both $\mathbf{A}$ and $\mathbf{B}$ and perpendicular to $\mathbf{C}$ is $\mathbf{D} = (\mathbf{A}\times\mathbf{B})\times\mathbf{C}$. Hence, computing, $\mathbf{D} = 17\mathbf{i} - 13\mathbf{j} + 8\mathbf{k}$ and the unit vector $= \mathbf{D}/|\mathbf{D}| = \frac{1}{\sqrt{522}}(17\mathbf{i} - 13\mathbf{j} + 8\mathbf{k})$.

6. (a) Parallel. (b) Neither. (c) Neither.

7. $-\frac{3}{\sqrt{10}}\mathbf{i} + \frac{1}{\sqrt{10}}\mathbf{j}$ and $\frac{3}{\sqrt{10}}\mathbf{i} - \frac{1}{\sqrt{10}}\mathbf{j}$.

8. (a) Scalar projection of $\mathbf{B}$ on $\mathbf{A} = (\mathbf{A}\cdot\mathbf{B})/|\mathbf{A}| = \frac{11}{\sqrt{13}}$; Vector projection of $\mathbf{B}$ on $\mathbf{A} = [(\mathbf{A}\cdot\mathbf{B})/|\mathbf{A}|](\mathbf{A}/|\mathbf{A}|) = \frac{11}{13}(2\mathbf{i} + 3\mathbf{j})$.
   (e) Scalar projection $= 1$; Vector projection $= \mathbf{i}$.

10. (a) The distance $= \frac{1}{\sqrt{5}}$. (b) The distance $= \frac{11}{5}$.

11. $\frac{9}{\sqrt{17}}$.

13. (a)   This formula can be established by two methods: (i) elementary calculus method and (ii) vector calculus method.

(i) *Elementary calculus method*:

Given the equation of the straight line as $ax + by + c = 0$. The equation of any line perpendicular to the given line from an external point (x_1, y_1) can be written at once $bx - ay = bx_1 - ay_1$. The point of intersection of these two lines can be obtained by Cramer's rule, and they are given as

$$x = \frac{b^2x_1 - aby_1 - ac}{a^2 + b^2} = x_2,$$

$$y = \frac{a^2y_1 - abx_1 - bc}{a^2 + b^2} = y_2.$$

Then if d is the distance between these two points (x_1, y_1) and (x_2, y_2), we have

$$d^2 = \left[x_1 - \frac{b^2x_1 - aby_1 - ac}{a^2 + b^2}\right]^2 + \left[y_1 - \frac{a^2y_1 - abx_1 - bc}{a^2 + b^2}\right]^2$$

$$= \frac{(a^2 + b^2)(ax_1 + by_1 + c)^2}{(a^2 + b^2)^2}$$

$$= \frac{(ax_1 + by_1 + c)^2}{(a^2 + b^2)}.$$

Thus $d = (|ax_1 + by_1 + c|)/\sqrt{a^2 + b^2}$. This is the desired result.

(ii) *Vector calculus method*:

Let us consider $P = (x_1, y_1)$ is the point in the $x - y$ plane, and $Q = (x, y)$ is any arbitrary point on the line $ax + by + c = 0$. Then the vector $\mathbf{QP} = \mathbf{i}(x - x_1) + \mathbf{j}(y - y_1)$. Let S is a point on the line such that the unit normal vector $\mathbf{SP} = \mathbf{i}a/\sqrt{a^2 + b^2} + \mathbf{j}b/\sqrt{a^2 + b^2}$. The perpendicular distance from $P = (x_1, y_1)$ to the point S is $|\mathbf{SP}|$ which is the scalar projection of $\mathbf{QP}$ on the $\mathbf{SP}$. Thus we have

$$\mathbf{SP} \cdot \mathbf{QP} = |\mathbf{SP}||\mathbf{QP}| \cos \theta$$

$$= |\mathbf{QP}| \cos \theta.$$

Hence, substituting the respective expressions yields,

$$\left\{\mathbf{i}\frac{a}{\sqrt{a^2 + b^2}} + \mathbf{j}\frac{b}{\sqrt{a^2 + b^2}}\right\} \cdot \{\mathbf{i}(x - x_1) + \mathbf{j}(y - y_1)\} = \text{projection of } \mathbf{QP} \text{ on } \mathbf{SP}.$$

Thus after reduction, we obtain

$$\frac{a(x - x_1) + b(y - y_1)}{\sqrt{a^2 + b^2}} = \text{Perpendicular distance,}$$

$$\frac{|ax_1 + by_1 + c|}{\sqrt{a^2 + b^2}} = \text{Perpendicular distance.}$$

This is the required perpendicular distance from P to the given straight line. Note that if $f(x, y) = ax + by = -c$ is a level curve then the unit normal vector at a point on the curve which passes through a given point is given by $\mathbf{n} = \nabla f / |\nabla f| = \mathbf{i}a/\sqrt{a^2 + b^2} + \mathbf{j}b/\sqrt{a^2 + b^2}$, where $\nabla = \mathbf{i}\partial/\partial x + \mathbf{j}\partial/\partial y$ (see Chapter 4 for more information).

    (b) $\frac{13}{5}$.

14. 30.
15. $\sqrt{4014}$.
16. The normal vector is $2\mathbf{i} - 10\mathbf{j} - 4\mathbf{k}$; the equation of the plane is $2x - 10y - 4z = -28$.
17. (a) 10. (b) 44.
18. (a) Length $= \sqrt{2}$; $\theta = \pi/4$. (b) Length $= \sqrt{2}$; $\theta = 3\pi/4$. (c) Length $= 2$; $\theta = \pi/3$. (d) Length $= 2$; $\theta = -\pi/3$.
19. (a) $3x - 2y + 4z - 11 = 0$.
    (b) $20x + 14y - 8z = 24$.
    (c) $4x + y + 2z = 12$.
21. $\mathbf{D} = \mathbf{i} + 2\mathbf{j} + 3\mathbf{k} = (-17\mathbf{A} + 14\mathbf{B} + 3\mathbf{C})/33 = -\mathbf{U} + 29\mathbf{V} + 13\mathbf{W}$.
22. $\mathbf{M} = -2\mathbf{i} + \mathbf{j} + \mathbf{k}$.

Chapter 2

1. (a) $\mathbf{R}'(t) = \cos t\,\mathbf{i} - \sin t\,\mathbf{j}$.
    (b) True since $\mathbf{k} \cdot \mathbf{R}'(t) = 0$.
    (c) $t = n\pi$, $(n = 0, \pm1, \pm2, \ldots)$.
    (d) Yes; $\sqrt{2}$.
    (e) Yes, 1.
    (f) $\mathbf{R}''(t) = -\sin t\,\mathbf{i} - \cos t\,\mathbf{j}$.
4. (a) Arc length $= s = 2\sqrt{5}\pi^2$. (b) $\mathbf{T}(t) = \frac{1}{\sqrt{5}}[\mathbf{i} \sin t + \mathbf{j} \cos t + 2\mathbf{k}]$.
    (c) $\mathbf{T}(\pi) = \frac{1}{\sqrt{5}}(-\mathbf{j} + 2\mathbf{k})$.
5. (a) Speed $= |\dot{\mathbf{R}}| = v = \sqrt{2}e^t$. (b) $a_T = \sqrt{2}e^t = a_N$.
    (c) $\mathbf{T}(t) = \frac{1}{\sqrt{2}}[\mathbf{i}(\cos t - \sin t) + \mathbf{j}(\cos t + \sin t)]$. (d) $\kappa = \frac{1}{\sqrt{2}}e^{-t}$.
9. (a) $v = 3/2$. (b) $\mathbf{a}(t) = -\cos t(\mathbf{i} - \mathbf{j}) - \sin t(\mathbf{i} + \mathbf{j})$.
    (c) $-2/3 \sin t(\mathbf{i} - \mathbf{j}) + 2/3 \cos t(\mathbf{i} + \mathbf{j}) + 1/3\mathbf{k}$. (d) $\kappa = \frac{8}{9\sqrt{2}}$.

11. (a) $\mathbf{T}(t) = [1/(1 + t^2)](1, \sqrt{2}t, t^2);$
         $\mathbf{N}(t) = [1/\sqrt{2}(1 + t^2)](-\sqrt{2}t, \sqrt{t}(1 - t^2), 2t).$
     (b) $\kappa = 1/3\sqrt{2}(1 + t^2)^2.$

13. (a) $\mathbf{T}(t) = [1/(2t^2 + 1)](2t^2, 2t, 1);$
         $\mathbf{N}(t) = [1/(2t^2 + 1)](2t, 1 - 2t^2, -2t).$
     (b) $\kappa = 2t/(2t^2 + 2)^2.$

15. (a) The normal plane is $y - 6x = \pi$; osculating plane is $x + 6y = 6\pi.$
     (b) The normal plane is $x + 2y + 3z = 6$; osculating plane is
         $3x - 3y + z = 1.$

17. The speed of the object is $|\mathbf{v}(t)| = \sqrt{25t^2 + 2}$ while its position is $\mathbf{r}(t) = ti - tj + 5/2t^2\mathbf{k}.$

19. (a) C intersects the xz-plane where $y = 0$, the point is $(15/8, 0, -\ln 2).$
     (b) The tangent line is $x = 1 - 3s, y = 1 + 2s, z = s.$
     (c) The equation of the normal plane is $3x - 2y - z = 1.$

24. $\mathbf{R}(t) = i(t^4/4 + t - 1) + j(t^2/2 + 2t) + \mathbf{k}(\frac{5}{4} - 3t - \cos 2t/4);$
     $\mathbf{V}(1) = 2i + 3j + (\sin 2/2 - 3)\mathbf{k}.$

Chapter 3

1. $\phi_y(1, 1, 0) = \frac{4}{3}2^{1/3}.$

3. $x\phi_x = [1/(x - y)][2x^4y - 3x^3y^2]; y\phi_y = x^4y/(x - y)^2 : x\phi_x + y\phi_y = 3\phi.$

6. $\partial\rho/\partial t + \nabla \cdot (\rho\mathbf{V}) = 0$ which can be written as $\partial\rho/\partial t + \partial(\rho u)/\partial x + \partial(\rho v)/\partial y + \partial(\rho w)/\partial z = 0.$ By calculating the partial derivatives we found that

$$\frac{\partial\rho}{\partial t} = z,$$

$$\frac{\partial(\rho u)}{\partial x} = 3x^2y^2 + yt + 2xyzt,$$

$$\frac{\partial(\rho v)}{\partial y} = 3xy^2z - 2xt^2 + 2yz^2t,$$

$$\frac{\partial(\rho w)}{\partial z} = 5xt + 6z^2t + 4xyz.$$

It can be easily seen that the continuity equation is not satisfied.

7. (a) Given that $u(x, y) = 1 + x/(x^2 + y^2)$ and $v(x, y) = -y/(x^2 + y^2).$

$$\frac{\partial u}{\partial x} = \frac{-x^2 + y^2}{(x^2 + y^2)^2},$$

$$\frac{\partial u}{\partial y} = \frac{-2xy}{(x^2 + y^2)^2},$$

$$\frac{\partial v}{\partial x} = \frac{2xy}{(x^2 + y^2)^2},$$

$$\frac{\partial v}{\partial y} = \frac{-x^2 + y^2}{(x^2 + y^2)^2}.$$

From these calculations it is easy to see that $u_x = v_y$ and $u_y = -v_x$. Hence, the Cauchy–Riemann conditions are satisfied.

(b) Given that $u(x, y) = e^x(x \cos y - y \sin y)$; $v(x, y) = e^x(x \sin y + y \cos y)$.

$$\frac{\partial u}{\partial x} = e^x(x \cos y - y \sin y + \cos y),$$

$$\frac{\partial u}{\partial y} = -e^x(x \sin y + \sin y + y \cos y),$$

$$\frac{\partial v}{\partial x} = e^x(x \sin y + \sin y + y \cos y),$$

$$\frac{\partial v}{\partial y} = e^x(x \cos y - y \sin y + \cos y).$$

It is obvious that the Cauchy–Riemann conditions are satisfied.

9. $u_t = -\alpha e^{-\alpha t} \sin x$; $u_{xx} = -\alpha e^{-\alpha t} \sin x$; $u_t = \alpha u_{xx}$.

11. $u_{xx} = -1/r^3 + 3(x - x_0)^2/r^5$; $u_{yy} = -1/r^3 + 3(y - y_0)^2/r^5$; $u_{zz} = -1/r^3 + 3(z - z_0)^2/r^5$; $u_{xx} + u_{yy} + u_{zz} = 0$.

13. (a) $J = -3y^2/x^4$.

  (b) $u_x = 2(y^2 - x^2)/(x^2 + y^2)^2$; $u_y = -4xy/(x^2 + y^2)^2$; $v_x = 4xy/(x^2 + y^2)^2$; $v_y = 2(y^2 - x^2)/(x^2 + y^2)^2$; $J = (u_x v_y - u_y v_x) = 4/(x^2 + y^2)^2$.

15. (a) $\partial P/\partial V = -\kappa T/V^2$. (b) $\partial V/\partial T = \kappa/P$. (c) $\partial T/\partial P = V/\kappa$.

Chapter 4

1. (a) $7\sqrt{3} - 16$. (b) $-672\sqrt{2}$. (c) 1. (d) $-1/2 - \sqrt{3}$.

3. (a) $\nabla f(x, y) = e^x \sin y\mathbf{i} + e^x \cos y\mathbf{j}$.

  (b) $\nabla f(1, \pi/4) = \frac{1}{\sqrt{2}}e(\mathbf{i} + \mathbf{j})$.

  (c) $\nabla f(1, \pi/4) \cdot \mathbf{u} = \frac{1}{\sqrt{10}}e$.

5. (a) $\nabla f(x, y, z) = (y + z^3, x + z^2, 2yz + 3xz^2)$.

  (b) $\nabla f(2, 0, 3) = (27, 11, 54)$.

  (c) $\nabla f(2, 0, 3) \cdot \mathbf{u} = 43/3$.

7. (a) $\nabla f(1, 0) = (1, 2)$ is the direction and the maximum rate is $|\nabla f(1, 0)| = \sqrt{5}$.

  (b) The maximum rate of change is $|\nabla f(1, 2)| = 2\sqrt{5}/5$ in the direction $(2/5, 4/5)$ or $(2, 4)$.

   (c) The maximum rate of change is $|\nabla f(4, 10)| = \sqrt{17}/6$ in the direction $(2/3, 1/6)$ or $(4, 1)$.
   (d) The maximum rate of change is $|\nabla f(4, 3, -1)| = \sqrt{11}$ in the direction $(1, -1, -3)$.
   (e) The maximum rate of change is $|\nabla f(\pi/6, -\pi/8)| = \sqrt{13/2}$ in the direction $(-3\sqrt{2}/2, -\sqrt{2})$ or $(-3, -2)$.
   (f) The maximum rate of change is $|\nabla f(4, 2, 1)| = \sqrt{17}/2$ in the direction $(1/2, 0, -2)$ or $(1, 0, -4)$.

9. (a) Proceed in the direction $\nabla f(60, 100) = (-1, 2, -4)$.
    (b) The maximum slope is equal to the maximum directional derivative, which is $|(1, 2, -4)| = \sqrt{17.44}$ and

$$\theta = \tan^{-1}\sqrt{17.44} \approx 76.5°.$$

11. (a) Satisfies. (b) Satisfies. (c) Satisfies provided $p^2 = q^2$.
13. $40[(z - y)\mathbf{i} + (x - z)\mathbf{j} + (y - x)\mathbf{k}]$.
15. (a) $2\mathbf{R} \cdot \mathbf{A}$. (b) $\mathbf{R} \times \mathbf{A}$. (c) $2r^2\mathbf{A}$. (d) $4(\mathbf{A} \cdot \mathbf{R})^3\mathbf{A}$.
     (e) $(\mathbf{A} \cdot \mathbf{R})/r$. (f) $(\mathbf{A} \cdot \mathbf{R})\mathbf{A}$. (g) 0. (h) $2\mathbf{A}$. (i) 6.

20.
$$\nabla \times \left[\frac{1}{r}(\mathbf{A} \times \mathbf{R})\right] = \frac{1}{r}\nabla \times (\mathbf{A} \times \mathbf{R}) + \nabla\left(\frac{1}{r}\right) \times (\mathbf{A} \times \mathbf{R})$$

$$= \frac{1}{r}[(\nabla \cdot \mathbf{R})\mathbf{A} - (\mathbf{A} \cdot \nabla)\mathbf{R}] - \frac{\mathbf{R}}{r^3}(\mathbf{A} \times \mathbf{R})$$

$$= \frac{2\mathbf{A}}{r} - \frac{1}{r^3}[(\mathbf{R} \cdot \mathbf{R})\mathbf{A} - (\mathbf{R} \cdot \mathbf{A})\mathbf{R}]$$

$$= \frac{\mathbf{A}}{r} + \frac{\mathbf{A} \cdot \mathbf{R}}{r^3}\mathbf{R}.$$

21. (a)
$$\nabla u \times \nabla v = \begin{vmatrix} \mathbf{i} & \mathbf{j} & \mathbf{k} \\ u_x & u_y & u_z \\ v_x & v_y & v_z \end{vmatrix}$$

$$= \mathbf{i}J\left(\frac{u, v}{y, z}\right) + \mathbf{j}J\left(\frac{u, v}{z, x}\right) + \mathbf{k}J\left(\frac{u, v}{x, y}\right).$$

$\nabla u \times \nabla v = 0$ implies that $J(u, v)/(y, z) = 0$; $J(u, v)/(z, x) = 0$; $J(u, v)/(x, y) = 0$. From these three Jacobians it is clear that $u(x, y, z)$ and $v(x, y, z)$ are related by some function $f(u, v) = 0$.

(b) $u = u(x, y)$, $v = v(x, y)$ and so $\nabla u \times \nabla v = \mathbf{k}J(u, v)/(x, y) = 0$ implies that $J(u, v)/(x, y) = 0$.

22. Given that $\mathbf{F} = (x^2 + y^2 + z^2)^n \mathbf{R} = r^{2n} \mathbf{R}$.

(a)
$$\nabla \cdot \mathbf{F} = \nabla \cdot (r^{2n} \mathbf{R}) = r^{2n} \nabla \cdot \mathbf{R} + \nabla(r^{2n}) \cdot \mathbf{R}$$
$$= 3r^{2n} + (2n)r^{2n-1} \nabla(r) \cdot \mathbf{R}$$
$$= 3r^{2n} + 2nr^{2n-1} \frac{\mathbf{R} \cdot \mathbf{R}}{r}$$
$$= (3 + 2n)r^{2n} = (2n + 3)(x^2 + y^2 + z^2)^n.$$

(b)
$$\nabla \times \mathbf{F} = \nabla \times (r^{2n} \mathbf{R})$$
$$= r^{2n} \nabla \mathbf{R} + \nabla(r^{2n}) \times \mathbf{R}$$
$$= 0.$$

(c) $\nabla \times \mathbf{F} = 0$ implies that there exists a scalar potential ϕ such that $\mathbf{F} = -\nabla \phi$. Hence,

$$\frac{\partial \phi}{\partial x} = -x(x^2 + y^2 + z^2)^n,$$

$$\frac{\partial \phi}{\partial y} = -y(x^2 + y^2 + z^2)^n,$$

$$\frac{\partial \phi}{\partial z} = -z(x^2 + y^2 + z^2)^n$$

and its solution can be written as $\phi(x, y, z) = -(x^2 + y^2 + z^2)^{n+1}/(2n + 2) + C$. The additive constant C can be taken as zero.

(d) $\phi(x, y, z) = -(r^{2n+2})/(2n + 2)$. ϕ diverges to infinity provided
   (i) $n = -1$ for any r.
   (ii) $n + 1 > 0$ for $r \to \infty$.
   (iii) $n + 1 < 0$ for $r = 0$.

23. The potential $= \phi = (GMm/2R^3)(x^2 + y^2 + z^2)$.

24. (a) Given that $\mathbf{B} = (\nabla u) \times (\nabla v)$. Then,

$$\nabla \cdot \mathbf{B} = \nabla \cdot (\nabla u \times \nabla v)$$
$$= \nabla v \cdot \nabla \times (\nabla u) - \nabla u \cdot \nabla(\nabla v)$$
$$= 0.$$

Therefore, $\mathbf{B}$ is solenoidal.

(b) $\nabla \cdot \mathbf{B} = 0$ implies that there exists a vector potential $\mathbf{A}$ such that

$$\mathbf{B} = \nabla \times \mathbf{A}$$
$$= \nabla \times \frac{1}{2}(u\nabla v - v\nabla u)$$
$$= \frac{1}{2}[u\nabla \times \nabla v + \nabla u \times \nabla v - v\nabla \times \nabla u - \nabla v \times \nabla u]$$
$$= \nabla u \times \nabla v.$$

This is the desired result.

25. The potential function is given by $\mathbf{F} = -\nabla\phi$. Hence, we have
     (i)   For $r \leq a$:

$$\frac{\partial\phi}{\partial x} = \frac{Mx}{a^3},$$

$$\frac{\partial\phi}{\partial y} = \frac{My}{a^3},$$

$$\frac{\partial\phi}{\partial z} = \frac{Mz}{a^3}.$$

The potential function ϕ for this case is obtained at once $\phi = (M/2a^3)(x^2 + y^2 + z^2) + C$.

(ii)   For $r \geq a$:

$$\frac{\partial\phi}{\partial x} = \frac{Mx}{r^3},$$

$$\frac{\partial\phi}{\partial y} = \frac{My}{r^3},$$

$$\frac{\partial\phi}{\partial z} = \frac{Mz}{r^3}.$$

The solution can be obtained as $\phi = -M/\sqrt{x^2 + y^2 + z^2} + C$. The additive constant can be assumed to be zero.

Chapter 5

1. (a) $\int_C x \, ds = \int_0^1 t^3 \sqrt{9t^4 + 1} \, dt = \frac{1}{54}(10^{3/2} - 1)$.
    (b) $\int_C xy^4 \, ds = \int_{-\pi/2}^{\pi/2} [(4)^5 \cos t \sin^4 t] 4 \, dt = 1638.4$.
    (c) $\int_C (x - 2y^2) \, dy = \int_{-2}^1 (2x^2 - 4x^5) \, dx = 48$.
    (d) $C = C_1 + C_2$. On C_1: $x = x$, $y = 0$, $0 \leq x \leq 2$. On C_2: $x = x$, $y = 2x - 4$, $2 \leq x \leq 3$. Then

$$\int_C xy \, dx + (x - y) \, dy$$

$$= \int_{C_1} xy \, dx + (x - y) \, dy + \int_{C_2} xy \, dx + (x - y) \, dy$$

$$= \int_0^2 0 \, dx + \int_2^3 [(2x^2 - 4x) + (4 - x)(2)] \, dx$$

$$= \int_2^3 (2x^2 - 6x + 8) \, dx = \frac{17}{3}.$$

(e)
$$\int_C x\sqrt{y}\,dx + 2y\sqrt{x}\,dy$$

$$= \int_{C_1} x\sqrt{y}\,dx + 2y\sqrt{x}\,dy + \int_{C_2} x\sqrt{y}\,dx + 2y\sqrt{x}\,dy$$

$$= \int_0^{\pi/2} [-\cos t\,(\sin t)^{3/2} + 2\sin t\,(\cos t)^{3/2}]dt$$

$$+ \int_0^1 [16t\sqrt{2t+1} + 8(2t+1)\sqrt{t}]dt$$

$$= \frac{32\sqrt{3} + 66}{5}.$$

3. 0.

5. (a) 2a. (b) 0.

7. $\int_c \mathbf{F} \cdot d\mathbf{r} = \int_0^1 (2te^{t^2-1} + 3t^7)dt = 11/8 - 1/e.$

9. $m = \int_0^{2\pi} (t^2 + 1)\sqrt{2}\,dt = \sqrt{2}(8/3\pi^3 + 2\pi);$
   $\bar{x} = 1/\sqrt{2}(8/3\pi^3 + 2\pi) \int_0^{2\pi} \sqrt{2}(t^3 + t)dt = 3\pi(2\pi^2 + 1)/(4\pi^2 + 3);$
   $\bar{y} = 0, \bar{z} = 0.$ Hence, we have $(\bar{x}, \bar{y}, \bar{z}) = (3\pi(2\pi^2 + 1)/(4\pi^2 + 3), 0, 0).$

11. Here $\mathbf{F}(x, y) = (2x \sin y)\mathbf{i} + (x^2 \cos y - 3y^2)\mathbf{j}.$ Then $f(x, y) = x^2 \sin y - y^3$ is a potential for $\mathbf{F}$, i.e. $\nabla f = \mathbf{F}$ so $\mathbf{F}$ is conservative and thus the line integral is independent of path. Therefore,

$$\int_C 2x \sin y\,dx + (x^2 \cos y - 3y^2)dy = f(5, 1) - f(-1, 0)$$

$$= 25\sin(1) - 1.$$

12. Given $\mathbf{F} = \mathbf{i}(2y^2 - 12x^3y^3) + \mathbf{j}(4xy - 9x^4y^2).$ It can be easily shown that $\nabla \times \mathbf{F} = 0$ which implies that $\mathbf{F} = \nabla\phi$ where ϕ is a scalar potential. Hence, $\partial\phi/\partial x = 2y^2 - 12x^3y^3$ and $\partial\phi/\partial y = 4xy - 9x^4y^2.$ By integration the potential is obtained as $\phi(x, y) = 2xy^2 - 3x^4y^3 + C.$ Thus the value of the integral $= [\phi(x, y)]_{(1,1)}^{(3,2)} = -1919.$

13. (a) $P = -y/(x^2 + y^2), \partial P/\partial y = (y^2 - x^2)/(x^2 + y^2)^2$ and $Q = x/(x^2 + y^2), \partial Q/\partial x = (y^2 - x^2)/(x^2 + y^2)^2.$ Thus, $\partial P/\partial y = \partial Q/\partial x.$

(b) $C_1: x = \cos t, y = \sin t, 0 \leq t \leq \pi; C_2: x = \cos t, y = \sin t, t = 2\pi$ to $t = \pi.$ Then

$$\int_{C_1} \mathbf{F} \cdot d\mathbf{r} = \int_0^\pi \frac{(-\sin t)(-\sin t) + (\cos t)(\cos t)}{\cos^2 t + \sin^2 t}dt = \pi$$

and

$$\int_{C_2} \mathbf{F} \cdot d\mathbf{r} = \int_{2\pi}^\pi dt = -\pi.$$

Since these are not equal, the line integral of $\mathbf{F}$ is not independent of path. This does not contradict Theorem 5.3 since the domain of $\mathbf{F}$ (which is $\mathcal{R}^2$ except the origin) is not simply connected.

15. (a) $\iint_S y ds = \int_0^3 \int_0^{(6-2y/3)} y\sqrt{14}\,dxdy = 3\sqrt{14}.$

    (b) $\iint_S xz ds = \int_0^1 \int_0^{1-y}(x(1-y)-x^2)\sqrt{3}\,dxdy = \frac{1}{8\sqrt{3}}.$

    (c) $\iint_S x ds = \int_0^2 \int_0^2 x\sqrt{4x^2+17}\,dxdz = \frac{3\sqrt{33}-17\sqrt{17}}{6}.$

    (d) $\iint_S (y^2+z^2)ds = \int_0^{2\pi}\int_0^2 r^2\sqrt{4r^2+1}r\,drd\theta$

             $= \frac{\pi}{60}(391\sqrt{17}+1).$

17. (a) $2e^2 - 10e^{-2}$. (b) 0. (c) $2\pi/3$. (d) -2π. (e) $10\pi/3$. (f) $\pi\sqrt{2}/4$.

19. $2\pi(1-e^{-2})$.

21. (a) $\pi(8\sqrt{2}-7)/6$. (b) 4π.

23. (a) $\pi(2-\sqrt{2})/3$. (b) $(4\tan^{-1}2-\pi)/12$. (c) $32\pi/3$.

24. (i) There are three surfaces in this cylindrical structure case. The total surface integrals $= (\pi a^3 h(2h+a))/4$; (ii) The total volume integral $= \iiint_V x^2 z dV = \int_0^h \int_0^{2\pi}\int_0^a (r\cos\theta)^2 z(rdrd\theta dz) = \frac{a^4}{4}\int_0^h \int_0^{2\pi} z\cos^2\theta d\theta dz = \pi a^4 h^2/8.$

25. (i) There are two surfaces in the upper hemispherical case: The surface integral $= \pi a^5/4$; (ii) The volume integral $= \pi a^6/24$.

28. The surface integral $= 3\pi a^3/2$.

Chapter 6

1. (a) (i) $\oint_C x^2 y dx + xy^3 dy = -\frac{1}{12}.$

       (ii) $\oint_C x^2 y dx + xy^3 dy = \int_0^1 \int_0^1 (y^3-x^2)dxdy = -\frac{1}{12}.$

    (b) (i) $\oint_C x dx - x^2 y^2 dy = -\frac{1}{5}.$

       (ii) $\oint_C x dx - x^2 y^2 dy = \int_0^1 \int_x^1 (-2xy^2)dydx = -\frac{1}{5}.$

    (c) (i) $\oint_C (x+2y)dx + (x-2y)dy = -\frac{1}{6}.$

       (ii) $\oint_C (x+2y)dx + (x-2y)dy = \int_0^1 \int_{x^2}^x (1-2)dydx = -\frac{1}{6}.$

    (d) (i) $\oint_C (x^2+y^2)dx + 2xy dy = 0.$

       (ii) $\oint_C (x^2+y^2)dx + 2xy dy = \iint_D (2y-2)dA = 0.$

3. By Green's theorem, $W = \int_C \mathbf{F}\cdot d\mathbf{r} = \int_C x(x+y)dx + xy^2 dy = \iint_D (y^2-x)dydx$ where C is the path described in the question and D is the triangle bounded by C. So $W = \int_0^1 \int_0^1 (y^2-x)dydx = -\frac{1}{12}.$

5. (a) $\iint_S \nabla\times\mathbf{F}\cdot d\mathbf{S} = \oint_C \mathbf{F}\cdot d\mathbf{r} = \int_0^{2\pi}\cos^2 t dt = \pi.$

    (b) $\iint_S \nabla\times\mathbf{F}\cdot d\mathbf{S} = \oint_C \mathbf{F}\cdot d\mathbf{r} = \int_0^{2\pi}(\cos^2 t - \sin^3 t)dt = \pi.$

    (c) $\iint_S \nabla\times\mathbf{F}\cdot d\mathbf{S} = \oint_C \mathbf{F}\cdot d\mathbf{r} = \int_0^{2\pi}\cos^4 t dt = \frac{3\pi}{4}.$

    (d) $\iint_S \nabla\times\mathbf{F}\cdot d\mathbf{S} = \oint_C \mathbf{F}\cdot d\mathbf{r} = 2\int_0^{2\pi}(-1+\cos 2t + \sin 2t)dt = -4\pi.$

7. $\int_C \mathbf{F}\cdot d\mathbf{r} = \int_S \nabla F\cdot \mathbf{n}dS = \int_0^1 \int_0^{1-x}(8-6x-10y)dydx = \frac{4}{3}.$

9. $\int_C \mathbf{F}\cdot d\mathbf{r} = \int_S \nabla F\cdot d\mathbf{S} = \iint_D 0 dA = 0.$

11. $\int_C \mathbf{F}\cdot d\mathbf{r} = \iint_S \nabla\times d\mathbf{S} = 2\int_0^{2\pi}\int_0^1 r^3 drd\theta = \pi.$

13. $\iint_S \nabla\times\mathbf{F}\cdot d\mathbf{S} = 16.$

15. $\iint_S \mathbf{F}\cdot d\mathbf{S} = \iiint_E (-z-z+2z)dV = 0.$

17. $\iint_S \mathbf{F} \cdot d\mathbf{S} = \iiint_E (1-1)dV = 0.$

19. $\iint_S \mathbf{F} \cdot d\mathbf{S} = \int_0^{2\pi} \int_0^{\pi/2} \int_1^2 3(\rho^2 \sin^2\phi + 1)\rho^2 \sin\phi \, d\rho d\phi d\theta = \frac{194}{5}\pi.$

21. $\iint_S \mathbf{F} \cdot d\mathbf{S} = \int_0^{2\pi} \int_0^{\pi} \int_2^3 (4\rho^2 \sin^2\phi \cos^2\theta)(\rho^2 \sin\phi) d\rho d\phi d\theta = \frac{3378}{15}\pi.$

22. From the given problem the surface integrals and the volume integral can be calculated easily to obtain

$$\iint_{S_1} \mathbf{F} \cdot \mathbf{n} dS_1 = -\frac{\pi}{4},$$

$$\iint_S \mathbf{F} \cdot \mathbf{n} dS = \frac{13\pi}{20},$$

$$\iint_{S_2} \mathbf{F} \cdot \mathbf{n} dS_2 = -\frac{5\pi}{20} + \frac{13\pi}{20} = \frac{2\pi}{5},$$

$$\iiint_V \nabla \cdot \mathbf{F} dV = \frac{2\pi}{5}.$$

Hence, $\iint_{S_2} \mathbf{F} \cdot \mathbf{n} dS_2 = \iiint_V \nabla \cdot \mathbf{F} dV$ which verifies the divergence theorem.

23. Here the same idea used in question 22 is pursued. $S_2 = S \cup S_1$, where S is the part of the paraboloid $x^2 + y^2 + z = 2$ that lies above the plane $z = 1$, and S_1 is the disk $x^2 + y^2 = 1$ on the plane $z = 1$ oriented downward, and then we apply the Divergence theorem. So we have $\iint_{S_1} \mathbf{F} \cdot d\mathbf{S} = \iint_{S_1} \mathbf{F} \cdot \mathbf{n} dS = \iint_{S_1} (-1)dS = -\pi.$ Then $\iint_{S_2} \mathbf{F} \cdot d\mathbf{S} = \int_0^1 \int_0^{2\pi} \int_1^{2-r^2} r dz d\theta dr = \pi/2.$ Thus, the flux of $\mathbf{F}$ across S is

$$\iint_S \mathbf{F} \cdot d\mathbf{S} = \iint_{S_2} \mathbf{F} \cdot d\mathbf{S} - \iint_{S_1} \mathbf{F} \cdot d\mathbf{S} = \frac{3\pi}{2}.$$

24. Given that $\mathbf{E} = \varepsilon Q \mathbf{R}/|\mathbf{R}|^3$, $|\mathbf{R}| = r$.

$$\nabla \cdot \mathbf{E} = \varepsilon Q \nabla \cdot \frac{\mathbf{R}}{|\mathbf{R}|^3}$$

$$= \varepsilon Q \left\{ \frac{1}{r^3} \nabla \cdot \mathbf{R} + \nabla(r^{-3}) \cdot \mathbf{R} \right\}$$

$$= \varepsilon Q \left[\frac{3}{r^3} - \frac{3}{r^3} \right] = 0.$$

This verifies that the given electric field is solenoidal.

25. $\iint_S (2x + 2y + z^2)dS = \iiint_B dV = \frac{4}{3}\pi.$

27. $\iint_S \nabla \times \mathbf{F} \cdot d\mathbf{S} = \oint_C \mathbf{F} \cdot d\mathbf{r} = \int_0^{2\pi} (16\cos^3 t \sin t + 4\sin^2 t)dt = 4\pi.$

29. $\iint_S \mathbf{F} \cdot d\mathbf{S} = \int_0^{2\pi} \int_0^1 \int_0^2 (3r^2 + 3z^2)r dz dr d\theta = 11\pi.$

31. Let S_1 be the sphere of radius 1 with centre at the origin and outer normal $\mathbf{n}_1$. Let S_2 be the surface of the ellipsoid with outer unit normal $\mathbf{n}_2$ and let E be

the solid region between S_1 and S_2. Then the outward flux of $\mathbf{F}$ through the ellipsoid is given by

$$\iint_{S_2} \mathbf{F} \cdot \mathbf{n}_2 dS = \iint_{S_1} dS + \iiint_E 0 dV$$

$$= (\text{Surface area of the unit sphere})$$

$$= 4\pi.$$

Chapter 7

1. $\mathbf{R} = \mathbf{i}x + \mathbf{j}y + \mathbf{k}z; \quad \partial\mathbf{R}/\partial x = \mathbf{i}, \quad \partial\mathbf{R}/\partial y = \mathbf{j}, \quad \partial\mathbf{R}/\partial z = \mathbf{k}: \quad |\partial\mathbf{R}/\partial x| = 1,$
    $|\partial\mathbf{R}/\partial y| = 1, |\partial\mathbf{R}/\partial z| = 1; \mathbf{e}_1 = \mathbf{i}, \mathbf{e}_2 = \mathbf{j}, \mathbf{e}_3 = \mathbf{k}.$
2. $\mathbf{e}_1 = (1/\sqrt{\sinh^2 u + \sin^2 v})[\mathbf{i} \sinh u \cos v + \mathbf{j} \cosh u \sin v],$
    $\mathbf{e}_2 = (1/\sqrt{\sinh^2 u + \sin^2 v})[-\mathbf{i} \cosh u \sin v + \mathbf{j} \sinh u \cos v], \mathbf{e}_3 = \mathbf{k}.$
5. (a) $2\mathbf{R}$. (b) $-\mathbf{R}/r^3$.
7. $V = 24\pi$.
12. Applying the divergence theorem to the vector $\mathbf{U} \times \mathbf{V}$ where $\mathbf{V}$ is a constant vector, we obtain

$$\iiint_V \nabla \cdot (\mathbf{U} \times \mathbf{V}) dV = \iint_S (\mathbf{U} \times \mathbf{V}) \cdot \mathbf{n} dS,$$

$$\iiint_V (\mathbf{V} \cdot \nabla \times \mathbf{U}) dV = \iint_S \mathbf{V} \cdot \mathbf{n} \times \mathbf{U} dS.$$

The last relation implies that $\iiint_V \nabla \times \mathbf{U} dV = \iint_S \mathbf{n} \times \mathbf{U} dS.$

Chapter 8

9. (a) Work done along the path $y = 6x - 4$ is given by $W = \int_1^2 (36x^2 - 16x) dx = 60$. (b) By the potential method, $\phi(x, y) = 2x^2 y + C$, and so $W = \phi(x, y)|_{(1,2)}^{(2,8)} = 60$. Hence, by these two methods we get the identical result.
17. We have

$$\frac{\partial\rho}{\partial t} = \varepsilon_0 \nabla \cdot \frac{\partial\mathbf{E}}{\partial t}$$

$$= \varepsilon_0 \nabla \cdot (\nabla \times \mathbf{B} - \mu_0 \mathbf{j})/\varepsilon_0 \mu_0$$

$$= -\nabla \cdot \mathbf{j},$$

since the combination div curl is always zero.

19. Taking the curl, we have

$$\nabla \times (\nabla \times \mathbf{B}) = \mu_0 \varepsilon_0 \frac{\partial \nabla \times \mathbf{E}}{\partial t}.$$

Expanding the right-hand side, we obtain

$$-\nabla^2 \mathbf{B} = -\mu_0 \varepsilon_0 \frac{\partial^2 \mathbf{B}}{\partial t^2},$$

so $\mathbf{B}$ obeys exactly the same wave equation as $\mathbf{E}$.

21. The energy can be written $w = \mathbf{B} \cdot \mathbf{B}/2 + \mathbf{E} \cdot \mathbf{E}/2c^2$, so the rate of change of energy is

$$\begin{aligned}
\frac{\partial w}{\partial t} &= \mathbf{B} \cdot \frac{\partial \mathbf{B}}{\partial t} + \frac{1}{c^2} \mathbf{E} \cdot \frac{\partial \mathbf{E}}{\partial t} \\
&= -\mathbf{B} \cdot \nabla \times \mathbf{E} + \mathbf{E} \cdot \nabla \times \mathbf{B} \\
&= -\nabla \cdot (\mathbf{E} \times \mathbf{B}).
\end{aligned}$$

This gives the conservation law

$$\frac{\partial w}{\partial t} + \nabla \cdot \mathbf{P} = 0,$$

where $\mathbf{P} = \mathbf{E} \times \mathbf{B}$ is known as the Poynting vector, representing the energy flux of the electromagnetic wave.

Appendix B

Vector formulae at a glance

B.1 Vector algebra

$$\mathbf{A} \cdot \mathbf{B} = \mathbf{B} \cdot \mathbf{A}$$

$$\mathbf{A} \times \mathbf{B} = -\mathbf{B} \times \mathbf{A}$$

$$\mathbf{A} \cdot (\mathbf{B} \times \mathbf{C}) = \mathbf{B} \cdot (\mathbf{C} \times \mathbf{A}) = \mathbf{C} \cdot (\mathbf{A} \times \mathbf{B})$$

$$\mathbf{A} \times (\mathbf{B} \times \mathbf{C}) = (\mathbf{A} \cdot \mathbf{C})\mathbf{B} - (\mathbf{A} \cdot \mathbf{B})\mathbf{C}$$

Gradient

$$\nabla(\Phi\theta) = \Phi\nabla\theta + \theta\nabla\Phi$$

$$\nabla(\mathbf{A} \cdot \mathbf{B}) = (\mathbf{A} \cdot \nabla)\mathbf{B} + (\mathbf{B} \cdot \nabla)\mathbf{A} + \mathbf{A} \times (\nabla \times \mathbf{B}) + \mathbf{B} \times (\nabla \times \mathbf{A})$$

Divergence

$$\nabla \cdot (\Phi\mathbf{A}) = \Phi\nabla \cdot \mathbf{A} + \mathbf{A} \cdot \nabla\Phi$$

$$\nabla \cdot (\mathbf{A} \times \mathbf{B}) = -\mathbf{A} \cdot (\nabla \times \mathbf{B}) + \mathbf{B} \cdot (\nabla \times \mathbf{A}),$$

where ϕ is a scalar function.

Curl

$$\nabla \cdot (\nabla\Phi) = \nabla^2\Phi$$

$$\nabla \cdot (\nabla \times \mathbf{A}) = 0$$

$$\nabla \times (\nabla\Phi) = 0$$

$$\nabla \times (\nabla \times \mathbf{A}) = \nabla(\nabla \cdot \mathbf{A}) - \nabla^2\mathbf{A}$$

B.2 Unit tangent, normal, binormal and the osculating plane

$$\mathbf{R} = x\mathbf{i} + y\mathbf{j} + z\mathbf{k} \quad \text{The position vector}$$

$$\mathbf{T} = \frac{\dot{\mathbf{R}}}{|\dot{\mathbf{R}}|} \quad \text{Unit tangent}$$

$$\mathbf{N} = \frac{\dot{\mathbf{T}}}{|\dot{\mathbf{T}}|} \quad \text{Unit normal}$$

$$\mathbf{B} = \mathbf{T} \times \mathbf{N} \quad \text{Binormal}$$

$$\mathbf{P_1P} \cdot \left\{ \frac{d\mathbf{R}}{dt} \times \frac{d^2\mathbf{R}}{dt^2} \right\}_{t_0} = 0 \quad \text{Osculating plane,}$$

where t_0 is a given point.

B.3 Curvature, tangential and normal acceleration

$$\mathbf{R} = x\mathbf{i} + y\mathbf{j} + z\mathbf{k} \quad \text{The position vector}$$

$$a_{\mathrm{T}} = \frac{\dot{\mathbf{R}} \cdot \ddot{\mathbf{R}}}{|\dot{\mathbf{R}}|} \quad \text{Tangential acceleration}$$

$$a_{\mathrm{N}} = \frac{|\dot{\mathbf{R}} \times \ddot{\mathbf{R}}|}{|\dot{\mathbf{R}}|} \quad \text{Normal acceleration}$$

$$\kappa = \frac{|\dot{\mathbf{R}} \times \ddot{\mathbf{R}}|}{|\dot{\mathbf{R}}|^3}$$

B.4 Serret–Frenet equations

$$\tau = \left| \frac{d\mathbf{B}}{ds} \right|, \quad \text{Torsion}$$

$$\frac{d\mathbf{T}}{ds} = \kappa\mathbf{N}$$

$$\frac{d\mathbf{B}}{ds} = -\tau\mathbf{N}$$

$$\frac{d\mathbf{N}}{ds} = \tau\mathbf{B} - \kappa\mathbf{T}$$

Divergence theorem

$$\iint_S \mathbf{A} \cdot \mathbf{n} \ dS = \iiint_V \nabla \cdot \mathbf{A} \ dV,$$

where V is a volume bounded by a closed surface S.

B.5 Stokes' theorem

$$\oint_C \mathbf{A} \cdot d\mathbf{R} = \iint_S (\nabla \times \mathbf{A}) \cdot \mathbf{n} \ dS,$$

where S is an open surface bounded by a closed contour C.

B.6 Cartesian coordinates (x, y, z)

$$\nabla G = \mathbf{i}\frac{\partial G}{\partial x} + \mathbf{j}\frac{\partial G}{\partial y} + \mathbf{k}\frac{\partial G}{\partial z}$$

$$\nabla \cdot \mathbf{A} = \frac{\partial A_x}{\partial x} + \frac{\partial A_y}{\partial y} + \frac{\partial A_z}{\partial z}$$

$$\nabla \times \mathbf{A} = \mathbf{i}\left[\frac{\partial A_z}{\partial y} - \frac{\partial A_y}{\partial z}\right] + \mathbf{j}\left[\frac{\partial A_x}{\partial z} - \frac{\partial A_z}{\partial x}\right] + \mathbf{k}\left[\frac{\partial A_y}{\partial x} - \frac{\partial A_x}{\partial y}\right]$$

$$\nabla^2 G = \frac{\partial^2 G}{\partial x^2} + \frac{\partial^2 G}{\partial y^2} + \frac{\partial^2 G}{\partial z^2}$$

B.7 Cylindrical coordinates (r, θ, z)

$$\nabla G = \mathbf{r}\frac{\partial G}{\partial r} + \frac{1}{r}\boldsymbol{\theta}\frac{\partial G}{\partial \phi} + \mathbf{k}\frac{\partial G}{\partial z}$$

$$\nabla \cdot \mathbf{A} = \frac{1}{r}\frac{\partial(rA_r)}{\partial r} + \frac{1}{r}\frac{\partial A_\theta}{\partial \theta} + \frac{\partial A_z}{\partial z}$$

$$\nabla \times \mathbf{A} = \mathbf{r}\left[\frac{1}{r}\frac{\partial A_z}{\partial \theta} - \frac{\partial A_\theta}{\partial z}\right] + \boldsymbol{\theta}\left[\frac{\partial A_r}{\partial z} - \frac{\partial A_z}{\partial r}\right] + \frac{1}{r}\mathbf{k}\left[\frac{\partial(rA_\theta)}{\partial r} - \frac{\partial A_r}{\partial \theta}\right]$$

$$\nabla^2 G = \frac{1}{r}\frac{\partial}{\partial r}\left(r\frac{\partial G}{\partial r}\right) + \frac{1}{r^2}\frac{\partial^2 G}{\partial \theta^2} + \frac{\partial^2 G}{\partial z^2}$$

B.8 Spherical coordinates (r, θ, ϕ)

$$\nabla G = \mathbf{r}\frac{\partial G}{\partial r} + \frac{1}{r}\boldsymbol{\theta}\frac{\partial G}{\partial \theta} + \frac{1}{r \sin \theta}\boldsymbol{\phi}\frac{\partial G}{\partial \phi}$$

$$\nabla \cdot \mathbf{A} = \frac{1}{r^2}\frac{\partial (r^2 A_r)}{\partial r} + \frac{1}{r \sin \theta}\frac{\partial (\sin \theta A_\theta)}{\partial \theta} + \frac{1}{r \sin \theta}\frac{\partial (A_\phi)}{\partial \phi}$$

$$\nabla \times \mathbf{A} = \frac{\mathbf{r}}{r \sin \theta}\left[\frac{\partial (A_\phi \sin \theta)}{\partial \theta} - \frac{\partial (A_\phi)}{\partial \phi}\right] + \boldsymbol{\theta}\left[\frac{1}{r \sin \theta}\frac{\partial (A_r)}{\partial \phi} - \frac{1}{r}\frac{\partial (r A_\phi)}{\partial r}\right]$$

$$+ \boldsymbol{\phi}\left[\frac{1}{r}\frac{\partial (r A_\theta)}{\partial r} - \frac{1}{r}\frac{\partial (A_r)}{\partial \theta}\right]$$

$$\nabla^2 G = \frac{1}{r^2}\frac{\partial}{\partial r}\left(r^2\frac{\partial G}{\partial r}\right) + \frac{1}{r^2 \sin \theta}\frac{\partial}{\partial \theta}\left(\sin \theta\frac{\partial G}{\partial \theta}\right) + \frac{1}{r^2 \sin^2 \theta}\frac{\partial^2 G}{\partial \phi^2}.$$

B.9 Maxwell's wave equations

$$\nabla \cdot \mathbf{E} = 0$$

$$\nabla \cdot \mathbf{H} = 0$$

$$\nabla \times \mathbf{E} = -\mu\frac{\partial \mathbf{H}}{\partial t}$$

$$\nabla \times \mathbf{H} = \epsilon\frac{\partial \mathbf{E}}{\partial t}$$

$$\frac{\partial^2 \mathbf{E}}{\partial t^2} = c^2 \nabla^2 \mathbf{E}$$

$$\frac{\partial^2 \mathbf{H}}{\partial t^2} = c^2 \nabla^2 \mathbf{H}$$

where $c = 1/\sqrt{\epsilon \mu}$.

B.10 Walli's formula

$$\int_0^{\pi/2} \sin^m \theta \cos^n \theta d\theta = \frac{[(m-1)(m-3)\cdots 2 \text{ or } 1][(n-1)(n-3)\cdots 2 \text{ or } 1]}{(m+n)(m+n-2)\cdots 2 \text{ or } 1}\alpha,$$

where

$$\alpha = \begin{cases} \pi/2, & \text{if } m \text{ and } n \text{ are both even} \\ 1, & \text{otherwise} \end{cases}$$

Example B.1

$$\int_0^{\pi/2} \cos^6 \theta \sin^4 \theta d\theta = \frac{(5 \times 3 \times 1)(3 \times 1)}{10 \times 8 \times 6 \times 4 \times 2} \times \frac{\pi}{2}$$

☐

Example B.2

$$\int_0^{\pi/2} \cos^5 \theta \sin^3 \theta d\theta = \frac{(4 \times 2)(2)}{8 \times 6 \times 4 \times 2} \times 1$$

☐

Example B.3

$$\int_0^{\pi/2} \cos^5 \theta \sin^4 \theta d\theta = \frac{(4 \times 2)(3 \times 1)}{9 \times 7 \times 5 \times 3 \times 1} \times 1$$

☐

Remark B.1   Walli's formula cited above is extremely important in evaluating the line, surface and volume integrals presented in Chapters 5–8. Many integrals are very much involved with difficult and complicated trigonometric functions involving high integral powers $n \geq 2$ as the integrand with the appropriate limits of integrations. Walli's formula will ease these calculations considerably. The intention of introducing this formula is to avoid frustration on the part of the reader in the evaluation of these highly complicated integrals.   ∎

Appendix C

Historical background of vector calculus

C.1 Introduction

In our natural world there are physical quantities that are associated with only a magnitude. These can be represented by real numbers. Some examples are temperature, density, volume, moment of inertia, speed and pressure. They are known as *scalars*. However, numerous other quantities have both magnitude and direction associated with them, and thus cannot be described by a single number. Velocity, acceleration, force, and electric current are the most obvious concepts in this category. Representation of such quantities mathematically necessitates the introduction of *vectors*.

Derived from the Latin for 'to carry', the word *vector* and the idea that a force is a vector quantity – in that the resultant of two forces can be found using the parallelogram law – was known to Aristotle. Galileo Galilei (1564–1642) stated the parallelogram law explicitly. While the roots of our subject lie far back in history, vector analysis as we know it today is largely a nineteenth century creation. As a result by work by Wessel in the late eighteenth century, the use of complex numbers to represent vectors in a plane was a well established result by the year 1830.

If several vectors act on a body these forces will not necessarily all lie in the same plane. This realization led to a search for a 'three dimensional complex number' and an associated algebra. There were two significant developments in this regard. The first was the invention of *quaternions* by W.R. Hamilton (1805–1865) in Ireland in 1843 and the second was the publication of the book *Die linearle Ausdehnungslehre [The Calculus of Extensions]* by H.G. Grassmann (1809–1877) in Germany in 1844.

These events were both independent and motivated by entirely different considerations even though they were almost simultaneous. Hamilton's desire was to develop mathematical tools for use in astronomy and physics while Grassmann's pursuits were more abstract and philosophical. Another of Grassmann's concerns was to develop a theoretical algebraic structure for multidimensional geometry.

C.2 Hamilton's quaternions

A quaternions as Hamilton defined it is an entity of the form

$$a + b\mathbf{i} + c\mathbf{j} + d\mathbf{k},$$

where a, b, c and d are scalars and $\mathbf{i}, \mathbf{j}$ and $\mathbf{k}$ are somewhat analogous to the quantity $\mathbf{i}$ in complex number theory. The quantity a is referred to as the scalar part of the quaternions and the remainder as the vector part. The three coefficients b, c and d are Cartesian coordinates of a point $\mathbf{P}$. The operation of adding two quaternions is defined in an obvious way, and the criterion for equality of two quaternions is as one would expect. The usual algebraic rules of multiplication are assumed to be valid, except that the rules for the products of $\mathbf{i}, \mathbf{j}$ and $\mathbf{k}$ are

$$\mathbf{i}^2 = \mathbf{j}^2 = \mathbf{k}^2 = -1 \quad \text{(hence the similarity to complex numbers),}$$
$$\mathbf{ij} = \mathbf{k}, \quad \mathbf{ji} = -\mathbf{k},$$
$$\mathbf{jk} = \mathbf{i}, \quad \mathbf{kj} = -\mathbf{i},$$
$$\mathbf{ki} = \mathbf{j}, \quad \mathbf{ik} = -\mathbf{j},$$

where the commutative law is abandoned. In these we can see the similarity with the familiar cross product results. Hence if,

$$\alpha = \alpha_0 + \alpha_1 \mathbf{i} + \alpha_2 \mathbf{j} + \alpha_3 \mathbf{k} \quad \text{and} \quad \beta = \beta_0 + \beta_1 \mathbf{i} + \beta_2 \mathbf{j} + \beta_3 \mathbf{k}$$

are two quaternions, then their product is

$$\begin{aligned}
\alpha\beta &= \alpha_0\beta_0 - (\alpha_1\beta_1 + \alpha_2\beta_2 + \alpha_3\beta_3) + (\alpha_2\beta_3 - \beta_2\alpha_3)\mathbf{i} \\
&\quad + (\alpha_3\beta_1 - \beta_3\alpha_1)\mathbf{j} + (\alpha_1\beta_2 - \beta_1\alpha_2)\mathbf{k} \\
&\quad + (\alpha_0\beta_1 + \beta_0\alpha_1)\mathbf{i} + (\alpha_0\beta_2 + \beta_0\alpha_2)\mathbf{j} \\
&\quad + (\alpha_0\beta_3 + \beta_0\alpha_3)\mathbf{k} \\
&= \alpha_0\beta_0 - (\alpha_1\beta_1 + \alpha_2\beta_2 + \alpha_3\beta_3) \\
&\quad + (\alpha_0\beta_1 + \beta_0\alpha_1 + \alpha_2\beta_3 - \beta_2\alpha_3)\mathbf{i} \\
&\quad + (\alpha_0\beta_2 + \beta_0\alpha_2 + \alpha_3\beta_1 - \beta_3\alpha_1)\mathbf{j} \\
&\quad + (\alpha_0\beta_3 + \beta_0\alpha_3 + \alpha_1\beta_2 - \beta_1\alpha_2)\mathbf{k}.
\end{aligned} \tag{C.1}$$

Hamilton also introduced the symbol ∇, to denote the differential operator

$$\mathbf{i}\frac{\partial}{\partial x} + \mathbf{j}\frac{\partial}{\partial y} + \mathbf{k}\frac{\partial}{\partial z},$$

which when applied to a scalar point function $\phi(x, y, z)$ produces a vector

$$\nabla\phi = \mathbf{i}\frac{\partial\phi}{\partial x} + \mathbf{j}\frac{\partial\phi}{\partial y} + \mathbf{k}\frac{\partial\phi}{\partial z}.$$

The application of this operator to a vector point function

$$\mathbf{B} = b_1\mathbf{i} + b_2\mathbf{j} + b_3\mathbf{k},$$

produced a quaternion of the form:

$$\begin{aligned}
\nabla\mathbf{B} &= \left(\mathbf{i}\frac{\partial}{\partial x} + \mathbf{j}\frac{\partial}{\partial y} + \mathbf{k}\frac{\partial}{\partial z}\right)(b_1\mathbf{i} + b_2\mathbf{j} + b_3\mathbf{k}) \\
&= -\left(\frac{\partial b_1}{\partial x} + \frac{\partial b_2}{\partial y} + \frac{\partial b_3}{\partial z}\right) + \left(\frac{\partial b_3}{\partial y} - \frac{\partial b_2}{\partial z}\right)\mathbf{i} \\
&\quad + \left(\frac{\partial b_1}{\partial z} - \frac{\partial b_3}{\partial x}\right)\mathbf{j} + \left(\frac{\partial b_2}{\partial x} - \frac{\partial b_1}{\partial y}\right)\mathbf{k},
\end{aligned}$$

following the multiplication rules. In our present understanding, we identify the scalar part of $\nabla\mathbf{B}$ as the negative of the divergence of $\mathbf{B}$ and the vector part as the *curl* of $\mathbf{B}$.

Hamilton's application of quaternions was in solving physical problems. For the remainder of his life he strongly encouraged their use to other physicists. P.G. Tait (1831–1901) of Scotland pressed for their use as a fundamental tool in both physics and geometry. Despite these efforts, quaternions were not extensively taken by either physicists or applied mathematicians. Conventional Cartesian coordinates continued to be the preferred tool.

C.3 Grassmann's calculus of extension

The extensive quantity of Grassmann's basic concept has n components,

$$\alpha_1 e_1 + \alpha_2 e_2 + \cdots + \alpha_n e_n,$$

where α_i are scalars and e_i are primary units. In the case of $n = 3$, we have geometric representations as directed line segments of unit length drawn from a common origin to determine an orthogonal right-handed system of axes. In addition to the obvious sum of two such quantities, Grassmann formulated two kinds of multiplication:

1. An inner product, such that $e_i/e_i = 1$ and $e_i/e_j = 0, i \neq j$. so that if $\alpha = \alpha_1 e_1 + \alpha_2 e_2 + \cdots + \alpha_n e_n$ and $\beta = \beta_1 e_1 + \beta_2 e_2 + \cdots + \beta_n e_n$, the inner product $[\alpha/\beta]$ is defined as,

$$[\alpha/\beta] = \alpha_1\beta_1 + \alpha_2\beta_2 + \cdots + \alpha_n\beta_n \qquad \text{(C.2)}$$

and $[\beta/\alpha] = [\alpha/\beta]$. Note that the usual notation '/' represents the inner product and not division here. The numerical value of one of Grassmann's extensive quantities was defined as

$$\sqrt{(\alpha/\alpha)} = \sqrt{(\alpha_1^2 + \alpha_2^2 + \cdots + \alpha_n^2)}.$$

It is to be noted here that in modern vector analysis, we define the inner product as the dot or scalar product with the usual notation $e_i \cdot e_i = 1$ and $e_i \cdot e_j = 0, i \neq j$.

2. An outer product $[\alpha/\beta]$, defined such that $[e_i e_i] = 0$ and $[e_i e_j] = -[e_j e_i]$, $i \neq j$, so that, in the $n = 3$ case,

$$[\alpha\beta] = (\alpha_2\beta_3 - \alpha_3\beta_2)[e_2 e_3] + (\alpha_3\beta_1 - \alpha_1\beta_3)[e_3 e_1]$$
$$+ (\alpha_1\beta_2 - \alpha_2\beta_1)[e_1 e_2]. \qquad \text{(C.3)}$$

It is to be noted here that in modern vector analysis, we define the outer product as the cross or vector product with the usual notation $e_i \times e_i = 0$ and $e_i \times e_j = -(e_j \times e_i), i \neq j$.

Comparison of Grassmann's inner product (C.2) in the case where $n = 3$ with Hamilton's quaternions product (C.1) reveals that the former is equivalent to the negative of the scalar part of Hamilton's quaternions product when the two quaternions have no scalar part(i.e. when they are vectors). In the three-dimensional case, if we replace $[e_2 e_3]$ by e_1, and so on, Grassmann's outer product (C.3) is equivalent to the vector part of the quaternions product of two vectors. An essential difference between the two is that in Grassmann's algebra the vector is the basic quantity while in quaternions theory it is just one part of the quaternions.

Grassmann's extension to the n-dimensional case introduced additional products and this thinking helped lead to more advanced concepts such as tensors. Despite the originality of his work, it remained little known for many years after its publication.

C.4 Maxwell's innovative ideas about vectors

Since Grassmann's work was in obscurity and the quaternions concept considered unsuitable for applications, other developments were required. J.C. Maxwell (1831–1879) of England provided the next significant development in vector analysis while working on his theoretical analysis of electricity and magnetism. Being familiar with Hamilton's quaternions concept, he separated its vector and scalar part. Maxwell denoted the scalar part of the quaternions by $S\nabla \mathbf{B}$ and called it the *convergence* of $\mathbf{B}$ since the expression was already in fluid dynamics. Shortly afterwards, W.L. Clifford (1845–1879) coined the term *divergence* and defined

$$\text{div } \mathbf{B} = -S\nabla\mathbf{B} = \frac{\partial b_1}{\partial x} + \frac{\partial b_2}{\partial y} + \frac{\partial b_3}{\partial z}.$$

Maxwell denoted the vector part by $V\nabla\mathbf{B}$ and called it the *curl* or *rotation* of $\mathbf{B}$. The choice of this term was attributed to the fact that it was already in use in fluid dynamics as twice the rate of rotation of the fluid at a point. Today, the term *rot* $\mathbf{B}$ still appears occasionally as a synonym for both *curl* $\mathbf{B}$ and $\nabla \times \mathbf{B}$. Maxwell also pointed out that the operator ∇ repeated gives

$$\nabla^2 = \nabla\nabla = \left(\mathbf{i}\frac{\partial}{\partial x} + \mathbf{j}\frac{\partial}{\partial y} + \mathbf{k}\frac{\partial}{\partial z}\right)\left(\mathbf{i}\frac{\partial}{\partial x} + \mathbf{j}\frac{\partial}{\partial y} + \mathbf{k}\frac{\partial}{\partial z}\right)$$

$$= -\left\{\frac{\partial^2}{\partial x^2} + \frac{\partial^2}{\partial y^2} + \frac{\partial^2}{\partial z^2}\right\} + \mathbf{i}\left(\frac{\partial^2}{\partial y\partial z} - \frac{\partial^2}{\partial z\partial y}\right)$$

$$+ \mathbf{j}\left(\frac{\partial^2}{\partial z\partial x} - \frac{\partial^2}{\partial x\partial z}\right) + \mathbf{k}\left(\frac{\partial^2}{\partial x\partial y} - \frac{\partial^2}{\partial y\partial x}\right)$$

$$= -\left\{\frac{\partial^2}{\partial x^2} + \frac{\partial^2}{\partial y^2} + \frac{\partial^2}{\partial z^2}\right\},$$

which he called the Laplacian operator. He went on to note the results

1. curl grad $A = 0$ for any scalar function A.
2. div curl $\mathbf{B} = 0$ for any vector field $\mathbf{B}$.

C.5 Josiah Willard Gibbs' modern vector analysis

J.W. Gibbs (1839–1903) in the USA and O. Heaviside (1850–1925), an electrical engineer in Britain, made the final break from quaternions theory into the beginning of modern vector analysis. Their work was independent but the end results were effectively the same, apart from differences in notation and terminology. Gibbs, a professor of mathematical physics, was motivated by the need for a simpler mathematical framework than quaternions, both for teaching and research purposes. In 1881 he produced a pamphlet *Elements of vector analysis* for private circulation among his students and other interested parties. Later, in 1901 he consented to the formal presentation of his work in book form, the actual writing of this book being done by one of his pupils, E.B. Wilson.

Most texts on vector analysis adhere to the form set down by Gibbs, although he himself was somewhat reluctant to publish his ideas. Heaviside, like Maxwell, was primarily interested in electricity and magnetism. He rejected quaternions as being unsuitable mathematical tools for this area and developed his own vector analysis, which he regarded as merely shorthand.

The correspondence between Gibbs and Heaviside on the one hand and Tait, as a 'quaternionsist', on the other makes for interesting and even amusing reading today. Heaviside was particularly acerbic in his remarks about quaternions. Some of the correspondence is quoted in the introduction to the classic text of C.E. Weatherburn – first published in 1921.

It is interesting to note that Maxwell's equations for the electromagnetic field were usually written out by Maxwell himself in component form: it was Heaviside who wrote them in vector form. The notation $\mathbf{A} \times \mathbf{B}$ for the vector product and $\nabla \times \mathbf{A}$ for the curl was used by the Italian vector analysts, particularly R. Marcolongo (1862–1943) and C. Burali-Forti (1861–1931). The problem of notation in vector analysis was a subject for heated debate in the early years of the twentieth century.

C.6 Other contributions

The two basic integral theorems — that of Stokes and the divergence theorem – were not originally developed in the concise language of vector analysis, indeed they predated the work of Gibbs and Heaviside by many years. In a paper published in 1831 on solving the heat conduction equation, the Russian mathematician Mikhail Vasilievich Ostrogradsky (1801-1862) converted a volume integral of the form

$$\iiint_V \left(\frac{\partial P}{\partial x} + \frac{\partial Q}{\partial y} + \frac{\partial R}{\partial z} \right) \, dx \, dy \, dz,$$

into a surface integral

$$\iint_S (P \cos \lambda + Q \cos \mu + R \cos \nu) \, dS,$$

where P, Q and R are scalar functions of x, y and z and are components of a vector and λ, μ and ν are the direction cosines of the normal to the surface S enclosing the volume V. However, the theorem, which in modern notation is of course

$$\iiint_V \nabla \cdot \mathbf{U} dV = \int \int_S \mathbf{U} \cdot \mathbf{n} dS$$

is also attributed to C.F. Gauss (1777–1855) in Germany, while George Green (1793–1841) in England also made systematic use of the integral identities equivalent to the divergence theorem in 1828. Stokes' theorem was probably first stated in a letter to Stokes by Lord Kelvin (Sir William Thomson) in England in 1850. Stokes (1819–1903) used the theorem as a question in a prize examination at Cambridge in 1854. Interestingly enough, one of the candidates for the prize was J.C. Maxwell.

Bibliography

Arfken, G., *Mathematical Methods for Physicists*, Academic Press, Inc.: San Diego, CA, 1985.

Curle, N. & Davies, H., *Modern Fluid Dynamics*, Vol. 1, Van Nostrand: London, 1968.

Hasselmann, K., On the non-linear energy transfer in a gravity-wave spectrum. Part 1: General theory. *J. Fluid Mech.*, **12**, pp. 481–500, 1962.

Hasselmann, K., On the non-linear energy transfer in a gravity-wave spectrum. Part 2: Conservation theorems; wave-particle analogy; irreversibity. *J. Fluid Mech.*, **15**, pp. 273–281; Part 3: Evaluation of the energy flux and swell-sea interaction for a Neumann spectrum, pp. 385–398, 1963.

Kellog, O.D., *Fundamentals of Potential Theory*, Dover: New York, 1929.

Kreyszig, E., *Advanced Engineering Mathematics*, John Wiley & Sons: New York, 1988.

Lamb, H., *Hydrodynamics*, 6th edn, Cambridge Mathematical Library, Cambridge University Press: Cambridge, 1993.

Lewis, P.E. & Ward, J.W., *Vector Analysis for Engineers and Scientists*, Addison-Wesley Publishing Co.: New York, 1989.

Lighthill, M.J., *Fourier Analysis and Generalised Functions*, Cambridge University Press: Cambridge, 1958.

Lighthill, J., *An Informal Introduction to Theoretical Fluid Mechanics*, Oxford University Press: Oxford, 1986.

Longuet-Higgins, M.S., On the nonlinear transfer of energy in the peak of a gravity wave spectrum: a simplified model. *Proc. Roy. Soc. London*, **A347**, pp. 311–328, 1976.

Love, A.E., *Treatise on the Mathematical Theory of Elasticity*, Cambridge University Press: Cambridge, 1927.

Marsden, J.E. & Tromba, A.J., *Vector Calculus*, W.H. Freeman and Company: New York, 1988.

Matthews, P.C., *Vector Calculus*, Springer-Verlag: London, 1998.

Milne-Thomson, L.M., *Theoretical Hydrodynamics*, 5th edn, Dover: New York, 1996.

Phillips, O.M., *The Dynamics of Upper Oceans*, Cambridge University Press: Cambridge, 1960.

Rahman, M., *Water Waves: Relating Modern Theory to Advanced Engineering Applications*, Oxford University Press: Oxford, 1995.

Russell, J.S., Experimental researches into the laws of certain hydrodynamical phenomena that accompany the motion of floating bodies and have not previously been reduced into conformity with the laws of resistance of fluids. *Transactions of the Royal Society of Edinburg*, **14**, pp. 47–109, 1840.

Sowerby, L., *Vector Field Theory*, Longmans: London, 1974.

Susilo, A., Perrie, W. & Rahman, M., Applied dominant transfer and fuzzy logic for nonlinear wave-wave interactions, *International Journal of Applied Mathematics and Engineering Sciences*, **1**, 1–21, 2007.

Tracy, B.A. & Resio, D.T., *Theory and Calculation of the Nonlinear Energy Transfer between Sea Waves in Deep Water*, WIS Report 11, 1982.

Trim, D., *Calculus for Engineers*, Prentice Hall Canada, Inc.: Scarborough, 1998.

Weatherburn, C.E., *Differential Geometry of Three Dimensions*, Vol. 1, Cambridge University Press: Cambridge, pp. 279–298, 1927.

Weatherburn, C.E., *Advanced Vector Analysis*, Bell: London, 1944.

Webb, D.J., Nonlinear transfer between sea waves. *Deep-Sea Res.*, **25**, pp. 279–298, 1978.

Wolfram, S., *The Mathematica*, Cambridge University Press: Cambridge, 1996.

Wylie, C.R., *Advanced Engineering Mathematics*, McGraw-Hill Book Co.: New York, 1975.

Zill, D.G. & Cullen, M.R., *Advanced Engineering Mathematics*, PWS-Kent Publishing Co.: Boston, 1992.

Index

Computational Methods and Experiments in Materials Characterisation III

Edited by: **A. MAMMOLI**, *The University of New Mexico, USA and* **C.A. BREBBIA**, *Wessex Institute of Technology, UK*

Until recently, engineering materials could be characterized successfully using relatively simple testing procedures. As materials technology advances, interest is growing in materials possessing complex meso-, micro- and nano-structures, which to a large extent determine their physical properties and behaviour. The purposes of materials modelling are many: optimization, investigation of failure, simulation of production processes, to name but a few. Modelling and characterisation are closely intertwined, increasingly so as the complexity of the material increases. Characterisation, in essence, is the connection between the abstract material model and the real-world behaviour of the material in question. Characterisation of complex materials therefore may require a combination of experimental techniques and computation. This book publishes papers presented at the Third International Conference on Computational Methods and Experiments in Material Characterisation.

WIT Transactions on Engineering Sciences, Vol 57

ISBN: 978-1-84564-080-4 2007
apx 400pp
apx £130.00/US$235.00/€195.00

The Art of Resisting Extreme Natural Forces

Edited by: **C.A. BREBBIA**, *Wessex Institute of Technology, UK*

According to the ancient Greeks, nature was composed of four elements: air, fire, water and earth. Engineers are continuously faced with the challenges imposed by those elements, when designing bridges and tall buildings to withstand high winds; constructing fire resistant structures, controlling flood and wave forces; minimizing earthquake damage; prevention and control of landslides and a whole range of other natural forces.

Natural disasters occurring in the last few years have highlighted the need to achieve more effective and safer designs against extreme natural forces. At the same time, structural projects have become more challenging.

Featuring contributions from the First International Conference on Engineering Nature, this book addresses the problems associated in this field and aims to provide solutions on how to resist extreme natural forces. Topics include: Hurricane, Tornadoes and High Winds; Aerodynamic Forces; Fire Induced Forces; Wave Forces and Tsunamis; Landslides and Avalanches; Earthquakes; Volcanic Activities; Bridges and Tall Buildings; Large Roofs and Communication Structures; Underground Structures; Dams and Embankments; Offshore Structures; Industrial Constructions; Coastal and Maritime Structures; Risk Prevention; Remediation and Retrofitting and Safety Based Design.

WIT Transactions on Engineering Sciences, Vol 58

ISBN: 978-1-84564-086-6 2007
apx 400pp
apx £130.00/US$235.00/€195.00

Plasticity of Cold Worked Metals

A Deductive Approach

A. PAGLIETTI, *University of Cagliari, Italy*

Plasticity Theory is a basic tool of Structural Engineering to evaluate the ultimate strength of structures and their behaviour under large loads. Any modern computing code in structural analysis makes use of it to model the response of a structure beyond the elastic range - often by exploiting very sophisticated computational analysis and advanced kinematics for large deformations. Yet all this valuable effort has its Achilles' heel: the evolution law of the yield surface of the material, also referred to as work-hardening rule.

The work-hardening rules that are usually introduced to model this phenomenon are the kinematic rule, the isotropic rule and some combinations -more or less linear- of them. Other work-hardening rules are also available. They are invariably more complicated than the former and possibly even less adequate to represent the post-elastic behaviour of real materials when it comes to general three-dimensional elastic-plastic processes. Still these processes are within the reach of many commercially available computing codes, which in principle should enable us to make general elastic-plastic analyses as needed. The lack of more realistic work-hardening rules, however, sets serious doubts about the validity of the results that can thus be obtained. As a consequence we often have to restrict the applications to particular classes of processes, reduce the range of admissible deformations and make frequent recourse to experimental validation.

ISBN: 978-1-84564-065-1   2007   192pp
£65.00/US$125.00/€97.50

High Performance Structures and Materials III

Edited by: **C.A. BREBBIA**, *Wessex Institute of Technology, UK*

This book contains the edited papers presented at the Third International Conference on High Performance Structures and Materials and it addresses issues involving advanced types of structures, particularly those based on new concepts or new types of materials. Particular emphasis is placed on intelligent 'smart structures' as well as the application of computational methods to model, control and manage these structures and materials.

The book covers topics such as: Damage and Fracture Mechanics; Composite Materials and Structures; Optimal Design; Adhesion and Adhesives; Natural Fibre Composites; Failure Criteria of FRP; Non-linear Behaviour of FRP Structures; Material Characterization; High Performance Materials; High Performance Concretes; Aerospace Structures; Reliability of Structures; Ceramics in Engineering.

WIT Transactions on The Built Environment, Vol 85

ISBN: 1-84564-162-0   2006   744pp
£230.00/US$425.00/€350.00

Find us at
http://www.witpress.com

WIT Press is a major publisher of engineering research. The company prides itself on producing books by leading researchers and scientists at the cutting edge of their specialities, thus enabling readers to remain at the forefront of scientific developments. Our list presently includes monographs, edited volumes, books on disk, and software in areas such as: Acoustics, Advanced Computing, Architecture and Structures, Biomedicine, Boundary Elements, Earthquake Engineering, Environmental Engineering, Fluid Mechanics, Fracture Mechanics, Heat Transfer, Marine and Offshore Engineering and Transport Engineering.